AGRICULTURE ISSUES AND POLICIES SERIES

FARM BILL OF 2008: MAJOR PROVISIONS AND LEGISLATIVE ACTION

AGRICULTURE ISSUES AND POLICIES SERIES

Agriculture Issues & Policies, Volume I
Alexander Berk (Editor)
2001. ISBN 1-56072-947-3

Agricultural Conservation
Anthony G. Hargis (Editor)
2009. ISBN 978-1-60692-273-6

Agricultural Conservation
Anthony G. Hargis (Editor)
2009. ISBN 978-1-60876-813-4
(Online Book)

Environmental Services and Agriculture
Karl T. Poston (Editor)
2009. ISBN: 978-1-60741-053-9

Weeds: Management, Economic Impacts and Biology
Rudolph V. Kingely (Editor)
2009. ISBN 978-1-60741-010-2

Soybean and Wheat Crops: Growth, Fertilization and Yield
Samuel Davies and George Evans
2009. ISBN: 978-1-60741-173-4

Effects of Liberalizing World Agricultural Trade
Henrik J. Ehrstrom (Editor)
2009. ISBN: 978-1-60741-198-7

Effects of Liberalizing World Agricultural Trade
Henrik J. Ehrstrom (Editor)
2009. ISBN: 978-1-60876-601-7
(Online Book)

Hired Farmworkers: Profile and Labor Issues
Rea S. Berube (Editor)
2009. ISBN 978-1-60741-232-8

Agricultural Wastes
Geoffrey S. Ashworth and Pablo Azevedo (Editors)
2009. ISBN: 978-1-60741-305-9

Sugar Beet Crops: Growth, Fertilization & Yield
Claus T. Hertsburg
2009. ISBN: 978-1-60741-491-9

Economic Impacts of Foreign-Source Animal Disease
Jace R. Corder (Editor)
2009. ISBN: 978-1-60741-601-2

Economic Impacts of Foreign-Source Animal Disease
Jace R. Corder (Editor)
2009. ISBN: 978-1-60876-602-4
(Online Book)

Essential Oils: Art, Agriculture, Science, Industry and Entrepreneurship (A Focus on the Asia-Pacific Region)
Murray Hunter
2009. ISBN: 978-1-60741-865-8

Corn Crop Production: Growth, Fertilization and Yield
Arn T. Danforth
2009. ISBN: 978-1-60741-955-6

Corn Crop Production: Growth, Fertilization and Yield
Arn T. Danforth
2009. ISBN: 978-1-60876-860-8
(Online Book)

Phosphate Solubilizing Microbes for Crop Improvement
Mohammad Saghir Khan and Almas Zaidi (Editors)
2009. ISBN: 978-1-60876-112-8

Organic Food - Economics and Issues
Earl D. Straub (Editor)
2009. ISBN: 978-1-60741-130-7

Global Beef Trade
Alessandro Ferrara (Editor)
2009. ISBN: 978-1-60741-121-5

Ecophysiology of Tropical Tree Crops
Fabio DeMatta (Editor)
2010. ISBN 978-1-60876-392-4

Assessing Disease Potential in U.S. Aquaculture Industry
Emilija Kalnins (Editor)
2010. ISBN: 978-1-60741-543-5

Inspection and Protection of U.S. Meat and Poultry
Finn J. Amundson (Editor)
2010. ISBN: 978-1-60741-120-8

Soil Phenols
A. Muscolo and M. Sidari (Editors)
2010. ISBN: 978-1-60876-264-4

No-Till Farming: Effects on Soil, Pros and Cons and Potential
Earl T. Nardali (Editor)
2010. ISBN: 978-1-60741-402-5

Drivers and Restraints for Economically Efficient Farm Production
Helena Hansson Karin Larsén Bo Öhlmér
2010. ISBN: 978-1-60876-171-5

Agriculture and Food
Claus Schäfer
2010. ISBN: 978-1-60692-038-1

Agricultural Economics: New Research
Tomas H. Lee (Editor)
2010. ISBN: 978-1-61668-077-0

Agricultural Economics: New Research
Tomas H. Lee (Editor)
2010. ISBN: 978-1-61668-445-7 (Online)

U.S. Biobased Products Market Potential and Projections through 2025
Meredith A. Williamson
2010 ISBN: 978-1-60741-033-1

The Sugar Industry and Cotton Crops
Peter T. Jenkins (Editor)
2010. ISBN: 978-1-61668-320-7

The Sugar Industry and Cotton Crops
Peter T. Jenkins (Editor)
2010. ISBN: 978-1-61668-439-6 (Online)

Agriculture Research and Technology
Kristian Bundgaard and Luke Isaksen (Editors)
2010 ISBN: 978-1-60741-850-4

Organic Farming and Peanut Crops
Darren C. Grossman and Terrance L. Barrios (Editors)
2010. ISBN: 978-1-60876-187-6

America's Family Farms
Efren J. Tamayo (Editor)
2010. ISBN: 978-1-60741-751-4

Sustainable Agriculture: Technology, Planning and Management
Augusto Salazar and Ismael Rios (Editors)
2010. ISBN: 978-1-60876-269-9

Pesticide Resistance, Population Dynamics and Invasive Species Management
Gregory J. McKee, Colln A. Carter, James A. Chalfant, Rachael E. Goodhue, and Frank G. Zalom
2010. ISBN: 978-1-60741-758-3

Tomatoes: Agricultural Procedures, Pathogen Interactions and Health Effects
Eric D. Aubé and Frederick H. Poole (Editors)
2010. ISBN: 978-1-60876-869-1

Rethinking Structural Reform in Turkish Agriculture: Beyond the World Bank's Strategy
Baris Karapinar, Fikret Adaman and Gokhan Ozertan (Editors)
2010. ISBN: 978-1-60876-718-2

Manure Use for Fertilizer and Energy
Connor D. Macias (Editor)
2010. ISBN: 978-1-60876-847-9

Price Dynamics behind Consumer Food Purchases
Morgan D. Fitzpatrick (Editor)
2010. ISBN: 978-1-60876-892-9

Governance of Agrarian Sustainability
Hrabrin Bachev
2010. ISBN: 978-1-60876-888-2

Transformation of U.S. Animal Agriculture
Justin M. Daigle (Editor)
2010. ISBN: 978-1-60876-938-4

The Peanut Plant and Light: Spermidines from Peanut Flowers
Victor S. Sobolev, James B. Gloe and Arlene A. Sy
2010. ISBN: 978-1-61668-028-2

The Peanut Plant and Light: Spermidines from Peanut Flowers
Victor S. Sobolev, James B. Gloer and Arlene A. Sy
2010. ISBN: 978-1-61668-371-9 (Online Book)

Agriculture and Environmental Security in Southern Ontario's Watersheds
Glen Filson, Bamidele Adekunle, and Katia Marzall
2010. ISBN: 978-1-61668-156-2

Agriculture and Environmental Security in Southern Ontario's Watersheds
Glen Filson, Bamidele Adekunle, and Katia Marzall
2010. ISBN: 978-1-61668-372-6

Million Dollar Farms in the New Century
Samuel D. Bosworth (Editor)
2010. ISBN: 978-1-60741-755-2

Million Dollar Farms in the New Century
Samuel D. Bosworth (Editor)
2010. ISBN: 978-1-61668-541-6 (Online Book)

Alignment-Free Models in Plant Genomics: Theoretical, Experimental, and Legal issues
Humberto González-Díaz, Guillermin Agüero-Chapin, Cristian Robert Munteanu, Francisco Prado-Prado, Kuo-Chen Chou, Aliuska Duardo-Sanchez, Grace Patlewicz and Antonio López-Diaz
2010. ISBN: 978-61668-333-7

Alignment-Free Models in Plant Genomics: Theoretical, Experimental, and Legal issues
Humberto González-Díaz, Guillermin Agüero-Chapin, Cristian Robert Munteanu, Francisco Prado-Prado, Kuo-Chen Chou, Aliuska Duardo-Sanchez, Grace Patlewicz and Antonio López-Diaz
2010. ISBN: 978-61668-603-1
(Online Book)

Manure: Management, Uses and Environmental Impacts
Carmen S. Dellaguardia (Editor)
2010. ISBN: 978-1-61668-424-2

Manure: Management, Uses and Environmental Impacts
Carmen S. Dellaguardia (Editor)
2010. ISBN: 978-1-61668-647-5
(Online Book)

Agroforestry Development on the Canadian Prairies
Suren N. Kulshreshtha, Ken Van Rees, Hayley Hesseln, Johnston, John Kort
2010. ISBN: 978-1-61668-266-8

Agroforestry Development on the Canadian Prairies
Suren N. Kulshreshtha, Ken Van Rees, Hayley Hesseln, Johnston, John Kort
2010. ISBN: 978-1-61668-266-8
(Online Book)

Farm Bill of 2008: Major Provisions and Legislative Action
Mary T. Conner (Editor)
2010. ISBN: 978-1-60741-750-7

AGRICULTURE ISSUES AND POLICIES SERIES

FARM BILL OF 2008: MAJOR PROVISIONS AND LEGISLATIVE ACTION

MARY T. CONNER
EDITOR

Nova Science Publishers, Inc.

New York

For permission to use material from this book please contact us:
Telephone 631-231-7269; Fax 631-231-8175
Web Site: http://www.novapublishers.com

LIBRARY OF CONGRESS CATALOGING-IN-PUBLICATION DATA

Farm Bill of 2008 : major provisions and legislative action / editor: Mary T. Conner.
p. cm.
Includes index.
ISBN 978-1-60741-750-7 (hardcover)
1. Agricultural laws and legislation--United States. I. Conner, Mary T.
KF1682.F3727 2009
343.73'076--dc22

2009038388

Published by Nova Science Publishers, Inc. ✢ New York

CONTENTS

PREFACE

Roughly every five years, Congress debates legislation governing federal farm and food policy. The Food, Conservation, and Energy Act of 2008 (also known as the "Farm Bill") was enacted into law on June 18, 2008. It contains 15 titles covering support for commodity crops, horticulture and livestock production, conservation, nutrition, trade and food aid, agricultural research, farm credit, rural development, energy, forestry and other related programs. The enacted bill succeeds the most recent 2002 Farm Bill and is to guide most federal farm and food policies through 2012. The Congressional Budget Office (CBO) estimates the total cost of the 2008 farm bill at just under $284 billion in total budget authority over five years. This book is a study of the new Farm Bill and its costs, provisions by title, programs and related topics.

Chapter 1 - The 2008 farm bill (P.L. 110-246, Food, Conservation, and Energy Act of 2008) was enacted into law on June 18, 2008. It contains 15 titles covering support for commodity crops, horticulture and livestock, conservation, nutrition, trade and food aid, agricultural research, farm credit, rural development, energy, forestry, and other related programs. It also includes tax-related provisions to offset some new spending initiatives in the rest of the bill. The bill succeeds the most recent 2002 farm bill (P.L. 107-171) and is to guide most federal farm and food policies through FY2012.

Chapter 2 - The Food, Conservation, and Energy Act of 2008 (P.L. 110-246, "2008 farm bill") was enacted into law on June 18, 2008. It contains 15 titles covering support for commodity crops, horticulture and livestock production, conservation, nutrition, trade and food aid, agricultural research, farm credit, rural development, energy, forestry, and other related programs. It also includes provisions that make certain changes to tax laws, in order to offset some new spending initiatives in the final bill. The enacted bill succeeds the most recent 2002 farm bill (P.L. 107-17 1) and is to guide most federal farm and food policies through 2012. Many provisions of the 2002 farm bill expired in September 2007, but were extended under a series of temporary extensions prior to final enactment of the 2008 bill.

The enacted 2008 farm bill continues and/or modifies most existing farm and commodity programs, and also creates new programs and provisions. For farm commodities, the bill generally continues the framework of the 2002 farm bill, revises payment limitations (tightening certain limits and relaxing others), adjusts support prices for some commodities, and creates a new revenue support program, in addition to the traditional direct, counter-cyclical, and marketing loan programs for major supported crops. The bill also adds new stand-alone titles containing provisions to address horticulture and livestock issues, including new mandatory funding for specialty crop block grants and to support organic production; and provisions to address meat and poultry inspection, country-of-origin labeling, and livestock competition. Other provisions include changes to the current crop insurance program, a new provision for ongoing disaster assistance, and expanded borrowing opportunities for beginning and socially disadvantaged farmers.

The bill's nutrition title increases food stamp benefits and sets new standards that will make more households eligible, and also raises funding for fresh fruits and vegetables in most domestic food programs. For research, the bill requires the reorganization of USDA's research, extension, and economic agencies. For most other titles — conservation, international trade and food aid, rural development, forestry, and energy — the enacted law reauthorizes, expands, and/or modifies many of the existing programs, creates new programs and initiatives, and allows some programs to expire.

The Congressional Budget Office (CBO) estimates the total cost of the 2008 farm bill (i.e., baseline plus new spending, using its March 2007 baseline) at just under $284 billion in total budget authority over five years (FY2008-FY2012). About $42 billion (15%) in projected spending will support commodity crops, $189 billion (67%) will support the cost of domestic nutrition programs, $24 billion (9%) will support conservation programs, and $22 billion (8%) will support crop insurance. Another $14 billion is expected to be spent on supplemental disaster assistance, trade, horticulture and livestock production, rural development, research, forestry and energy, and other programs. Offsets from tax provisions and proceeds from the credit, crop insurance, and commodity program titles are estimated at $10 billion (FY2008-FY201 2).

Chapter 3 - The 2008 enacted farm bill (Food, Conservation, and Energy Act of 2008, P.L. 110-246) reauthorizes almost all existing conservation programs, modifies several programs, and creates various new conservation programs. A new Conservation Stewardship program replaces the existing Conservation Security Program and a new Agricultural Water Enhancement Program under the Environmental Quality Incentives Program is also authorized with mandatory funding. Other new programs include the Chesapeake Bay Watershed Program and a "Sodsaver" provision to help preserve native sod, including virgin prairie in the Prairie Pothole National Priority Area. Significant modifications to existing programs include a reduction of the maximum enrolled acreage under the Conservation Reserve Program to 32 million acres and an increase in the cap for the Wetlands Reserve Program to over 3 million acres.

Other changes in the enacted bill include modifications to address eligibility requirements, program definitions, enrollment and payment limits, contract terms, evaluation and application ranking criteria, among other administrative issues. Eligibility is expanded for many programs and technical assistance under most programs is broadened to cover forested and managed lands, pollinator habitat and protection, and identified natural resource areas. Beginning, limited resource, and socially disadvantaged producers, specialty crop producers, and producers transitioning to organic production are also targeted for special consideration in many existing programs.

Estimated new spending on the conservation title—not including estimated conservation-related revenue and cost-offset provisions in the bill—is projected to increase by $2.7 billion over 5 years and $4.0 billion over 10 years. Total mandatory spending for the conservation title is projected at $24.3 billion over 5 years (FY2008-FY2012) and $55.2 billion over 10 years (FY2008-FY2017).

A comparison of conservation provisions in the enacted 2008 farm bill with existing law and the House and Senate farm bills is provided in the **Appendix**.

This report will not be updated.

The Food, Conservation, and Energy Act of 2008 (P.L. 110-246), the 2008 farm bill, reauthorizes almost all existing conservation programs, modifies several programs, and creates various new conservation programs. These changes address eligibility requirements, program definitions, enrollment and payment limits, contract terms, evaluation and application ranking criteria, among other administrative issues. In general, the conservation title includes specific changes that expand eligibility and delivery of technical assistance under most programs to cover more broadly, for example, forested and managed lands, pollinator habitat and protection, and identified natural resource areas, among other expansions. Producer coverage across most programs is also expanded to include beginning, limited resource, and socially disadvantaged producers; speciality crop producers; and producers transitioning to organic production. The enacted bill also creates new conservation programs to address emerging issues and priority resource areas, and also new subprograms under existing programs.

Estimated new spending on the 2008 farm bill's conservation title—not including estimated conservation-related revenue and cost-offset provisions in the bill—is projected to increase by $2.7 billion over 5 years and $4.0 billion over 10 years. Total mandatory spending for the title is projected at $24.3 billion over 5 years (FY2008-FY2012) and $55.2 billion over 10 years (FY2008-FY2017).

Chapter 4 - The Food, Conservation, and Energy Act of 2008 (P.L. 110-246, the 2008 farm bill) extends and expands many of the renewable energy programs originally authorized in the 2002 farm bill. The bill also continues the emphasis on the research and development of advanced and cellulosic bioenergy authorized in the 2007 Energy Independence and Security Act (P.L. 110-140).

Farm bill debate over U.S. biomass-based renewable energy production policy focused mainly on the continuation of subsidies for ethanol blenders, continuation of the import tariff for ethanol, and the impact of corn-based ethanol on agriculture. The enacted bill requires reports on the economic impacts of ethanol production, reflecting concerns that the increasing share of corn production being used for ethanol is contributing to high commodity prices and food price inflation.

Title VII, the research title of the 2008 farm bill, contains numerous renewable energy related provisions that promote research, development, and demonstration of biomass-based renewable energy and biofuels. The Sun Grant Initiative coordinates and funds research at land grant institutions on biobased energy technologies. The Agricultural Bioenergy Feedstock and Energy Efficiency Research and Extension Initiative provides support for on-farm biomass energy crop production research and demonstration.

Title IX, the energy title of the farm bill, authorizes mandatory funds of $1.1 billion and discretionary funds (subject to appropriations) totaling $1.0 billion for the FY2008-FY2012 period. Energy grants and loans provided through initiatives such as the Bioenergy Program for Advanced Biofuels promote the development of cellulosic biorefinery capacity. The Repowering Assistance Program supports increasing efficiencies in existing refineries. Programs such as the Rural Energy for America Program (REAP) assist rural communities and businesses in becoming more energy-efficient and self-sufficient, with an emphasis on small operations. The Biomass Crop Assistance and Development Program, the Biorefinery

Assistance Program, and the Forest Biomass for Energy Program provide support to develop alternative feedstock resources and the infrastructure to support the production, harvest, storage, and processing of cellulosic biomass feedstocks. Cellulosic feedstocks—for example, switchgrass and woody biomass—are given high priority both in research and funding.

Title XV of the 2008 farm bill contains tax and trade provisions. It continues current biofuels tax incentives, reducing those for corn-based ethanol but expanding tax credits for cellulosic ethanol. The tariff on ethanol imports is extended.

In: Farm Bill of 2008: Major Provisions and Legislative Action ISBN: 978-1-60741-750-7
Editor: Mary T. Conner

Chapter 1

WHAT IS THE "FARM BILL"?

Renée Johnson
Agricultural Policy Resources, Science, and Industry Division

SUMMARY

The 2008 farm bill (P.L. 110-246, Food, Conservation, and Energy Act of 2008) was enacted into law on June 18, 2008. It contains 15 titles covering support for commodity crops, horticulture and livestock, conservation, nutrition, trade and food aid, agricultural research, farm credit, rural development, energy, forestry, and other related programs. It also includes tax-related provisions to offset some new spending initiatives in the rest of the bill. The bill succeeds the most recent 2002 farm bill (P.L. 107-171) and is to guide most federal farm and food policies through FY2012.

WHAT IS THE "FARM BILL"?

Federal farm support, food assistance, agricultural trade, marketing, and rural development policies are governed by a variety of separate laws. Although many of these policies can be and sometimes are modified through freestanding authorizing legislation, or as part of other laws, the omnibus, multi-year farm bill provides an opportunity for policymakers to address agricultural and food issues more comprehensively. The omnibus farm bill is renewed about every five years.[1] The omnibus nature of the farm bill can create a broader coalition of support among sometimes conflicting interests for policies that, individually, might not survive the legislative process. This same climate can also stir fierce competition for available funds.

The Food, Conservation, and Energy Act of 2008 (P.L. 110-246, "2008 farm bill") is the most recent omnibus farm bill. It was enacted into law on June 18, 2008, and succeeded the 2002 farm bill.[2] The farm bill governs federal farm and food policy, covering a wide range of programs and provisions, and, as noted above, undergoes review and renewal roughly every five years. The 2008 farm bill contains 15 titles encompassing commodity price and income supports, farm credit, trade, agricultural conservation, research, rural development, energy, and foreign and domestic food programs such as food stamps and other nutrition programs, among other programs. More information on individual titles and programs in the 2008 farm bill is in CRS Report RL33934, *The 2008 Farm Bill: A Summary of Major Provisions and Legislative Action.*

WHAT IS THE COST?

The Congressional Budget Office (CBO) estimates the total cost of the 2008 bill (i.e., baseline plus new funding, using the March 2007 baseline) at $284 billion over FY2008-FY2012 and $604 billion over FY2008-FY2017 (Table 1). These costs are mandatory outlays that do not require appropriations actions. The farm bill also authorizes discretionary programs that require appropriators to allocate funds not accounted for here.

Table 1. CBO Estimated Costs for the 2008 Conference Agreement on the Farm Bill (FY2008-FY2012) (outlays in million $)

	FY2008-FY2012			FY2008-FY2017		
	Baseline	**CBO Score (change)**	**Total**	**Baseline**	**CBO Score (change)**	**Total**
Commodities (Title I)	43,354	(1,726)	41,628	87,179	(1,658)	85,521
Conservation (Title II)	21,392	2,720	24,112	50,699	4,000	54,699
Trade/Food Aid (Title III)	1,823	30	1,853	3,715	(78)	3,637
Nutrition (Title IV)[a]	186,005	2,897	188,902	397,131	9,218	406,349
Credit (Title V)	(1,046)	(378)	(1,424)	(2,321)	(306)	(2,627)
Rural Development (Title VI)	72	122	194	72	149	221
Research (Title VII)	290	31	321	1,290	(907)	383
Forestry (Title VIII)	0	38	38	0	45	45
Energy (Title IX)	41	602	643	43	836	879
Horticulture/Organic (Title X)	—	402	402	—	938	938
Livestock (Title XI)	—	1	1	—	1	1
Crop Insurance (Title XII)	25,718	(3,860)	21,858	52,743	(5,591)	47,152
Commodity Futures (Title XIII)	—	0	0	—	0	0
Miscellaneous (Title XIV)[b]	6,338	44	6,382	13,668	(138)	13,530
Disaster Assistance (Title XV)	—	3,807	3,807	—	3,807	3,807
Tax/Other (Title XV)	—	(4,798)	(4,798)	—	(10,429)	(10,429)
	283,987	**(66)**	**283,921**	**604,218**	**(107)**	**604,111**

Source: Compiled by CRS using the Congressional Budget Office (CBO) March 2007 baseline and CBO score of the conference agreement for H.R. 2419, the Food, Conservation, and Energy Act of 2008; also Senate Finance Committee, E*stimated Revenue Effects of the Conference Agreement for Title XV of H.R. 2419, Fiscal Years 2008-2018*, 08-2 068 R10 (Preliminary), May 13, 2008. May not add due to rounding.

a. New outlays for the expanded Fresh Fruit and Vegetable program required in the nutrition title, $274 million (FY2008-FY2012) and $1.020 billion (FY2008-FY2017), are not reflected in this table because they are effectively offset with money from permanent appropriations under Section 32, mandated in Title XIV.

b. Excludes estimates for crop insurance provisions previously included as part of the farm bill's miscellaneous provisions.

Of the $284 billion in projected total five-year net outlays for programs under the farm bill — including revenue and cost-offset provisions in the bill — about $42 (14%) in projected spending will support commodity crops, $189 billion (67%) will support the cost of domestic nutrition programs, $24 billion (9%) will support conservation programs, and $22 billion (8%) will support crop insurance. For FY2008-FY2012, the enacted bill also includes nearly $4 billion in new spending for supplemental disaster assistance (included under Title XV). Another $10 billion is expected to be spent on trade, horticulture and livestock production, rural development, research, forestry and energy, and other programs. Tax-related provisions and cost savings from some farm bill programs are expected to generate additional funding to offset any new spending. CBO estimates that offsets in the bill total more than $10 billion over five years (FY2008-FY2012).

What Are the Key Farm Bill Provisions by Title?

Below is a summary of the types of provisions and programs of individual titles in the 2008 farm bill. More information is in CRS Report RL33934, *The 2008 Farm Bill: A Summary of Major Provisions and Legislative Action*, and other CRS reports.

Title I: Commodity Programs

For the major commodity crops — grains, oilseeds, and cotton — the 2008 farm bill generally continues the farm commodity price and income support framework of the 2002 farm bill. It revises payment limitations by tightening some annual limits and relaxing others, and adjusts target prices and loan rates for some commodities. It continues the direct payment, counter-cyclical payment, and marketing loan programs for the 2008-2012 crop years. The bill creates a pilot revenue-based counter-cyclical program — the Average Crop Revenue Election (ACRE) program — beginning with the 2009 crop year. It also has a pilot program for planting flexibility, restricts base acres developed for residential use, and eliminates benefits to farms with less than 10 acres.

For dairy, the 2008 farm bill extends, with modifications, two federal programs that support milk prices and dairy farm income — the dairy price support program (DPSP) and the Milk Income Loss Contract (MILC) program. It also authorizes farmers to voluntarily enter into forward price contracts as part of the federal milk marketing order program, among other dairy-related provisions. The bill also continues the sugar program that supports prices for domestic producers and processors. To address the possibility of increased sugar imports from Mexico under theNorth American Free Trade Agreement, the enactedbill mandates an 85% market share for U.S. sugar producers and creates a sugar-for-ethanol program to sell surplus sugar to ethanol producers. Across all commodities, CBO data show estimated total five-year outlays for the title at $41.6 billion (FY2008-FY2012, **Table 1**).

For more detailed information, see CRS Report RL34594, *Farm Commodity Programs in the 2008 Farm Bill*, CRS Report RL34036, *Dairy Policy and the 2008 Farm Bill,* and CRS Report RL34103, *Sugar Policy and the 2008 Farm Bill.*

Title II: Conservation

The 2008 farm bill reauthorizes almost all 2002 farm bill conservation programs, modifies several programs, and creates several new conservation programs. The bill makes changes to and/or expands both working lands programs, such as the Environmental Quality Incentives Program and the (renamed) Conservation Stewardship Program, and land retirement programs, such as the Conservation Reserve Program and the Farmland Protection Program. Program changes address eligibility requirements, program definitions, enrollment and payment limits, contract terms, evaluation and ranking criteria, and other administrative issues, among other program conditions. Producer coverage across most programs is also expanded to include beginning, limited-resource, and socially disadvantaged producers; specialty crop producers; and producers transitioning to organic production. The enacted bill also creates new conservation programs to address emerging issues and priority resource areas, and also new subprograms under existing programs. CBO data show estimated total outlays at $24.1 billion (FY2008-FY2012). See CRS Report RL34060, *Conservation and the 2008 Farm Bill.*

Title III: Trade

The 2008 farm bill reauthorizes and amends USDA's food aid, export market development, and export credit guarantee programs. The bill reauthorizes the largest U.S. food aid program, the P.L. 480 food aid program, along with other smaller programs that provide food aid to countries that are promoting the development of market-oriented agricultural sectors (Food for Progress) or school feeding and nutrition programs (the McGovern-Dole International School Feeding and Child Nutrition Program). It also establishes a pilot program for local and regional purchase of commodities for famine prevention. The farm bill also terminates some export programs, while selected others receive increased funding. CBO data show estimated total five-year outlays for this title at nearly $1.9 billion (FY2008-FY2012). See also CRS Report RL33553, *Agricultural Export and Food Aid Programs*, CRS Report RL34145, *International Food Aid and the 2008 Farm Bill*, and CRS Report RL34227, *Agricultural Exports and the 2008 Farm Bill.*

Title IV: Nutrition

The 2008 farm bill's nutrition title accounts for well over half of all spending covered by the bill, with the overwhelming majority financing the Food Stamp program. The most significant issues in this title deal with administration of, eligibility for, and benefits under the Food Stamp program, funding for The Emergency Food Assistance Program (TEFAP), and support for a program making free fresh fruits and vegetables available in schools. The enacted 2008 farm bill includes provisions that extend expiring authorities in covered programs (generally through FY2012) and increase spending for most programs above what would have been expected under prior law (above the "baseline"). CBO data show estimated total five-year outlays for nutrition programs at $188.9 billion (FY2008-FY2012). See CRS Report RL33829, *Domestic Food Assistance: The Farm Bill and Other Legislation in the 110th Congress.*

Title V: Credit

The farm bill enacted relatively minor changes to the permanent statutes for two government-related farm lenders: the USDA Farm Service Agency (FSA) and the Farm Credit System (FCS). CBO estimates thesechanges will result in total cost savings of about $1.4 billion over the next five years. See CRS Report RS21977, *Agricultural Credit: Institutions and Issues*.

Title VI: Rural Development

The 2008 farm bill reauthorizes and/or amends rural development loan and grant programs and authorizes several new provisions, including rural infrastructure, economic development, and broadband and telecommunications development, among other programs. The bill creates several new programs intended to assist with regional development strategies and to provide technical and financial assistance for rural businesses. CBO data show estimated total five-year outlays for this title at $0.2 billion (FY2008-FY2012). See CRS Report RL34126, *Rural Development Provisions of the 2008 Farm Bill*.

Title VII: Research

The 2008 farm bill reorganizes the administration of USDA's research, extension, and economic agencies to coordinate the mission area's intramural and extramural activities across the department through a new Research, Extension, and Economics Office (REEO). Intramural research is carried out by the Agricultural Research Service, Economic Research Service, and National Agricultural Statistics Service. Extramural research, both formula-funded and competitively awarded, has been administered through the Cooperative State Research, Education, and Extension Service. As of October 2009, this agency becomes the new National Institute of Food and Agriculture. The bill establishes new and expands existing research initiatives, providing more support with mandatory funds for this mission area. CBO data show estimated five-year outlays for this title at $0.3 billion (FY2008-FY2012). For more information see CRS Report RL34352, *Agricultural Research, Extension, and Education: Farm Bill Issues*.

Title VIII: Forestry

The enacted farm bill makes changes to existing forestry programs, allows one to expire, and creates some new programs to assist local entities to protect forests threatened with conversion to non-forest uses, and to restore forests damaged by natural disaster, among other programs. The bill also establishes priorities for forestry assistance funding, requires statewide forest resource assessments and plans and creates a new coordinating committee to oversee state assistance funding. The bill also amends existing law to restrict imports of illegally logged wood and modifies income tax deductions for qualified timber gains. CBO data show estimated total five-year outlays for this title at less than $40 million (FY2008-FY2012). See also CRS Report RL33917, *Forestry in the 2008 Farm Bill*.

Title IX: Energy

The 2008 farm bill reauthorizes, expands, and/or modifies existing programs, and creates new programs and initiatives to promote biofuels and cellulosic ethanol production. The bill supports farm and community renewable energy systems; promotes production, marketing, and processing of biofuel feedstocks other than corn starch; and expands research, education, and demonstration programs for advanced biofuels. It also expands programs for federal procurement of biofuels and bio-refinery repowering projects and establishes USDA coordination of federal biobased energy efforts. CBO data show estimated total five-year outlays at $0.6 billion (FY2008-FY2012). For more information, see CRS Report RL34130, *Renewable Energy Policy in the 2008 Farm Bill.*

Title X: Horticulture and Organic Agriculture

The 2008 farm bill includes new provisions for horticultureand organic production under a new bill title, providing nearly $1 billion in funding over the next ten years. About half of this spending will be used to expand the Specialty Crop Block Grant Program, which provides funds to state agriculture departments for U.S. specialty crop marketing, promotion, research, and other activities. The bill also provides new mandatory funding for growth of farmers' markets and for transitioning producers to organic production, authorizes funding for a new federal-state cooperative pest and disease early detection program, and provides for price reportingand organic data collection, among other provisions. CBO data show estimated total five-year outlays for this title at $0.4 billion (FY2008-FY2012). See also CRS Report RL33520, *Specialty Crops: 2008 Farm Bill Issues.*

Title XI: Livestock

The 2008 farm bill includes new livestock-related provisions under a new bill title. The bill makes changes to existing laws governing livestock and poultry marketing and competition, including specifying that producers may not be forced into mandatory arbitration in livestock or poultry contracts, allowing producers to decline arbitration prior to entering into the contract, enabling producers to litigate a contract dispute where the principal part of their production occurs, and requiring additional reporting and tracking of enforcement action under the Packers and Stockyards Act. The bill also modifies country-of-origin labeling (COOL) requirements for retailers, opens the way for state-inspected meat and poultry to enter interstate commerce, and extends mandatory safety inspection to catfish. CBO data show total five-year outlays at $1 million (FY2008-FY2012). See CRS Report RL33958, *Animal Agriculture: 2008 Farm Bill Issues.*

Title XII: Crop Insurance and Disaster Assistance Programs

The 2008 farm bill provides for changes to the crop insurance program, along with other disaster assistance provisions, under a new bill title. The enacted bill contains several revisions to the crop insurance program, many of which are designed to reduce program costs. CBO estimates net savings of $3.9 billion over five years (FY2008-FY2012), mostly through changes in the timing of premium receipts from farmers, and payments to the companies. The title also includes other disaster assistance provisions, including the addition of the Small Business Disaster Response and Loan Improvements Act of 2008, which makes significant changes to the Small Business Administration's (SBA's) response to disaster. CBO data show estimated total five-year outlays for this title at $21.9 billion (FY2008-FY2012). See CRS Report RL34207, *Crop Insurance and Disaster Assistance: 2008 Farm Bill Issues.*

Title XIII: Commodity Futures

The enacted 2008 farm bill has a new title that reauthorizes appropriations for the Commodity Futures Trading Commission (CFTC) through FY2013, and makes amendments to the Commodity Exchange Act.

Title XIV: Miscellaneous

The miscellaneous title in the 2008 farm bill covers various provisions affecting research, energy, and rural development titles, as well as provisions covering socially disadvantaged and limited-resource producers and agricultural security, among other miscellaneous provisions. CBO data show estimated total outlays for provisions in this title at $6.4 billion (FY2008-FY2012).

Title XV: Trade and Tax Provisions

The enacted bill contains a new farm bill title that provides for a new permanent Supplemental Agricultural Disaster Assistance program, as well as various trade and tax provisions. The supplemental revenue assistance payment program for crop producers is designed to compensate eligible producers for a portion of crop losses that are not eligible for an indemnity payment under the crop insurance program (i.e., the portion of losses that is part of the deductible on the policy). CBO data show estimated total five-year outlays for the permanent disaster program at $3.8 billion (FY2008-FY2012). See CRS Report RL34207, *Crop Insurance and Disaster Assistance in the 2008 Farm Bill.*

The enacted bill's tax provisions consist of six groups, respectively containing provisions for revenue, an agriculture disaster reserve fund, conservation, energy, agriculture, and other provisions. The single largest revenue-raising provision involves a change in the estimated tax payment of corporations. The single largest revenue-losing provision in the enacted bill pertains to the agriculture disaster reserve fund, described above. CBO data show these provisions will generate a $10 billion offset over the next ten years.

End Notes

[1] There have been seven omnibus farm bills since the 1970s (2008, 2002, 1996, 1990, 1985, 1981, 1977). Prior farm legislation was in 1973, 1970, 1965, 1956, 1954, 1949, 1948, 1938, and 1933.

[2] Farm Security and Rural Investment Act of 2002 (P.L. 107-171).

In: Farm Bill of 2008: Major Provisions and Legislative Action ISBN: 978-1-60741-750-7
Editor: Mary T. Conner

Chapter 2

THE 2008 FARM BILL: MAJOR PROVISIONS AND LEGISLATIVE ACTION

Renée Johnson*[1], *Geoffrey S. Becker*[1], *Tom Capehart*[1], *Ralph M. Chite*[1], *Tadlock Cowan*[1], *Ross W. Gorte*[1], *Charles E. Hanrahan*[1], *Remy Jurenas*[1], *Jim Monke*[1], *Jean M. Rawson*[1], *Randy Schnepf*[1], *Joe Richardson*[2], *Donald J. Marples*[3], *Mark Jickling*[3] and *N. Eric Weiss*[3]

[1]Resources, Science, and Industry Division
[2]Domestic Social Policy Division
[3]Government and Finance Division

SUMMARY

The Food, Conservation, and Energy Act of 2008 (P.L. 110-246, "2008 farm bill") was enacted into law on June 18, 2008. It contains 15 titles covering support for commodity crops, horticulture and livestock production, conservation, nutrition, trade and food aid, agricultural research, farm credit, rural development, energy, forestry, and other related programs. It also includes provisions that make certain changes to tax laws, in order to offset some new spending initiatives in the final bill. The enacted bill succeeds the most recent 2002 farm bill (P.L. 107-17 1) and is to guide most federal farm and food policies through 2012. Many provisions of the 2002 farm bill expired in September 2007, but were extended under a series of temporary extensions prior to final enactment of the 2008 bill.

The enacted 2008 farm bill continues and/or modifies most existing farm and commodity programs, and also creates new programs and provisions. For farm commodities, the bill generally continues the framework of the 2002 farm bill, revises payment limitations (tightening certain limits and relaxing others), adjusts support prices for some commodities, and creates a new revenue support program, in addition to the traditional direct, counter-cyclical, and marketing loan programs for major supported crops. The bill also adds new stand-alone titles containing provisions to address horticulture and livestock issues, including new mandatory funding for specialty crop block grants and to support organic production; and provisions to address meat and poultry inspection, country-of-origin labeling, and livestock competition. Other provisions include changes to the current crop insurance program, a new provision for ongoing disaster assistance, and expanded borrowing opportunities for beginning and socially disadvantaged farmers.

The bill's nutrition title increases food stamp benefits and sets new standards that will make more households eligible, and also raises funding for fresh fruits and vegetables in most domestic food programs. For research, the bill requires the reorganization of USDA's research, extension, and economic agencies. For most other titles — conservation, international trade and food aid, rural development, forestry, and energy — the enacted law reauthorizes, expands, and/or modifies many of the existing programs, creates new programs and initiatives, and allows some programs to expire.

The Congressional Budget Office (CBO) estimates the total cost of the 2008 farm bill (i.e., baseline plus new spending, using its March 2007 baseline) at just under $284 billion in total budget authority over five years (FY2008-FY2012). About $42 billion (15%) in projected spending will support commodity crops, $189 billion (67%) will support the cost of domestic nutrition programs, $24 billion (9%) will support conservation programs, and $22 billion (8%) will support crop insurance. Another $14 billion is expected to be spent on supplemental disaster assistance, trade, horticulture and livestock production, rural development, research, forestry and energy, and other programs. Offsets from tax provisions and proceeds from the credit, crop insurance, and commodity program titles are estimated at $10 billion (FY2008-FY201 2).

Key CRS Policy Staff

Area of Expertise	Name	Telephone
Report Coordinator/Overview	Renée Johnson	7-9588
Food and Feed Grain Support	Jim Monke	7-9664
Other Commodity Support Programs	Ralph M. Chite (dairy) Remy Jurenas (sugar) Randy Schnepf (cotton)	7-7296 7-728 1 7-4277
Payment Limits Planting Flexibility Farm Credit	Jim Monke	7-9664
Crop Insurance; Disaster Assistance	Ralph M. Chite	7-7296
Conservation	Renée Johnson Tadlock Cowan	7-9588 7-7600
Energy	Tom Capehart	7-2425
Agricultural Trade and Food Aid	Charles E. Hanrahan	7-7235
Specialty Crops; also Agricultural Research, Extension, & Economics	Jean M. Rawson	7-7283
Meat and Poultry Inspection; Marketing and Regulatory Programs; Food Safety	Geoffrey S. Becker	7-7287
Rural Development	Tadlock Cowan	7-7600

(Continued)

Area of Expertise	Name	Telephone
Forestry	Ross Gorte	7-7266
Domestic Food Assistance and Nutrition Programs	Joe Richardson	7-7325
Revenue-Raising Tax Provisions	Donald J. Marples	7-3739
Small Business Disaster Response	N. Eric Weiss	7-6209
Commodity Futures Trading Commission (CFTC) reauthorization	Mark Jickling	7-7784

ACKNOWLEDGMENTS

Portions of this report were originally written by retired CRS specialists Jasper Womach, Jeffrey Zinn, and David Brumbaugh.

OVERVIEW

Roughly every five years, Congress debates legislation governing federal farm and food policy. The 2002 farm bill (P.L. 107-17 1) covered a wide range of programs including commodity price and income support, farm credit, agricultural conservation, research, rural development, and foreign and domestic food programs, among others. In 2007, both the House and Senate completed committee and floor action on their respective versions of the new farm bill. However, conference negotiations were initially delayed because of differences between committee leadership and the Administration's position, largely over tax-related provisions needed to offset new spending in the bill. Many provisions of the 2002 farm bill expired in September 2007, but were continued under a series of temporary extensions to allow more time to resolve differences between the House- and Senate- passed bills.

On May 8, 2008, House and Senate farm bill conferees announced the details of a completed conference agreement (H.R. 2419, the Food, Conservation, and Energy Act of 2008). The following week, both chambers completed floor action and approved the final conference agreement on the 2008 farm bill. The Bush Administration vetoed the legislation, but both the House and Senate voted to override the veto. On May 22, the 2008 farm bill was enacted into law (P.L. 110- 234). However, the newly enacted law contained only 14 of the 15 farm bill titles because an enrolling error resulted in one title of the bill being omitted from the version that was sent to the White House. To resolve this issue, both the House and Senate passed a version of the 2008 farm bill with all 15 original bill titles (H.R. 6124). The President vetoed H.R. 6124 on June 18. That same day, both the House and Senate voted to override the veto and the bill became law (P.L. 110-246), replacing P.L. 110-234.

Congressional Action

2008 Farm Bill Policy Setting

The 2007/2008 farm bill debate differed from the 2002 debate in some important ways. First, the 2008 farm bill faced potentially significant budgetary and spending constraints, while the 2002 farm bill was crafted during a period of budget surplus. Second, the 2008 farm bill debate was set against a backdrop of an ongoing multilateral negotiation, which targeted farm subsidies and challenged U.S. support and legal programs under existing trade rules. Third, the Bush Administration submitted its own detailed proposal for the new farm bill, whereas for other recent farm bills the Administration had not issued specific recommendations. Fourth, many other groups, including both traditional and non-agricultural interests, also submitted specific recommendations that ranged from maintaining the status quo to making dramatic policy changes.

Budgetary Considerations

As with all federal programs, the farm bill debate was influenced by budgetary constraints imposed by Congress. Recent federal deficits raised concerns with respect to reauthorization or expansion of current farm programs. Prior to congressional consideration of a new farm bill, budget projections showed a lower baseline budget for agriculture programs, mainly because high commodity prices had caused projections of future farm program spending to fall sharply below previous projections. The Congressional Budget Office's (CBO's) March 2007 baseline budget served as the official benchmark for the FY2008 budget resolution and for scoring the budgetary impacts of the new farm bill. The CBO baseline assumed continuation of current farm bill policies under expected economic conditions. The budget resolution set the actual spending constraints for the agriculture committees as they drafted a new farm bill.

In May 2007, Congress approved the FY2008 budget resolution, which adopted the baseline budget as the fiscal parameter for the next farm bill. It also included a $20 billion reserve fund (above baseline) for new farm bill spending over five years.[1] However, any new spending was required to be deficit-neutral, meaning that it would have to be offset with equivalent reductions in other federal spending for existing mandatory programs, or by raising revenues. Large increases in the market prices of corn and other commodities since the summer of 2006 contributed to a lower March 2007 baseline for farm program spending. For example, the March 2007 baseline projected spending for commodity support payments under current law to be $42.4 billion for the FY2008-FY201 3 period, about $30 billion lower than actual spending in the previous six years (**Table 1**). Baseline estimates for mandatory conservation and nutrition programs for the next six years were higher compared to the previous six years.

Trade Negotiations and Commitments

The farm bill debate was also influenced by obligations concerning the design and size of farm subsidies under the World Trade Organization (WTO) Agreement on Agriculture, as well as by the U.S. position in the Doha Round of multilateral negotiations.

Table 1. 2002 Farm Bill Actual Spending (FY2002-FY2007 est.) and the March 2007 CBO Baseline (FY2008-FY2013) (outlays in $ millions)

	Commodity Support	Conservation	Exports	Food Stamps	Total
Baseline (FY08-FY13)	42,446	26,496	2,005	225,845	**296,792**
Actual (FY02-FY07)	72,934	18,323	1,648	178,158	**271,063**
Baseline vs. Actual	-30,488	+8,173	+357	+47,687	**+25,729**

Source: Compiled by CRS from various Congressional Budget Office (CBO) baselines for 2007. The FY2008-2013 baseline is as of March 2007 and does not include the cost of the 2008 farm bill. The report section titled "Projected Cost" provides cost estimates of the 2008 farm bill.

Agreement in the Doha Round was expected to converge in 2007 with the expiration of the 2002 farm bill, and to occur well before the June 30, 2007, expiration of Trade Promotion Authority (TPA), which provides for expedited congressional consideration of trade agreements. Some policymakers wanted a Doha Round agreement so that the next farm bill could be made consistent with new farm trade rules; others argued that the United States should not unilaterally change its own subsidy programs ahead of any multilateral trade agreement. Progress on the Doha Round stalled in 2006 and again in 2008, and criticisms and legal challenges by some WTO member countries of current U.S. farm programs have continued.[2]

The Administration's Policy Recommendations

In January 2007, the Bush Administration released its own detailed recommendations for the farm bill that aimed to substantially alter many aspects of the current commodity support system and other existing farm bill programs.[3] The Administration's stated approach for designing its 2007 farm bill recommendations was to take a "reform-minded and fiscally responsible approach to making farm policy more equitable, predictable and protected from challenge."[4] In part, this referred to the perceived need to more evenly distribute federal program spending and benefits across a larger share of the U.S. farm community, as well as the perceived need to modify current farm programs to better comply with WTO obligations and limit future legal challenges from other countries. Some of these same concerns were voiced in recommendations and proposals by other organizations and interest groups.

The Administration's proposed changes would have also faced potential funding obstacles, given budget constraints for the 2008 farm bill. Nevertheless, President Bush repeatedly threatened to veto any legislation that included certain revenue and tax-related provisions being considered by Congress, as well as legislation that failed to implement certain policy changes, including cuts in farm income subsidies, among other policy issues and concerns.[5] Also during the debate, the Administration recommended that Congress consider a one- or two-year extension of the current farm bill and take up a new farm bill in the next session.[6]

On May 21, President Bush vetoed the conference agreement on H.R. 2419. A second bill containing all 15 original farm bill titles, H.R. 6124, was again vetoed on June 18. Among the reasons cited by the Administration for the veto were Congress' inclusion of certain revenue and tax-related provisions in both bills. The Administration was also concerned that the legislation did not include certain policy reforms in farm income subsidies, and was concerned about possible incompatibility with U.S. obligations under the WTO, among other policy issues. Doha Round negotiations are continuing, however. If agreement is reached, then implementing legislation could require changes in some farm bill programs and limit spending on farm support.

Other Recommendations/Proposals

The 2007/2008 farm bill debate differed from the 2002 debate in the number and scope of proposals seeking a range of changes to existing programs, as well as the addition of new ones. Many of these recommendations gained support within and outside Congress. In addition to the Administration proposal, several organizations and interest groups released their own farm bill recommendations. These included state organizations, national farm groups, commodity associations, conservation and rural development organizations, nutrition program advocates, and several non-traditional interest groups. These policy recommendations represented diverse interests seeking objectives ranging from maintaining current programs to substantially altering or eliminating them. Some recommendations were incorporated into bills introduced by some Members of Congress during the debate, who sought to directly challenge the existing farm legislation and programs through comprehensive and broad-based proposed legislative changes. Others in Congress were reluctant to change current programs because they are strongly supported by the long-time beneficiaries.

Legislative Development

In anticipation of the 2007 farm bill, both the House and Senate Agriculture Committees conducted hearings in Washington DC and across the country during 2006, and continued to hold hearings early in 2007.[7] Early in 2007, the chairmen of both the House and Senate Agriculture Committees indicated their intention to complete work on a new farm bill prior to the August 2007 recess, with full congressional action by September. The House Agriculture Committee conducted its markup of its version of the farm bill (H.R. 2419) in mid-July, and completed House floor action on July 27, 2007. The Senate Agriculture Committee approved its version (S. 2302) in October and, on December 14, the Senate completed floor action on its bill, which was offered as a substitute amendment to the House bill (H.R. 2419).

Conference negotiations were initially delayed because of differences between committee leadership and the Administration, and also differences between the House and Senate on how to resolve approaches to finance new spending above baseline using tax provisions not usually associated with farm bills. During this time, many provisions in the 2002 farm bill were set to expire in September 2007. Certain provisions were extended until March 15, 2008, under the Consolidated Appropriations Act for FY2008 (P.L. 110-161). Following this extension, Congress approved a series of additional temporary extensions, including a one-month extension and four consecutive short-term extensions lasting through May 23, 2008.[8] Conferees began official meetings in April 2008.

On May 8, 2008, House and Senate farm bill conferees announced the details of a completed conference agreement (H.R. 2419, the Food, Conservation, and Energy Act of 2008). On May 14, 2008, the House passed the conference agreement on the 2008 farm bill by a vote of 318-106. On May 15, the Senate passed the same bill by a vote of 81-15. Concurrently, on May 14, both the House and Senate passed, by voice vote, the final temporary extension of current law lasting until the earlier of May 23, 2008, or the date the 2008 farm bill was signed into law.

On May 21, 2008, the Bush Administration vetoed the legislation. The House voted to override the veto by a vote of 316-108 also on May 21, followed by a Senate veto override by a vote of 82-13 the next day. On May 22, the 2008 farm bill was enacted into law (P.L. 110-234). However, an enrolling error resulted in one title of the bill (Title III, Trade) being omitted from the version that was sent to the White House. The newly enacted law contained 14 of 15 farm bill titles. To resolve this issue, both the House and Senate passed a version of the 2008 farm bill with all 15 original bill titles (H.R. 6124). The President vetoed H.R. 6124 on June 18, 2008. That same day both the House (80-14) and the Senate (317-109) voted to override the veto and the bill became law (P.L. 110-246), replacing P.L. 110-234. A timeline showing a chronology of major events is provided at the end of this report.

Brief Overview of Provisions

The enacted 2008 farm bill contains 15 titles covering support for commodity crops, horticulture and livestock production, conservation, nutrition, trade and food aid, agricultural research, farm credit, rural development, energy, forestry, and other related programs. It also includes tax-related provisions that make certain changes to tax laws in order to offset new spending initiatives in the rest of the bill. The 2008 farm bill replaces many of the provisions of the 2002 farm bill (P.L. 107-171) and guides most federal farm and food policies through 2012.

The 2008 farm bill includes five new titles that were not in the 2002 farm bill, covering horticulture and livestock products, crop insurance and disaster assistance, commodity futures, and various tax and trade provisions. The text box shows the titles of the 2008 farm bill and briefly describes some title provisions.

The 2008 Farm Bill: Titles and Respective Programs and Policies

- **Title I, Commodities:** Income support to growers of selected commodities, including wheat, feed grains, cotton, rice, oilseeds, peanuts, sugar, and dairy. Support is largely through direct payments, counter-cyclical payments, and marketing loans. Other support mechanisms include government purchases for dairy, and marketing quotas and import barriers for sugar.
- **Title II, Conservation:** Environmental stewardship of farmlands and improved management practices through land retirement and working lands programs, among other programs geared to farmland conservation, preservation, and resource protection.
- **Title III, Agricultural Trade and Food Aid:** U.S. agriculture export and international food assistance programs, and various World Trade Organization (WTO) obligations.

- **Title IV, Nutrition:** Domestic food and nutrition and commodity distribution programs, such as food stamps and supplemental nutrition assistance.
- **Title V, Farm Credit:** Federal direct and guaranteed farm loan programs. Also specifies loan eligibility rules and other policies.
- **Title VI, Rural Development:** Business and community programs for planning, feasibility assessments, and coordination activities with other local, state, and federal programs, including rural broadband access.
- **Title VII, Research:** Agricultural research and extension programs, including bio security and response, biotechnology, and organic production.
- **Title VIII, Forestry:** USDA Forest Service programs, including forestry management, enhancement, and agroforestry programs.
- **Title IX, Energy:** Bioenergy programs and grants for procurement of biobased products to support development of biorefineries and assist eligible farmers, ranchers, and rural small businesses in purchasing renewable energy systems, as well as user education programs.
- **Title X, Horticulture and Organic Agriculture:** A new farm bill title covering fruits, vegetables, and specialty crops and organic agriculture.
- **Title XI, Livestock:** A new farm bill title covering livestock and poultry production, including provisions that amend existing laws governing livestock and poultry marketing and competition, country-of-origin labeling requirements for retailers, and meat and poultry state inspections, among other provisions.
- **Title XII, Crop Insurance and Disaster Assistance:** A new farm bill title covering crop insurance and assistance previously included in the miscellaneous title (not including the supplemental disaster assistance provisions in the Trade and Tax title).
- **Title XIII, Commodity Futures:** A new farm bill title covering reauthorization of the Commodity Futures Trading Commission (CFTC) and other changes to current law.
- **Title XIV, Miscellaneous:** Other types of farm programs and assistance not covered in other bill titles, including provisions to assist limited-resource and socially disadvantaged farmers, agricultural security, and other provisions.
- **Title XV, Trade and Tax Provisions:** A new title covering tax-related provisions intended to offset spending initiatives for some programs, including those in the nutrition, conservation, and energy titles. The title also contains other provisions, including the new supplemental disaster assistance and disaster relief trust fund.

For commodities (Title I), the enacted 2008 farm bill generally continues the framework of the 2002 farm bill, but with changes to program eligibility criteria and payment limitations, and adjustments to target prices and loan rates for some commodities, covering the 2008 through 2012 crop years. The bill creates a new Average Crop Revenue Election (ACRE) program beginning in crop year 2009. It also adds new provisions to address horticulture and livestock issues, and creates two new titles to address these sectors (Title X and Title XI). The bill provides mandatory funding for specialty crop block grants and adds new provisions supporting pest and disease programs, new funding for growth of farmers' markets and for transitioning producers to organic production, and price reporting and organic data collection, among other provisions. New animal agriculture provisions include changes to existing laws governing livestock and poultry marketing and competition, and changes in country-of-origin labeling requirements and meat and poultry inspections.

The nutrition title (Title IV) reauthorizes and increases funding for most farm bill authorized programs. It increases benefits and makes more households eligible under in the Food Stamp program, which the farm bill conferees have renamed the Supplemental Nutrition Assistance program. The 2008 farm bill also provides new spending to increase purchases of commodities for The Emergency Food Assistance Program (TEFAP), expands the Fresh Fruit and Vegetable program, and adds funding for the Senior Farmers' Market Nutrition program (SFMNP). The bills' international food aid and trade provisions (Title III) reauthorize funding for USDA's international food aid export market development, export credits, and export guarantees, and also address barriers to U.S. agriculture exports.

Under the conservation (Title II), energy (Title IX), rural development (Title VI), and forestry titles (Title VIII), the 2008 farm bill reauthorizes, expands, and modifies many existing programs, creates new programs and initiatives, and allows some programs to expire. The bill also reauthorizes, expands, and modifies many of the existing provisions under the research title (Title VII) by requiring the reorganization of USDA's research, extension, and economic agencies.

The 2008 farm bill expands borrowing opportunities under USDA's Farm Service Agency loan program (Title V). It also creates a new farm bill title to modify crop insurance (Title XII), which provides significant savings to offset the cost of new spending in other parts of the bill. Provisions in the bill also provide ongoing disaster assistance (Title XV) and address agricultural security and animal quarantine inspections (Title XIV).

The bill includes revenue and offsetting cost provisions that are outside the jurisdiction of the agriculture committees. These provisions make certain changes to tax laws that are intended to offset additional spending in the farm bill, and were added by both chambers to comply with current pay-go budget rules (Title XV). The 2008 farm bill also includes the reauthorization of and modifications to the Commodity Futures Trading Commission (CFTC).

The report section titled "Summary of the 2008 Farm Bill Provisions" provides more detailed discussions of the major provisions in the final conference agreement and in the House and Senate versions of the farm bill.

Projected Cost

The Congressional Budget Office (CBO) estimates the total cost of the 2008 bill (i.e., baseline plus new funding, using the March 2007 baseline) at $284 billion over FY2008-FY2012 and $604 billion over FY2008-FY2017 (**Table 2**).[9] The costs discussed in this report are mandatory outlays that do not require appropriations actions. The farm bill also authorizes discretionary programs that require appropriators to allocate funds and thus are not reflected in the table.

Table 2 provides a title-by-title breakdown of CBO spending estimates for the enacted 2008 farm bill, covering both FY2008-FY2012 and FY2008-FY2017. The overwhelming share (more than 95%) of estimated total net outlays for programs included in the farm bill is anticipated to be spent on programs and activities covered by the nutrition (67%), the commodities (15%), the conservation (9%), and crop insurance titles (8%). Of the $284 billion in projected total five-year net outlays for programs under the farm bill — including revenue and cost-offset provisions in the bill — about $42 billion in projected spending will support commodity crops, $189 billion will support the cost of food stamps and commodity assistance, $24 billion will support conservation programs, and $22 billion will support crop insurance. For FY2008-FY2012, the enacted bill also includes nearly $4 billion in new spending for supplemental disaster assistance (included under Title XV). Another $10 billion is expected to be spent on trade, horticulture and livestock production, rural development, research, forestry and energy, and other programs.

Tax-related provisions — particularly from customs user fees and corporate estimated tax payments in the bill — along with cost savings from some farm bill programs, are expected to generate additional funding to offset any new spending. CBO estimates that offsets in the bill total more than $10 billion over five years (FY2008-2012) (**Table 2**), which includes tax-related provisions, and proceeds from the credit, crop insurance, and commodity program titles. Disaster assistance and programs under the nutrition and conservation titles account for the majority of the new spending in the bill.

2008 Farm Bill Implementation

Many of the program changes included in the 2008 farm bill will require action by USDA to develop regulations and guidelines. How USDA interprets these changes and follows through with modifications at the program level will likely be subject to additional scrutiny over the coming year, especially since many of these legislative changes must be made by dates specified in the farm bill.

Table 2. CBO Estimated Costs for the 2008 Conference Agreement on the Farm Bill (FY2008-FY2012) (outlays in million $)

	FY2008-FY2012			FY2008-FY2017		
	Baseline	CBO Score (change)	Total	Baseline	CBO Score (change)	Total
Commodities (Title I)	43,354	(1,726)	41,628	87,179	(1,658)	85,521
Conservation (Title II)	21,392	2,720	24,112	50,699	4,000	54,699
Trade/Food Aid (Title III)	1,823	30	1,853	3,715	(78)	3,637
Nutrition (Title IV)[a]	186,005	2,897	188,902	397,131	9,218	406,349
Credit (Title V)	(1,046)	(378)	(1,424)	(2,321)	(306)	(2,627)
Rural Development (Title VI)	72	122	194	72	149	221
Research (Title VII)	290	31	321	1,290	(907)	383
Forestry (Title VIII)	0	38	38	0	45	45
Energy (Title IX)	41	602	643	43	836	879
Horticulture/Organic (Title X)	—	402	402	—	938	938
Livestock (Title XI)	—	1	1	—	1	1

Table 2. (Continued)

	FY2008-FY2012			FY2008-FY2017		
	Baseline	CBO Score (change)	Total	Baseline	CBO Score (change)	Total
Crop Insurance (Title XII)	25,718	(3,860)	21,858	52,743	(5,591)	47,152
Commodity Futures (Title XIII)	—	0	0	—	0	0
Miscellaneous (Title XIV)[b]	6,338	44	6,382	13,668	(138)	13,530
Disaster Assistance (Title XV)	—	3,807	3,807	—	3,807	3,807
Tax/Other (Title XV)	—	(4,798)	(4,798)	—	(10,429)	(10,429)
	283,987	**(66)**	**283,921**	**604,218**	**(107)**	**604,111**

Source: Compiled by CRS using the Congressional Budget Office (CBO) March 2007 baseline and CBO score of the conference agreement for H.R. 2419, the Food, Conservation, and Energy Act of 2008; also Senate Finance Committee, *Estimated Revenue Effects of the Conference Agreement for Title XV of H.R. 2419, Fiscal Years 2008-2018*, 08-2 068 R10 (Preliminary), May 13, 2008. May not add due to rounding.

[a] New outlays for the expanded Fresh Fruit and Vegetable program required in the nutrition title, $274 million (FY2008-FY2012) and $1 .020 billion (FY2008-FY2017), are not reflected in this table because they are effectively offset with money from permanent appropriations under Section 32, mandated in Title XIV.

[b] Excludes estimates for crop insurance provisions previously included as part of the farm bill's miscellaneous provisions.

Requirements and/or regulatory guidelines will be needed to implement various newly enacted or modified program provisions. For example, these include the new revenue-based counter-cyclical program and new requirements on producer payment limits (Title I); new mandatory country-of-origin labeling for certain livestock, and poultry, and other products, as well as new state meat and seafood inspections (Title XI); and new mandatory provisions for fruit and vegetable crops and organic agriculture (Title X). The farm bill's new Supplemental Agricultural Assistance programs (Title XV) will require regulations to implement, with the biggest implementation issue involving the timeliness of payments under the crop disaster assistance portion of the programs. In addition, the farm bill's research title (Title VII) takes major step in consolidating programs and planning in USDA's agricultural research, education, and economics mission area, which will merge three existing USDA agencies.

The farm bill's nutrition title (Title IV) makes changes to several existing nutrition programs but requires little or no federal guidance as to their implementation because they are "self-executing" (e.g., change or remove a specific dollar amount specified in law), call for minimal revisions to existing rules, or are state/local options. However, some initiatives, like the fresh fruit and vegetable program, may be revisited in the 111th Congress when it takes up a scheduled reauthorization of child nutrition laws. Other programs and provisions throughout the farm bill — covering conservation, forestry, energy, and rural development, among others — will also require new requirements and regulatory guidelines to implement.

In June, 2008, USDA announced its plans for developing implementation structures and procedures for the 2008 farm bill.[10] As part of this plan, USDA began working on a timeline for various priority items and regulatory development, identifying responsible implementing agencies and also cross-cutting issues, working with the Office of Management and Budget

(OMB) on the rulemaking process, and identifying action items for immediate action, among other steps. Earlier media reports, however, have USDA asserting it might need more money beyond the $50 million that Congress provided to implement the new farm bill;[11] other reports suggest that technology needs at USDA could slow delivery of these enacted program changes.[12]

In October 2008, USDA representatives told key congressional committee staff that implementation of the 2008 farm bill would involve the development of more than 150 individual regulations.[13] An estimated 40 regulations are will be developed by USDA's Farm Service Agency (FSA); an estimated 24 regulations would be required within a short turnaround, given that Congress had specified that regulations be developed within 90 days of bill enactment. USDA representatives indicated that there could be delays in the rulemaking process because of the "massive workload" required and possible backlog in *Federal Register* publications.[14]

SUMMARY OF THE 2008 FARM BILL PROVISIONS

The following is a discussion of the major provisions in the 2008 farm bill. As with any omnibus bill covering many issues and laws, the 2008 farm bill may contain technical inconsistencies that are likely to be rectified in a later law. The effects of these inconsistencies are not reflected in the following side-by-side.

Following the discussion is a detailed side-by-side comparison summarizing the major provisions in the 2008 farm bill and comparing the enacted bill provisions with previous law and provisions in the House- and Senate-passed bills. USDA's Economic Research Service (ERS) also has published a side-by-side comparison of the 2008 and the 2002 farm bills, as enacted.[15]

Title I: Commodity Programs

Grains, Oilseeds, and Cotton Support

The enacted 2008 farm bill generally continues the farm commodity price and income support framework of the 2002 farm bill. It revises payment limitations by tightening some limits and relaxing others, and adjusts target prices and loan rates for some commodities. It continues the direct payment, counter-cyclical payment, and marketing loan programs for the 2008 through 2012 crop years. The enacted bill also creates a pilot revenue-based counter-cyclical program beginning with the 2009 crop year. It also has a pilot program for planting flexibility, restrictions on base acres developed for residential use, and a new provision that eliminates benefits to farms with fewer than 10 acres.

For direct payments, the payment rates per commodity remain the same as under the 2002 farm bill, but the overall formula contains a 2% reduction in direct payments for crop years 2009-2011. This is accomplished by changing the ratio of base acres on which direct payments are made from 85% to 83.3% of base acres. The 85% ratio is restored for the 2012 crop year to maintain baseline for the next farm bill at a higher level. The reduction to 83.3% does not affect the counter-cyclical payment formula. The final bill also adopts House and

Senate provisions that eliminate making advance direct payments in the 2012 crop year and thereafter. This provision delays payment of 22% of the direct payment amount from December to the following October, thus into a new fiscal year and allowing the farm bill to score budget savings of about $1.1 billion in FY2012. Farmers will have to wait longer, but will receive their full payment.

Support levels for counter-cyclical payments and marketing loans are adjusted with many crops receiving notable increases, and support for cotton being reduced slightly. Several commodity groups felt that their support levels were not high enough relative to other commodities in the 2002 farm bill, and did not receive counter cyclical support ever or as often (e.g., wheat, soybeans). For counter-cyclical payments, six out of 10 commodities have an increase in their target price (wheat, sorghum, barley, oats, soybeans and minor oilseeds), one has a small decrease (cotton), and four are new in 2009 (dry peas, lentils, small chickpeas, large chickpeas). For marketing loans, eight out of 20 commodities have an increase in their loan rate (wheat, barley, oats, minor oilseeds, graded wool, honey, cane sugar, beet sugar), two have a decrease (dry peas, lentils), and one is new in 2009 (large chickpeas).

The 2008 farm bill does not change the "beneficial interest" rules, and thus continues to allow farmers to lock in their loan deficiency payments (LDP) when market prices are low, continue to own the commodity, and sell it at a future and possibly higher market price. Policy makers want farmers to continue to have the flexibility to market their commodities in response to market signals and benefit from the program, but advocates for change point out that if farmers can sell their crop for more than the support price, then government support should be unnecessary. The Bush Administration had identified this as one of its priorities for commodity title reform.

For the new revenue counter-cyclical option — the Average Crop Revenue Election (ACRE) program — the final bill adopted the Senate approach, but with significant modifications. Compared to the Senate-passed bill, the ACRE program starts a year earlier in 2009 with less change to its interaction with direct payments and marketing loans. The House-passed farm bill offered a pilot revenue counter-cyclical program based on national-level revenues, while the Senate-passed bill offered a state-level revenue counter-cyclical pilot program beginning in 2010 that replaced direct payments with a "fixed payment" and offered only recourse loans.

Farmers will choose either the traditional price-triggered counter-cyclical program or the new revenue-based ACRE option. Participants in the ACRE program will continue to receive direct payments, but at a 20% reduced rate. Participants will also continue to be eligible for nonrecourse marketing loans, but with a 30% lower loan rate. To receive an ACRE payment, two triggers need to be met. First, the actual state revenue for a supported crop during the crop year must be less than the state-level revenue guarantee amount. Second, an individual farm's actual revenue for a supported crop during the crop year must be less than the farm's benchmark revenue. Benchmark yields at the state and farm levels are Olympic averages (dropping the highest and lowest price) of the most recent five years. Price guarantees are averages of the marketing year price (or the marketing loan rate reduced by 30%, if greater) for the most recent two years. If both triggers are met, an individual farm will receive an ACRE payment that is based on the state-level difference between actual revenue and the ACRE guarantee per acre multiplied by a percentage (83.3% or 85% depending on the crop year) of the farm's planted acreage, but pro-rated based on the individual farm's yield history compared to the state's yield history.

The White House has criticized the ACRE program because its two-year price guarantee feature will incorporate the historically high recent market prices into the guarantee, and consequently possibly require large payments to farmers if market prices decline from their currently record high levels in 2007 and 2008. The White House has argued that the CBO score does not reflect the magnitude of this possibility because market prices in the baseline are assumed to remain high.

The 2008 farm bill also includes a pilot planting flexibility program for fruits and vegetables for processing, while continuing the overall restriction on planting fruits and vegetables on base acreage. The pilot program begins in 2009, and allows farmers in seven midwestern states to plant base acres to cucumbers, green peas, lima beans, pumpkins, snap beans, sweet corn, and tomatoes grown for processing. Their base acres are temporarily reduced for the year (resulting in lower direct and counter-cyclical payments), but restored for the next crop year. The states include Minnesota (34,000 acres), Wisconsin (9,000 acres), Michigan (9,000 acres), Illinois (9,000 acres), Indiana (9,000 acres), Ohio (4,000 acres), and Iowa (1,000 acres).

The farm bill includes a Senate provision that eliminates base acres on land that had been subdivided into multiple residential units or other non-farming uses. Prior rules have eliminated base only for land developed for nonagricultural commercial or industrial use.

The farm bill also eliminates payments to farms with fewer than 10 base acres of all crops, except for farms owned by socially disadvantaged or limited-resource farmers and ranchers. The acreage approach is different than a House provision which set a minimum threshold of $25 per type of payment. The Senate had no similar provision.

The bill requires USDA to reconcile social security numbers of program recipients with a Social Security database twice a year to assure program beneficiaries are alive, and to issue regulations describing how long a deceased person's estate may continue to qualify for program benefits. This is less specific than the Senate provision, which specified a two-year period. This provision is in response to a 2007 GAO report showing that farm commodity payments continue to be paid to deceased farmers or their estates beyond the two-year regulation.

Payment Limits

Two types of payment limits exist. One sets the maximum amount of farm program payments that a person can receive per year. The other sets the maximum amount of income that an individual can earn and still remain eligible for program benefits (a means test).

Regarding the limit on the amount of payments, the enacted 2008 farm bill continues the $40,000 limit on direct payments and $65,000 limit on counter-cyclical payments. The counter-cyclical limit will apply to both traditional and revenue counter-cyclical payments. The final bill does not place any limit on marketing loan benefits, and thus they are now unlimited. In the 2002 farm bill, marketing loan benefits were limited to $75,000 per farm per year, but use of commodity certificates and forfeiture were unlimited, thus creating equity issues.

The 2008 farm bill still allows doubling of those limits by having a spouse, but eliminates the "three-entity rule" that formerly allowed an alternative means of doubling by having multiple farms with different ownership arrangements. Along with elimination of the three-entity rule, the conference agreement requires "direct attribution" of program benefits to a living person. If a program payment cannot be traced to a living person within four levels of ownership, the payment to the original entity owning the farm is reduced proportionately.

Under the 2002 farm bill, the limit on payments was commonly regarded as $360,000 per farm per year.[16] Given the elimination of limits on the marketing loan program, an equivalent comparison to the 2008 farm bill is difficult. The limit on direct and counter-cyclical payments continues to be $210,000 per farmer couple per year.[17]

Regarding the adjusted gross income (AGI) limit, the 2008 farm bill adopts a slightly different approach than the 2002 farm bill or the House or Senate bills. Formerly, the AGI limit had an exception if a certain proportion of AGI (e.g., 75%) was earned from farming sources. The 2008 farm bill eliminates the exception and creates two new measures of AGI: adjusted gross nonfarm income, and adjusted gross farm income.

First, if a three-year average of nonfarm AGI exceeds $500,000, then no program benefits are allowed (direct, counter-cyclical and marketing loan). Second, if a three-year average of farm AGI exceeds $750,000, then no direct payments are allowed (but counter-cyclical and marketing loan benefits are allowed for these higher-income farmers). Program participants can have income from both sources, but the caps for each type are "hard" caps (that is, there are no exceptions to the cap as with "soft" caps, except that the cap on farm AGI applies only to direct payments). For example, if a full-time farmer has nonfarm AGI over $500,000, their program payments are eliminated regardless of their farm income. A taxpayer's AGI may also be between $750,000 and $1.25 million and still receive program benefits if the income is split in such a way to remain below the caps on farm and nonfarm income. Moreover, the 2008 farm bill adopts a Senate provision that allows AGI of a married couple to be divided as if separate tax returns were filed. While in principle this provision could allow doubling of the AGI limits, the income needs to be legitimately allocated between the spouses, likely by Social Security numbers or equivalent identifiers.

For more information on the commodity programs above and payment limits, see CRS Report RL34594, *Farm Commodity Programs in the 2008 Farm Bill.*

Dairy

Two federal programs that support milk prices and dairy farm income were among the farm bill programs set to expire in 2007 — the dairy price support program (DPSP) and the Milk Income Loss Contract (MILC) program. In the past under the DPSP, USDA had been required to indirectly support the farm price of milk, most recently at $9.90 per cwt. (100 pounds), which it has done by purchasing surplus butter, nonfat dry milk, and cheese at specified minimum prices. The 2008 farm bill continues the DPSP through December 2012, but modifies the program by specifying, in the law itself, the minimum purchase prices for these manufactured dairy products. If net removals (essentially, USDA's surplus purchases) for 12 consecutive months exceed statutory limits, USDA may reduce product purchase prices, under the final law.

Under expiring law, the MILC program had paid participating farmers 34% of the difference between a target price of $16.94 per cwt. and the monthly market price for farm fluid milk in New England, when the market price is below the target. Per farm payments had been limited to the first 2.4 million lbs. of annual milk production. The enacted 2008 farm bill extends the program, but generally increases the payment factor to 45% of the price differential for the period from October 1, 2008, through August 31, 2012, as proposed by the

Senate. Conferees also increased the production limit for payments to 2.985 million pounds, compared with a Senate proposal that would have raised the limit to 4.15 million pounds. Furthermore, the $16.94 per cwt. payment rate must be adjusted to reflect feed cost increases above trigger levels specified in the enacted farm bill. CBO has estimated the total net five-year increase in outlays for the bill's key dairy provisions at $386 million.

A third ongoing federal dairy policy tool, federal milk marketing orders, require dairy processors to pay a minimum price for farm milk depending on its end use (i.e., the type of product produced). Federal orders are permanently authorized, but a number of issues were brought to the attention of Congress for the farm bill debate. Dairy processors had been seeking a change in statute to exempt them from paying the federal milk marketing order minimum price whenever they forward contract prices with dairy farmers. The 2008 farm bill authorizes farmers to voluntarily enter into forward price contracts until September 30, 2012, with none to extend beyond September 30, 2015. The legislation contains safeguards designed to ensure that dairy farmers are not compelled by processors to participate in the program. The final bill also establishes, subject to the availability of appropriations, a commission to review and evaluate federal milk marketing order policies and procedures, and in the meantime revises the formal hearing procedures used to consider amendments to the orders.

Other dairy-related provisions in the enacted bill bring Alaska, Hawaii, the District of Columbia, and Puerto Rico into the dairy research and promotion (checkoff) program; lower the promotion program's assessment rate for imported products to 7.5 cents per cwt.; extend the dairy indemnity program; and provide for new USDA directives related to dairy product price reporting. For more information, see CRS Report RL34036, *Dairy Policy and the 2008 Farm Bill.*

Sugar

The sugar program is designed to guarantee the price received by growers and processors of sugar beets and sugarcane, but at no cost to the U.S. Treasury. To accomplish this, USDA limits the amount of sugar that processors can sell domestically under "marketing allotments" and restricts imports, in order to keep market prices above support levels. This way, the incentive exists for sugar cane processors and beet refiners to repay nonrecourse price support loans[18] extended by USDA rather than hand over processed sugar as payment (commonly referred to as loan forfeitures).

To address the potential for a U.S. sugar surplus caused by unrestricted imports from Mexico under the North American Free Trade Agreement (NAFTA) and from other countries under other free trade agreements, and the resulting loan forfeitures, the 2008 farm bill mandates a sugar-for-ethanol program. USDA is now required to purchase U.S.-produced sugar roughly equal to excess imports, if necessary to maintain market prices above support levels. The sugar purchased must then be sold to bioenergy producers for processing into ethanol. USDA's Commodity Credit Corporation will provide open-ended funding for this program. Other provisions increase the raw sugar and refined beet loan rates by 4%-5% by FY201 2, mandate an 85% domestic market share for the U.S. sugar producing sector, and remove some of the discretionary authority that USDA exercises to administer import quotas. Though CBO scored some savings for the ethanol program, it projects the sugar program will cost about $650 million over five years. The Bush Administration opposed the program, arguing that instead of reform, its provisions "actually increase government intervention to drive up sugar prices." It asserted that the sugar-forethanol component will operate at a "huge loss" as excess sugar supply is auctioned off to ethanol processors.[19]

For more background, see CRS Report RL34103, *Sugar Policy and the 2008 Farm Bill.*

Title II: Conservation

The 2008 farm bill reauthorizes almost all 2002 farm bill conservation programs, modifies several programs, and creates several new conservation programs. CBO-estimated new spending on the conservation title — not including estimated conservation-related revenue and cost-offset provisions in the bill — is projected to increase by $2.7 billion over 5 years and $4.0 billion over 10 years. Total mandatory spending (new spending plus baseline) for the conservation title is projected at $24.3 billion over 5 years (FY2008-FY2012) and $55.2 billion over 10 years (FY2008-FY2017).

Conservation programs administered by USDA can be broadly grouped into land retirement programs and so-called "working lands" programs. In general, land retirement and easement programs take land out of crop production and provide for program rental payments and cost-sharing to establish longer term conservation coverage, in order to convert the land back into forests, grasslands, or wetlands. Working lands programs provide technical and financial assistance to assist farmers to improve land management practices. Major land retirement and easement programs include the Conservation Reserve Program (CRP) and the Conservation Reserve Enhancement Program (CREP), the Wetlands Reserve Program (WRP), the Grasslands Reserve Program (GRP), and the Farmland Protection Program (FPP), among other programs. Major working lands programs include the Environmental Quality Incentives Program (EQIP), the (renamed) Conservation Stewardship Program (CSP), the Agricultural Management Assistance (AMA) program, and the Wildlife Habitat Incentives Program (WHIP), among others.

Changes to existing programs address eligibility requirements, program definitions, enrollment and payment limits, contract terms, evaluation and ranking criteria, and other administrative issues. In general, the conservation title includes certain changes that expand eligibility and the delivery of technical assistance under most programs to cover more broadly, for example, forested and managed lands, pollinator habitat and protection, and identified natural resource areas, among other expansions. Producer coverage across most programs is also expanded to include beginning, limited resource, and socially disadvantaged producers; speciality crop producers; and producers transitioning to organic production. The enacted bill also creates new conservation programs to address emerging issues and priority resource areas, and also new subprograms under existing programs.

The majority of the agriculture and farmland conservation groups have responded favorably to the expanded provisions and increased funding for programs in the conservation title of the 2008 farm bill. However, a few wildlife groups have expressed concern about changes that were made to some provisions during the conference negotiations, which they perceive as providing fewer benefits for the protection of wildlife and wildlife habitat. Among the concerns expressed by these groups are the reduction in the CRP acreage enrollment cap, easing of the requirements under the so-called sodsaver provision, limitations on the types of lands eligible under WHIP, and the new permanent disaster fund, which could encourage marginal land plantings, among other concerns.

Land Retirement/Easement Programs

The largest conservation program in terms of total annual funding is the CRP. The enacted bill caps CRP enrollment at 32 million acres, down from its previous cap of 39.2 million acres. The managers report on the conference agreement states this reduction is "not ... an indicator of declining or reduced support for CRP;" however, in other sections of the report USDA is encouraged to assist producers who are transitioning from land retirement to working lands conservation. The bill makes certain program changes, including allowing for USDA to address state, regional, and national conservation initiatives; expanding the program to cover beginning and socially disadvantaged farmers/ ranchers; allowing for certain types of managed grazing and installation of wind turbines on enrolled lands (but at reduced rates); requiring that program participants manage lands according to a conservation plan; requiring USDA to survey annually the per-acre estimates of county cash rents paid to CRP contract holders; clarifying the status of alfalfa grown as part of a rotation practice; and establishing cost-sharing rates for certain types of conservation structures. The bill also modifies the pilot program that allows for wetland and buffer acreage to enroll in CRP, subject to state acreage and maximum size limitations.[20]

The bill increases the WRP maximum enrollment cap to over 3.014 million acres (up from a previous cap of 2.275 million acres), and expands eligible lands to include certain types of private and tribal wetlands, croplands, and grasslands, as well as lands that meet the habitat needs of specific wildlife species. The 2008 farm bill authorizes a new Wetlands Reserve Enhancement Program, to establish agreements with states similar to that for CREP, which includes a Reserved Rights Pilot program to explore whether reserving grazing rights is compatible with WRP. The bill makes certain program changes, including changing the payment schedule for easements; specifying criteria for ranking program applications; requiring that USDA conduct an annual survey starting with FY2008 of the Prairie Pothole Region in the northern Great Plains area; and requiring USDA to submit a report to Congress on long-term conservation easements under the program.

For GRP, the enacted bill adopts a new acreage enrollment goal of an additional 1.2 million acres by 2012, with 40% of funds for rental contracts (10-, 15-, and 20- year duration) and 60% for permanent easements. The bill modifies terms and conditions of GRP contracts and easements to permit fire pre suppression and addition of grazing-related activities, such as fencing and livestock watering. It does not include a Grassland Reserve Enhancement provision, as proposed in the House. For FPP, the bill makes several technical changes to the program, covering the program's administrative requirements, appraisal methodology, and terms and conditions, among other issues. The farm bill conferees did not to rename FPP as the "Farm and Ranchland Protection Program," as the program is often referred to by USDA. The bill provides additional budget authority for FPP of $773 million (FY2008-2012).

Working Lands Programs

EQIP and CSP are the two largest USDA working lands programs, and received additional budget authority under the 2008 farm bill. The enacted bill did not adopt a Senate proposal that would have closely coordinated CSP and EQIP under the so-called Comprehensive Stewardship Incentives Program.

For EQIP, the enacted bill expands the program to cover practices that enhance soil, surface and ground water, and air quality, and conserve energy; it also covers grazing land, forestland, wetland, and other types of land and natural resources that support wildlife. The bill sets aside 5% of EQIP spending for beginning farmers and ranchers and 5% for socially disadvantaged farmers and ranchers, providing up to 90% of the costs of implementing an EQIP plan for these farmers. It also provides payments to assist tribal or native corporation members, and producers transitioning to organic production. The bill lowers the EQIP payment limit to $300,000 (down from $450,000) in any six-year period per entity, except in cases of special environmental significance, including projects involving methane digesters, as determined by USDA. Projects with organic production benefits are capped at $20,000 annually or $80,000 in any six-year period. The enacted bill retains the requirement that 60% of funds be made available for cost-sharing to livestock producers, including incentive payments for producers who develop a comprehensive nutrient management plan.

The bill modifies EQIP's Conservation Innovation Grants program to cover air quality concerns associated with agriculture (including greenhouse gas emissions). It also replaces the Ground and Surface Water Conservation Program within EQIP with a new Agricultural Water Enhancement Program (AWEP) to address water quality and quantity concerns on agricultural land, highlighting certain priority areas and providing additional mandatory funds for the program. The bill provides additional budget authority for EQIP of $3.4 billion (FY2008-2012).

For CSP, the enacted bill replaces the Conservation Security Program with a new and renamed Conservation Stewardship Program (CSP).[21] The new CSP, beginning in 2009, will continue to encourage conservation practices on working lands, but will be different than the former program in that it eliminates the three-tier approach, removes 10-year contracts, and requires direct attribution of payments, among other changes, thus requiring that USDA promulgate new rules for the program. The bill sets a target of enrolling more than 13 million acres annually in the new program, with individual producer payments limited to $200,000 per entity in any five-year period. The types of eligible lands are expanded to include priority resource concerns, as identified by states; certain private agricultural and forested lands; and also some nonindustrial private forest lands (limited to not more than 10% of total annual acres under the program). Technical assistance will also be provided to specialty crop and organic producers, along with pilot testing of producers who engage in innovative new technologies. Supplemental payments may be available to producers who engage in certain types of crop rotations. Program payments may not be used for the design, construction, or maintenance of animal waste storage or treatment facilities or associated waste transport or transfer devices.

Among other programs, the 2008 farm bill reauthorizes WHIP at current funding levels, but limits program eligibility to focus on lands "for the development of wildlife habitat on private agricultural land, nonindustrial private forest land, and tribal lands," thus potentially excluding some previously covered areas. It also allows USDA to provide priority to projects that address issues raised by state, regional, and national conservation initiatives. The bill raises the limit on cost-share payments to 25% for long-term projects under WHIP and limits total payments to $50,000 per year. The bill also authorizes increased funding for several programs, including the Grassroots Source Water Protection Program and the Small

Watershed Rehabilitation Program; it also provides additional mandatory funding for AMA and includes Hawaii as an eligible state under that program.

New Conservation Programs

The 2008 farm bill expands the range of USDA conservation activities and creates several new programs, including a program expanding conservation activities in the Chesapeake Bay region, a new state grants program, a provision to limit production on native sod, and provisions promoting market-based approaches to conservation.

The new Chesapeake Bay Watershed Program applies to all tributaries, backwaters, and side channels, including watersheds, draining into the Chesapeake Bay, but gives priority to the Susquehanna, Shenandoah, Potomac, and Patuxent Rivers. The bill authorizes $438 million in total mandatory funding for FY2009- FY20 12. The Voluntary Public Access and Habitat Incentives Program (also referred to as the "Open Fields" program) authorizes state grants to encourage landowners to provide public access for wildlife-dependent recreation, subject to a 25% reduction for the total grant amount if the opening dates for migratory bird hunting in the state are not consistent for residents and non-residents. The bill provides $50 million in mandatory funds for FY2009-FY2012 for the program. The so-called sodsaver provision makes producers that plant an insurable crop (over 5 acres) on native sod ineligible for crop insurance and the noninsured crop disaster assistance program for the first five years of planting. The enacted 2008 farm bill states that this provision may apply to virgin prairie converted to cropland in the Prairie Pothole National Priority Area, if elected by the state. Finally, the bill includes a provision intended to facilitate the participation of farmers and landowners in emerging environmental services markets, such as water and air quality, habitat protection, and carbon storage. The farm bill directs USDA to establish a framework for developing consistent standards and processes for quantifying environmental services from the agriculture and forestry sectors, but does not authorize funding for this effort.

For more detailed information, see CRS Report RL34557, *Conservation Provisions of the 2008 Farm Bill*. For more information on individual conservation programs and past conservation funding, see CRS Report RL33556, *Soil and Water Conservation: An Overview*. For more information on conservation programs and funding, see CRS Report RL32940, *Agricultural Conservation Programs: A Scorecard*.

Title III: Trade

The 2008 farm bill reauthorizes, in Title III of the farm bill, programs that provide international food aid and that promote U.S. commercial agricultural exports. A relatively few export programs are terminated, while selected others receive increased funding.

Food Aid

The United States is the world's largest provider of food aid, accounting for about 60% of total global food aid over the last decade. The bill extends P.L. 480 food aid programs through 2012 and changes the title of the underlying act from Agricultural Trade Development and Assistance Act to Food for Peace Act. The bill also removes export market development as an objective of the programs under the statute. P.L. 480 Title II is the largest

U.S. food aid program. The bill authorizes $2.5 billion to be appropriated annually for P.L. 480 Title II, which provides U.S. commodities for emergency relief and non-emergency (development) projects overseas. Were Congress to make the full appropriation, that amount would represent a substantial increase over the average annual appropriation of $1.2 billion in recent years. Although authorized in the farm bill, P.L. 480 Title II is administered by the U.S. Agency for International Development (USAID).

The 2008 farm bill increases the amounts of P.L. 480 funds that can be allocated to various food aid program activities. It increases the amount available for administrative and distribution expenses of food aid project implementing organizations from between 5% to 10% to between 7.5% and 13% of the funds available for Title III. The bill also provides $4.5 million for FY2009-2001 to study and improve food aid quality issues. The limit on funding available for pre- positioning of commodities overseas to help expedite delivery is increased from its 2002 farm bill level of $2 million to $10 million each fiscal year. The bill also reauthorizes a program of assistance for stockpiling and rapid transportation, delivery, and distribution of shelf-stable, prepackaged foods and increases the program's funding from $3 million to $8 million each fiscal year. For monitoring and evaluation of Title II non-emergency programs, the bill provides up to $22 million annually, not more than $8 million of which could be used for USAID's Famine Early Warning System (FEWS).

Both the House- and Senate-passed versions of the farm bill had contained hard earmarks for non-emergency or development food aid. The farm bill retains an earmark for development food aid (termed a "safe box") beginning at $375 million in FY2009 and ending at $450 million in FY2012. The safe box designation can only be waived if the President determines that an extraordinary food emergency exists and that resources available from the Bill Emerson Humanitarian Trust (see below) have been exhausted, and if the President has submitted a request to Congress for additional appropriations equal to the reduction in safe box and Emerson Trust levels. Private voluntary organizations (PVOs) argued for the safe box, maintaining that it would give them assurances of supply of the commodities they rely on to carry out development projects. The Administration opposed the safe box concept, saying that it would deprive it of the flexibility needed to respond to emergency food aid needs.

The 2008 farm bill reauthorizes other smaller programs that provide food aid to countries promoting the development of market-oriented agricultural sectors (Food for Progress, FFP) or for school feeding and nutrition programs (the McGovern-Dole International School Feeding and Child Nutrition Program). USDA administers these smaller food aid programs. The bill reauthorizes FFP without lifting the cap on CCC-funded transportation of commodities (an action that the Senate farm bill had recommended), which effectively determines the volume of commodities that can be provided. For the McGovern-Dole program, the bill provides an additional $84 million of CCC funds to remain available until expended. The House-passed bill, however, had proposed changing the funding basis of McGovern-Dole from discretionary to mandatory and increasing spending from $140 million in FY2009 to $300 million in FY2012. The final enacted bill also reauthorizes the Bill Emerson Humanitarian Trust, providing a reserve of commodities and cash that can be used to provide food aid in the event of unanticipated emergency food needs.

The final bill authorizes $60 million of CCC funds to carry out a pilot program for local or regional purchase of agricultural commodities for food aid programs for FY2009-2012. Under current law, the United States can use P.L. 480 funds only to purchase U.S.

commodities. The Administration's proposal for local/regional purchase, its only farm bill food aid proposal, would have provided for up to 25% of the funds available for P.L. 480 Title II to be allocated to this purpose. In FY2007, that would have amounted to up to $447 million. Local or regional purchases, the Administration argued, would make the U.S. response to emergencies more timely and cost-effective. Opponents of the proposal, however, maintained that it would undermine the coalition of producers, shippers, and charitable organizations that support U.S. food aid and would result in less U.S. food aid being provided. Congress's rejection of the local/regional purchase proposal is one of the reasons listed by the Administration for its veto of the farm bill.

Trade

The 2008 farm bill extends USDA's export market development programs through FY20 12. Although both the House- and Senate-passed farm bills had proposed increased funding for the Market Access Program (MAP), the bill maintains funding at the FY2007 level — $200 million annually. Similarly, the bill maintains funding for the Foreign Market Development Program (FMDP) — $34.5 million annually — over the life of the bill. MAP promotes mainly high value farm exports, while FMDP promotes mainly bulk or generic commodity exports. The bill revises the export credit guarantee programs to bring them into compliance with a WTO dispute settlement decision in the U.S.-Brazil cotton case that the United States lost. Changes include elimination of the 1% cap on origination fees for export credit guarantees and repeal of legislative authority for the supplier credit program (a short-term credit guarantee) and the intermediate export credit guarantee program (3-10 years). The 2008 farm bill repeals authority for the Export Enhancement Program (EEP), a direct export subsidy. The Administration requested repeal of EEP because, it argued, the program had been inactive since 1995 and repealing it would be in line with the U.S. effort to eliminate all export subsidies in ongoing multilateral trade negotiations. Additional funding is provided for the Technical Assistance for Specialty Crops (TASC) program, which focuses on eliminating sanitary and phytosanitary (food safety) barriers to U.S. agricultural exports.

Finally, the 2008 farm bill includes a provision requiring USDA to establish a softwood lumber importer declaration program. Importers will report their lumber imports, allowing data to be collected, verified, and reconciled, to assure implementation of the U.S.-Canada Softwood Lumber Agreement. The Senate had included a sense-of-the-Senate resolution encouraging the President to ensure lumber imports consistent with that bilateral agreement.

For background information on farm bill trade and food aid programs, see CRS Report RL33553, *Agricultural Export and Food Aid Programs*; for a discussion of food aid and the farm bill, see CRS Report RL34145, *International Food Aid and the 2008 Farm Bill*; for a discussion of export programs and the farm bill, see CRS Report RL34227, *Agricultural Exports and the 2008 Farm Bill.*

Title IV: Nutrition

The farm bill's nutrition title accounts for well over half of all spending on programs and activities covered by the bill, with the overwhelming majority financing the Food Stamp program. The most significant issues in (and provisions of) this title address the administration of, eligibility for, and benefits under the Food Stamp program, funding for The Emergency Food Assistance Program (TEFAP), and support for a program of making free fresh fruits and vegetables available in schools.

The enacted 2008 farm bill includes provisions that extend expiring authorities in covered programs (generally, through FY2012) and increase spending for most programs above what would be expected under prior law (above the "baseline"). The nutrition title covers the Food Stamp program, TEFAP, the fresh fruit and vegetable program in schools, the Senior Farmers' Market Nutrition program, programs in lieu of food stamps in Puerto Rico and American Samoa and on Indian reservations, rules governing procurement of food for school meal programs, and various special nutrition projects. Under previous law, these programs would cost nearly $40 billion a year and be expected to grow to almost $50 billion in FY2017.

Total spending on these and other nutrition programs (including amounts offset by other savings) is boosted by an estimated $3.2 billion (outlays) over 5 years (FY2008-FY2012) and $10.2 billion over 10 years (FY2008-FY2017).

Supplemental Nutrition Assistance Program

The largest share of the new spending mandated by the nutrition title results from changes that increase food stamp benefits and establish new standards to make more households eligible under the Food Stamp program. The farm bill also renames the existing program as the Supplemental Nutrition Assistance Program (SNAP). Added food stamp spending is estimated to total $2.3 billion over 5 years and $7.82 billion over 10 years — 73% and 77%, respectively — of the title's total new spending. The major food stamp revisions:

- boost the minimum amount of income that is disregarded when benefits are calculated by increasing and then indexing the "standard deduction,"resulting in a small, but growing, general benefit increase in addition to regular increases for food-price inflation (around $4 a month in FY2009);
- increase and then index the minimum monthly benefit guarantee, setting at it at 8% of the indexed maximum benefit for one person (raising it from the current $10 to at least $14 in FY2009);
- disregard all income spent on dependent care when calculating benefits (removing existing caps on this disregard); and
- substantially loosen eligibility rules relating to assets by indexing the dollar limits on allowable liquid assets and disregarding all retirement savings/plans and education savings.

Other provisions (1) continue inflation-indexed funding for nutrition assistance grants (in lieu of food stamps) to Puerto Rico and American Samoa, (2) extend the authority to operate a Food Distribution Program on Indian reservations, (3) simplify some administrative processes (like reporting requirements), (4) expand the availability of "transitional" benefits for those leaving public assistance programs, (5) give the federal government a great deal more flexibility in imposing penalties on retail food stores that violate food stamp rules, (6) add disqualification penalties for those selling food bought with food stamp benefits and those using benefits to obtain cash for container deposits, and (7) require greater federal scrutiny and oversight of state efforts to "privatize" and expand the use of computers in their administration of food stamps. The overwhelming majority of food stamp provisions represent policy changes that were included in both the House and Senate bills (although budget limits forced changes), or that were non-controversial. However, a number of significant initiatives were not adopted, generally for policy rather than cost reasons. These notably included a House proposal to place major limits on state privatization of food stamp administration, and the Senate's provisions loosening eligibility rules for able-bodied adults without dependents and eventually permitting the use of food stamp benefits to buy dietary supplements.

The Emergency Food Assistance Program

The nutrition title's second-largest share of new spending is for TEFAP, with estimated additional outlays of $526 million over FY2008-FY2012 and $1.26 billion over FY2008-FY2017 (17% and 12%, respectively, of the title's total estimated cost). Closely following the provisions of both the House and Senate bills, this provision in the final law greatly increases mandatory funding of food purchases for the program to levels well above the current requirement to acquire $140 million a year. Required commodity buys are expanded by (1) an immediate infusion of $50 million in FY2008 and (2) raising annual mandatory purchases to $250 million in FY2009 (indexed annually for food- price inflation in later years).

Other Programs

Another major initiative in the nutrition title is a dramatic increase in funding for the fresh fruit and vegetable program in schools. In FY2008, approximately $20 million is available for this effort. The bill boosts mandatory outlays by $274 million (FY2008-FY2012) and $1 billion (FY2008-FY2017), representing some 10% of total new spending.

In addition to the changes in major programs noted above, the 2008 farm bill (1) includes limited authority for schools in school meal programs to use geographic preference for locally grown and raised agricultural products when procuring food, (2) increases mandatory funding for the Senior Farmers' Market Nutrition program (from $15 million a year to $20.6 million a year), (3) continues and expands support for community food projects, (4) provides money for an initiative to use the (renamed) Food Stamp program to promote health and nutrition, and (5) authorizes (and, in some cases, funds) several projects related to food distribution efforts, school gardens, "hunger-free community" initiatives, provision of whole grain products to schools, and an urban food enterprise center.

In all but a very few cases (e.g., privatization and dietary supplement provisions), there were no important policy differences between the House- and Senate-passed versions of the farm bill. However, the bills diverged greatly in the amount of new spending they proposed. For example, the FY2008-FY2012 estimated cost (outlays) of the House bill's nutrition title was $4.2 billion versus the Senate's $5.3 billion. The bills also differed in funding priorities. The House devoted 78% of new funding to new food stamp spending, 14% to extra funding for TEFAP, and 7% to expanding the fresh fruit and vegetable program. In contrast, the Senate bill provided 66% of its projected funding to new food stamp spending, along with 10% for TEFAP and 21% for the fruit and vegetable initiative. The House and Senate measures further differed in another matter. The House would have made its policy amendments part of permanent law, producing a 10-year (through FY2017) cost estimate of

$11.5 billion. On the other hand, most of the Senate's significant revisions (e.g., increased food stamp benefits) were scheduled to terminate after FY2012, resulting in a much lower 10-year cost estimate than the House or than would have been the case with permanent changes (outlays totaling $6.7 billion). Finally, the House and Senate bills provided for different extensions of expiring authorities (like the authorization of appropriations for food stamps). The House extended these authorities through FY2012, while the Senate opted for indefinite extension in most cases. The final 2008 farm bill deals with funding level issues and issues of allocation among programs as discussed earlier in this section, makes all policy changes permanent law (as in the House version), and generally extends expiring authorities through FY2012 (as in the House version).

For more information, see CRS Report RL33829, *Domestic Food Assistance: The Farm Bill and Other Legislation in the 110th Congress*.

Title V: Credit

Farm bills usually contain provisions that modify the permanent statutes for two government-related farm lenders. First, the USDA Farm Service Agency (FSA) is a federal government lender of last resort that makes direct loans or guarantees loans made by commercial lenders to farmers who cannot qualify for commercial loans. Second, the Farm Credit System (FCS) is a private lender with a statutory requirement, and limitation, to lend to farmers and certain farm-related businesses. For more information, see CRS Report RS2 1977, *Agricultural Credit: Institutions and Issues*.

Farm Service Agency

The 2008 farm bill (1) further prioritizes and subsidizes Farm Service Agency lending for beginning and socially disadvantaged farmers, (2) increases lending limits per individual to $300,000 (up from $200,000) for each of the direct farm ownership and direct operating loan programs, and (3) extends and expands the guarantee program for seller-financed land loans. It creates a conservation loan guarantee program for conservation projects. Regarding "term limits" on guaranteed operating loans, which require farmers to graduate from FSA credit to commercial lenders, the enacted 2008 farm bill extends the suspension of the enforcement of "term limits" until December 31, 2010. It also creates a pilot program of "individual development accounts" for beginning farmers and ranchers.

The Pigford Decision

The 2008 farm bill adopts a Senate provision that would permit any claimant in the Pigford decision (a 1999 suit based on past discrimination against minority farmers applying for USDA loans) who has not received compensation to petition in civil court to obtain such compensation. The total amount of payment and debt relief would be limited to $100 million. USDA would be restricted from beginning a foreclosure if the borrower can show foreclosure is related to a Pigford claim. A similar provision is also included in the House-passed bill. See CRS Report RS20430, *The Pigford Case: USDA Settlement of a Discrimination Suit by Black Farmers*.

Farm Credit System

In recent years, FCS has sought to expand its lending authority beyond traditional farm loans and into more rural housing and non-farm businesses. Commercial banks oppose expanding FCS lending authority, saying that the availability of commercial credit in rural areas is not constrained, and that FCS's government-sponsored enterprise (GSE) status provides an unfair competitive advantage. The enacted bill, like the House and Senate bills, does not allow any expansion of Farm Credit System lending authority. It does address technical changes in the payment of insurance premiums by FCS banks to the FCS Insurance Corporation, and expands the list of borrowers eligible to own Bank for Cooperatives voting stock. For more information, see CRS Report RS2 1977, *Agricultural Credit: Institutions and Issues*.

Title VI: Rural Development

More than 88 programs administered by 16 different federal agencies target rural economic development. The Rural Development Policy Act of 1980 (P.L. 96-355) named USDA the lead federal agency for rural development. USDA administers most of the existing rural development programs and has the highest average of program funds going directly to rural counties (approximately 50%). Three mission agencies, Rural Housing Service, Rural Business-Cooperative Service, and Rural Utilities Service, administer the various loan and grant programs. More information on these programs is in CRS Report RL3 1837, *An Overview of USDA Rural Development Programs*.

The enacted 2008 farm bill reauthorizes and/or amends rural development loan and grant programs and authorizes several new provisions. The bill adopts the Senate measure to redefine "rural" with certain modifications, most notably, striking the housing density criterion; however, it also directs USDA to conduct rulemaking to develop restrictions on areas where housing density is greater than 200 units per square mile. The bill does not change current law with respect to rural eligibility for water and waste water disposal loans and grants and the community facility program. The bill requires USDA to prepare a report assessing the various definitions of "rural" and the effect these various definitions have on programs administered by USDA Rural Development.

Although both the House and Senate farm bills included mandatory funding for several programs, the enacted 2008 farm bill reduced that spending, while in some cases adding discretionary authorization. The enacted bill provides \$194 million (FY2008-FY2012) in mandatory spending for rural development programs. This is a reduction from up to \$550 million proposed in the House- and Senate-passed bills. The bill provides mandatory spending for a one-time funding of backlogged water and wastewater applications (\$120 million); a Rural Microentrepreneur Assistance program (\$13 million in mandatory and \$40 million annually in discretionary spending); and Value-Added Product Grants (\$15 million in mandatory spending). The bill provides no mandatory funding for rural health care facilities or the construction of child care facilities, although such funds were proposed by the Senate.

Other provisions authorized in the enacted bill include support for locally produced agricultural food products[22] and grants for assisting employment opportunities for disabled individuals in rural areas. The enacted bill also establishes a new Rural Collaborative

Investment Program and provides support for several water and wastewater programs, and adopts the Senate proposed measure to authorize a new interest rate structure for water and wastewater projects based on an index of outstanding municipal obligations. The bill also adopts other Senate provisions that authorize assistance to the Housing Assistance Council and reauthorize the Rural Business Investment Program, as well as the appropriate technology transfer to rural areas program (ATTRA). The bill deletes the House measure reauthorizing the Rural Strategic Investment Program, and deletes both House and Senate proposals providing guaranteed loans and grants to improve rural health care facilities.

The enacted bill adopted several House and Senate provisions to assist rural broadband development. The bill reauthorizes the Access to Broadband Telecommunications Services in Rural Areas and grants to broadcasting systems; directs the Secretary to develop a comprehensive national broadband strategy for rural areas; and authorizes a new National Center for Telecommunications Assessment. The bill also reauthorizes the Distance Learning and Telemedicine program and includes a provision to make library connectivity a feature of the program. However, the farm bill deleted various House- and Senate-passed proposals and grant programs targeting broadband service.

The 2008 farm bill reauthorizes several rural development and grant programs, including the Rural Economic Area Partnership; the National Rural Development Partnership; SEARCH grants; and Rural Business Opportunity Grants. The bill also reauthorizes the Delta Regional Authority and the Northern Great Plains Regional Authority. In addition to these two reauthorized commissions, the bill creates three new regional development authorities: the Southeast Crescent Regional Commission, the Southwest Border Regional Commission, and the Northern Borders Economic Development Commission (included under the Title XIV, Miscellaneous). The bill also includes a provision directing USDA to conduct studies on rural transportation issues and on rural electric power generation.

For more information, see CRS Report RL34126, *Rural Development and Provisions of the 2008 Farm Bill.*

Title VII: Agricultural Research

Under the mission area called Research, Extension, and Economics (REE), the USDA is responsible for conducting agricultural research at the federal level, and for providing partial support for cooperative research, extension, and post-secondary agricultural education programs in the states. The USDA's intramural activities are carried out by the Agricultural Research Service (ARS), Economic Research Service (ERS), National Agricultural Statistics Service (NASS), and National Agriculture Library (NAL). The federally funded extramural activities are managed by the Cooperative State Research, Education, and Extension Service (CSREES). For more information on these agencies' activities, see CRS Report RL33327, *Agricultural Research, Education, and Extension: Issues and Background.*

The issues confronting Congress concerning federal agricultural research can be generally categorized under two topics: the structure of the management organization and the level of research funding, both of which are long-standing issues. Congress addressed the management issue in the 2002 farm bill by directing USDA to examine and report on the structure of Agricultural Research Service (ARS) management and the merits of establishing a National Institute of Food and Agriculture (possibly modeled after the National Institutes of Health). With respect to funding, there has long been a struggle under persistent budget constraints to obtain increased appropriations even sufficient to keep up with inflation. With farm commodity support as a model, the research community has attempted to obtain a portion of its money in mandatory funds, with less reliance on discretionary appropriations.

The USDA task force report, *National Institute for Food and Agriculture: A Proposal*, was issued July 2004.[23] The proposal was presented to Congress in USDA's *2007 Farm Bill Proposals*.[24] While the USDA task force was conducting its review, the National Association of State Universities and Land-Grant Colleges (NASULGC) developed a proposal called *Create Research, Extension, and Teaching Excellence for the 21st Century* (CREATE-21).[25] CREATE-21 was presented to Congress as H.R. 2398 and S. 1094.

The research provisions in the individual House and Senate farm bills drew heavily on the recommendations of the USDA and NASULGC.[26]

Research Management

The enacted 2008 farm bill represents an amalgam of the research reorganization provisions in the House and Senate bills. The bill creates an umbrella coordinating entity known as the Research, Extension, and Education Office (REEO) in Office of the Under Secretary for Research, Education, and Economics, and designates the Under Secretary as the Chief Scientist of USDA. The new REEO will contain six divisions, each with its own director, representing the broad range of subject areas addressed by agricultural research, extension, and education programs.

The division directors are expected to work with the National Agricultural Research, Extension, Education, and Economics Advisory Board to coordinate all of the mission area's activities across the Department, including intramural research (ARS, ERS, NASS) and extramural research. CSREES, which currently is responsible for managing extramural research, will be eliminated as an agency and will become the National Institute of Food and Agriculture (NIFA). The conferees intend NIFA to be an independent, scientific, policy-setting agency for the food and agricultural sciences whose primary role is to administer competitive grants.

The enacted bill ends the National Research Initiative (NRI) and the Initiative for Future Agriculture and Food Systems (IFAFS) as distinct competitive grant programs, and establishes within NIFA an Agriculture and Food Research Initiative (AFRI) to award competitive grants for fundamental and applied research, extension, and education. The farm bill authorizes annual appropriations of $700 million for AFRI, representing the combined level of authorized and mandatory funding that the NRI and IFAFS were authorized to receive in previous years (appropriators have prohibited the use of mandatory funds for IFAFS since 2002). The Under Secretary (chief scientist) is required to submit a unified annual budget covering all activities of the REEO and NIFA. The budget is to represent the balance of several factors, including fundamental and applied research, funding for research capacity and infrastructure, and increased support for Hispanic-serving agricultural colleges and universities, for non-land grant colleges of agriculture, and for the University of the District of Columbia.

Funding

Apart from mandatory funding of $230 million over five years for a Specialty Crop Research Initiative, and $78 million in mandatory funding for the Organic Research and Extension Initiative (included in the Horticulture and Organic Agriculture title), the 2008 farm bill authorizes annual appropriation of such sums as necessary for research, extension, and education programs, much the same as in the previous farm bill. The House bill would have preserved mandatory funding of $200 million for competitive grants under a merged NRI/IFAFS program for FY2010-FY2012.

For more information, see CRS Report RL34352, *Agricultural Research, Extension, and Education: Farm Bill Issues*.

Title VIII: Forestry

Farm bills typically deal with forestry both directly (usually in a forestry title) and indirectly (for example, by including forests and forestry practices in more general conservation programs). For a description of existing programs, see CRS Report RL3 1065, *Forestry Assistance Programs*.

The enacted 2008 farm bill includes a forestry title (Title VIII) with several sections addressing statewide forest resource planning. One section establishes "national private forest conservation priorities" as (1) conserving and managing working forest landscapes for multiple values and uses; (2) protecting forests from threats and restoring appropriate forest types; and (3) enhancing public benefits from private forests. Other sections require statewide assessments and strategies for forest resources (with periodic revision). The bill creates a new Forest Resource Coordinating Committee, requires the competitive allocation of a portion of state assistance funding (based on how the statewide assessments and strategies fulfill the national priorities), and allows up to 5% of state assistance funding for competitively allocated innovative projects to address the national priorities. The bill also creates a new community forest and open space conservation grant program for local entities to protect forests threatened with conversion to non-forest uses, and creates an Emergency Forest Restoration Program to provide assistance for restoration efforts for forests damaged by natural disaster.

The 2008 farm bill extends, through 2012, the authorizations for the Office of International Forestry, the Rural Revitalization Technologies Program, the Renewable Resources Extension Act, and the Healthy Forest Reserves Program (with minor changes) under the Healthy Forests Restoration Act of 2003 (P.L. 108-148, 16 U.S.C. Sec. 501, et seq.). The bill also amends existing law to restrict imports of illegally logged wood. A separate subtitle — Cultural and Heritage Cooperation Authority — provides for tribal-Forest Service cooperative relations and assistance. The bill authorizes a competitive grant program to Hispanic-serving institutions to increase diversity in forestry and related fields, and allows contract modification options for certain Forest Service timber sales.

The enacted bill includes forestry-related provisions other than in the forestry title. In Title II (Conservation), numerous programs were modified to include forestry among approved conservation activities. In Title III (Trade), it requires USDA to required to establish a softwood lumber importer declaration program, for import data to be collected, verified, and reconciled to assure implementation of the U.S.-Canada Softwood Lumber Agreement. In Title IX (Energy), two programs were created to use woody biomass for energy production. In Title XV (Tax Provisions), the agreement authorizes a new type of tax-exempt private bond whose proceeds are used to finance forest conservation. It also modifies income tax deductions for qualified timber gains, and includes several provisions to modernize and clarify the tax treatment of timber real estate investment trusts (REITs).

See also CRS Report RL33917, *Forestry in the 2008 Farm Bill.*

Title IX: Energy

Interest in renewable energy has grown rapidly since late 2005 due, in large part, to a strong rise in domestic and international fuel prices and a dramatic acceleration in domestic biofuel production (mostly ethanol). Many policymakers view agriculture-based biofuels as both a catalyst for rural economic development and a response to growing energy import dependence. Renewable energy's role in the 2002 farm bill was contained in the farm bill's energy title (Title IX), which concentrated on grants, loan, and loan guarantees to foster research on agriculture-based renewable energy, to share development risk, and to promote the adoption of renewable energy systems. USDA's Bioenergy Program (Sec. 9006 of P.L. 107-171) — funding for which expired in FY2006 — had been the primary exception in that it provided incentives to expand actual production of bioenergy.

The enacted 2008 farm bill expands and extends the provisions in the energy section of the 2002 farm bill, and provides additional funding. The bill makes numerous changes to the programs in the energy title. For example, the bill combines the so-called Section 9006 program with the Energy Audit and Renewable Energy Development Program under the new Renewable Energy for America Program. The bill also creates new programs, including a Biomass Crop Assistance Program to provide financial assistance to producers for growing biomass crops and developing conversion facilities, and the Agricultural Bioenergy Feedstock and Energy Efficiency Research and Extension Initiative to provide for competitive grants to fund projects with a focus on supporting on-farm biomass crop research and extension. This latter initiative is under the bill's research title (Title VII) and includes other bioenergy research programs.

The enacted bill continues programs for federal purchase of biobased products under the Biobased Markets Program. The bill includes a Senate-proposed provision, Biorefinery Assistance, which provides grants and loan guarantees for construction and retrofitting of biorefineries for the production of advanced biofuels. The bill also provides for grants for constructing demonstration-scale biorefineries, and loan guarantees for the development and construction of commercial-scale biorefineries that use technologies that are either pre-commercial or commercially available. The bill provides for the repowering of existing biorefineries. It incorporates the Biomass Research and Development Act of 2000 as part of the bill's energy title and will fund projects that address the critical need for integrated research and technology development in the area of biofuels. It continues the Biodiesel Fuel Education Program with an expanded list of entities targeted. Also, among the miscellaneous provisions (Title XV), the ethanol production tax credit is lowered from 51 to 46 cents per gallon beginning in the first year following that in which ethanol production of 7.5 billion gallons is achieved. Finally, the 2008 farm bill establishes the cellulosic biofuel producer credit of $1.01 per gallon with special provisions for small cellulosic ethanol producers.

Mandatory spending for the enacted bill's agriculture-based energy programs are projected at about $600 million (FY2008-FY2012) and $900 million (FY2008- FY2017). This reflects a reduction compared to funding levels proposed in the House- and Senate-passed bills. The House bill had authorized more than $3 billion in new mandatory funding and more than $1 billion in discretionary funding for provisions of the energy title; the Senate bill authorized more than $1 billion in new mandatory funding and more than $2 billion in discretionary funding. Both the House and the Senate sought to fund these provisions through various revenue and cost offset provisions in both bills, although in very different ways.

For more information on agriculture and bioenergy, see CRS Report RL34130, *Renewable Energy Policy in the 2008 Farm Bill*; CRS Report RL32712, *Agriculture-based Renewable Energy Production*; CRS Report RL33290, *Fuel Ethanol: Background and Public Policy Issues*; and CRS Report RL33572, *Biofuels Incentives: A Summary of Federal Programs*.

Many of the federal programs that currently support renewable energy production in general, and agriculture-based energy production in particular, are outside the purview of USDA and have legislative origins outside of the farm bill. For example, the energy act signed into law in December 2007 (P.L. 110-140) covers a wide range of topics with extensive attention to biofuels. In particular, it includes a dramatic expansion of the renewable fuels mandate to 36 billion gallons by 2022 with carve-outs for biodiesel (1 billion gallons by 2012), cellulosic ethanol (16 billion gallons by 2022), and corn-starch ethanol (15 billion gallons by 2015). Legislative proposals focused on renewable energy are summarized in CRS Report RL33831, *Energy Efficiency and Renewable Energy Legislation in the 110th Congress*.

Title X: Horticulture and Organic Agriculture

Sales of specialty crops, such as fruits, vegetables, and tree nuts, account for nearly one-third of U.S. crop cash receipts and one-fifth of U.S. agricultural exports, according to USDA. When floriculture, greenhouse, and nursery crop sales are included, total specialty crops account for nearly 50% of all U.S. farm crop cash receipts. However, specialty crop producers are not eligible for commodity income support programs; also, few provisions in the farm bill's 2002 conservation, trade, research, and nutrition titles specifically addressed the specialty crop industry. For more information, see CRS Report RL33520, *Specialty Crops: 2008 Farm Bill Issues*.

The enacted 2008 farm bill contains a new Horticulture and Organic Agriculture title (Title X). Among the key provisions is reauthorization of the specialty crop block grant program established by the Specialty Crops Competitiveness Act of 2004 (P.L. 108-465). Under this program, each state receives a grant to support marketing research and promotion to enhance the competitiveness of specialty crops grown in the state. The 2004 act authorized appropriations to support the program (it received $7 million in each of FY2006-FY2008). The enacted 2008 farm bill provides mandatory funding in the amounts of $10 million for FY2008, $49 million for FY2009, and $55 million annually in FY2010 through FY2012.

Another key provision affecting specialty crops includes mandatory funding of $207 million through FY2012 to (1) establish cooperative agreements with state departments of agriculture for early plant pest detection activities; (2) establish a threat identification and mitigation program for foreign pests and diseases; and (3) provide funds and technical assistance to specialty crop producers to develop audit-based certification systems to lessen the risks of pest emergence and movement. The 2008 farm bill also provides $20 million over four years to establish centers where the specialty crop industry can obtain pest- and disease-free plant source material, and authorizes appropriations to establish a Pest and Disease Revolving Loan Fund to help local governments purchase equipment for the speedy removal of trees that must be destroyed to stop the spread of a pest or disease infestation.

The enacted bill also contains several provisions to encourage the consumption of fresh fruits and vegetables. One provides $33 million through FY2012 to expand the existing Farmers' Market Promotion Program, with a requirement that 10% of the funds be used to make it possible for beneficiaries of federal nutrition programs to use their electronic benefits cards at farmers' markets. Another provides $5 million to establish a Healthy Urban Food Enterprise Development Center designed to help make affordable, nutritious fresh foods more readily available in low-income communities and neighborhoods. Finally, the bill contains provisions to increase the amount of fresh fruits and vegetables to be purchased for USDA nutrition programs (see Title IV, Nutrition programs).

Another major component of Title X of the 2008 farm bill includes provisions supporting organic agriculture. The bill reauthorizes the National Organic Cost-share Program[27] and provides a one-time transfer of $22 million in FY2008 (available until expended) to help defray farmers' costs for obtaining certification under the National Organic Program; the amount that an individual farmer can receive in cost-share assistance is raised from $500 to $750. The conference report also includes $5 million in mandatory funding for data collection on organically grown crops, authorizes the appropriation of an additional $25 million over five years for that purpose, and requires USDA to spend $3.5 million of the total to collect and distribute up-to-date price data for the organic market. Another provision authorizes increased appropriations to support the administrative work of the National Organic Program office. Title II (Conservation) contains a provision making producers who want to convert from conventional to organic farming eligible for cost-share and technical assistance under the Environmental Quality Incentives Program.

See also CRS Report RL3 1595, *Organic Agriculture in the United States: Program and Policy Issues*.

Title XI: Livestock

Competition and Marketing

Rapid changes have occurred in recent decades in the structure and business methods of agriculture in general and of animal agriculture in particular. Production and marketing have been moving toward fewer and larger operations (sometimes referred to as consolidation or concentration), and toward vertical integration, although the pace of these changes has varied widely across the sectors. Debate has revolved around the impact of such changes on farm prices, on the traditional system of independent, family-based agriculture, on consumers, and on global competitiveness. Inherent in these questions is the role government should play in monitoring and regulating agricultural markets. For more information see CRS Report RL3 3958, *Animal Agriculture: 2008 Farm Bill Issues.*

The enacted 2008 farm bill contains a new title on Livestock (Title XI) that scales back much of the language in the Senate-passed bill aimed at more closely regulating livestock and poultry markets. For example, conferees deleted Senate language that would have prohibited most major packers from owning, feeding, or controlling livestock except within 14 days of slaughter. Also deleted was a Senate provision to establish at USDA a new Special Counsel for Agricultural Competition to investigate and prosecute violations of competition laws. Title XI of the enacted 2008 farm bill changes the Agricultural Fair Practices Act to alter the definitions of associations and handlers. Not included in the enacted bill were various Senate provisions intended to strengthen USDA's oversight and enforcement of the act. Also not included in the enacted bill were Senate provisions to give USDA stronger enforcement authorities over live poultry dealers, among other provisions, under the Packers and Stockyards Act (P&SA), which governs market competition in the meat packing sectors. In their place, conferees added language requiring an annual report detailing investigations into possible violations of the P&SA.

Also narrowed was Senate language governing contractual arrangements between producers and integrators. Under the final farm bill, a poultry or swine grower — a more limited definition of a contract producer than in the original Senate bill — has the right to cancel a production contract within 3 business days of execution, unless a later date is specified in the contract. In lieu of Senate language limiting the conditions under which a contractor could require a producer to make additional capital investments, the 2008 farm bill stipulates that the possibility of such an investment be conspicuously stated in the contract. Several other provisions retained, in somewhat modified form, in the enacted bill are intended to give producers additional protections when disputing contract terms.

The enacted bill contains provisions intended to improve electronic reporting under the Livestock Mandatory Price Reporting program administered by USDA's Agricultural Marketing Service (AMS), and to study the effects of requiring pork processing plants to report wholesale pork price information.

Country-of-Origin Labeling

The 2002 farm bill (Sec. 10816 of P.L. 107-171) required retailers to provide country-of-origin labeling (COOL) for fresh produce, red meats, peanuts, and seafood by September 30, 2004. Congress twice postponed implementation for all but seafood; COOL now must be implemented by September 30, 2008. Exempted from COOL are COOL-processed versions of these products, and dining-out establishments. There has been vigorous debate over whether this new program is desirable and necessary, its purposes, and its likely impacts on farmers, processors, retailers, and consumers. Opponents of mandatory COOL prefer a voluntary or market-driven program or at least some relaxation of the COOL law's compliance language. Supporters have continued to seek Congress's and USDA's assurance that the mandatory program will be implemented expeditiously. For a more detailed description of current law, requirements and issues see CRS Report RS22955, *Country-of-Origin Labeling for Foods.*

The 2008 farm bill implements the mandatory program on its current schedule, and adds goat meat, chicken (which competes with red meats in the market and which, unlike red meats, primarily are domestically produced), ginseng, pecans, and macadamia nuts as covered commodities. However, for red meats, it creates several new types of label categories that are intended to facilitate and simplify compliance in specifying the country or countries of the products. For all covered commodities, the bill also seeks to ease recordkeeping and verification requirements, and lower noncompliance penalties.

Inspection, Registries, and Grading

The 2008 farm bill includes provisions covering state-inspected meat and poultry, reportable meat and poultry registries, and catfish grading and inspection, among other provisions.

Federal law has prohibited state-inspected meat and poultry plants from shipping their products across state lines, a ban that many states and small plants have long sought to overturn. Limiting state-inspected products to intrastate commerce is unfair, many state agencies and state-inspected plants have argued, because the 27 current state-operated programs by law already must be, and are, "at least equal" to the federal system. Those who have opposed allowing state-inspected products in interstate commerce argued that state programs are not required to have, and do not have, the same level of safety oversight as the federal, or even the foreign, plants. Both the House and Senate farm bills contained language to enable state-inspected plants to sell products in interstate commerce, but under divergent policy approaches.

Conferees adopted the Senate's version, whereby state-inspected plants with 25 employees or fewer may opt into a new program that subjects them to federal laws and oversight, for which they may gain the federal mark of inspection and the ability to ship interstate. They would still be inspected by state employees, but these employees would be under the supervision of a federal official who will oversee training, inspection, compliance, and other activities. States would receive at least 60% reimbursement of their costs (compared with 50% under the existing federal-state program provisions, which also continues). The Senate language was a compromise package acceptable to both supporters and opponents of the House-passed language, which among other things could have enabled many plants currently under federal inspection to apply for state inspection and continue to ship interstate. Opponents of the House option feared that many would seek to leave the federal system if they believed they could receive more lenient oversight by the states. (For background, see CRS Report RL34202, *State-Inspected Meat and Poultry: Issues for Congress.*)

Conferees modified a provision in the Senate but not the House bill to require USDA to establish "reportable food registries" for meat and poultry and their products, whereby establishments would have to report whenever there is a probability of such foods causing adverse health consequences. (The FDA amendments legislation passed in 2007 (P.L. 110-85) establishes a similar registry for FDA-regulated foods.) The enacted farm bill amends the meat and poultry laws to require an establishment to promptly notify USDA if it has reason to believe that an adulterated or misbranded product has entered commerce. Another adopted provision requires meat and poultry establishments to prepare and maintain written recall plans.

Conferees modified Senate bill language to provide for two new USDA initiatives affecting farm-raised domestic catfish: a voluntary grading program administered through the Agricultural Marketing Service (AMS), and mandated safety inspection of such products by Food Safety and Inspection Service (i.e., making catfish an amenable species along with the major meat and poultry species). The House bill lacked this language. The final version provides for catfish grading as a voluntary fee-based program, with producers of other seafood species eligible to petition USDA for a similar service. Conferees agreed to extend mandatory inspection to catfish processors, further authorizing FSIS to take into account the conditions under which catfish are raised and transported. Although other fish and shellfish are not covered by the final amendment, conferees noted in their accompanying report that the Secretary of Agriculture has standing authority to add species if appropriate.

The conference report states the intent of Congress "that catfish be subject to continuous inspection and that imported catfish inspection programs be found to be equivalent under USDA regulations before foreign catfish may be imported into the United States." Language in the bill itself instructs the Secretary to define the term "catfish." However, the 2002 farm bill (P.L. 107-171, Sec. 10806) had amended the Federal Food, Drug, and Cosmetic Act to limit the acceptable definition to one family of catfish ("Ictaluridae"), effectively prohibiting the labeling of certain Asian-grown fish as catfish. So, the scope of the regulatory definition developed by the Secretary of Agriculture could be of some interest.

Other Provisions

Conferees deleted two provisions in the Senate bill. One provision would have established a Congressional Bipartisan Food Safety Commission that would have been required to report, within one year, on recommendations for modernizing food programs. The provision was intended to be in response to recent food safety incidents linked to both imported and domestic foods, which have brought into focus the question of whether there is a need for changes in federal food safety oversight. At issue is whether the current system has the statutory authorities, resources, and structural organization to protect consumers from unsafe food. The second deleted provision would have prohibited FDA from issuing a final risk assessment or from lifting the voluntary moratorium until completion of newly mandated studies on the safety and market impacts of introducing products from cloned animals. The provision was intended to be in response to FDA's request issued in late 2008 that companies refrain voluntarily from marketing meat and milk from cloned animals or their progeny until it can complete a final assessment of their safety (see CRS Report RL33334, *Biotechnology in Animal Agriculture: Status and Current Issues*).

See also CRS Report RL33958, *Animal Agriculture: 2008 Farm Bill Issues*.

Title XII: Crop Insurance and Disaster Assistance Programs

The enacted 2008 farm bill contains a new title covering crop insurance and disaster assistance programs (Title X). In addition, the enacted bill contains other provisions authorizing "Supplemental Agricultural Disaster Assistance," provided for under the bill's Title XV (Trade and Tax Provisions).

Crop Insurance Program

The federal crop insurance program is designed to protect crop producers from unavoidable risks associated with adverse weather, weather-related plant diseases, and insect infestations. Although the scope of the crop insurance program has widened significantly over the past 25 years, the anticipated goal that it would replace *ad hoc* disaster payments has not been achieved.

The crop insurance program is permanently authorized and hence does not require consideration in the farm bill. Some policymakers expressed interest in expanding the crop insurance program in the context of the farm bill and/or complementing it with a permanent disaster payment program. However, many viewed the crop insurance program as a potential target for program cost reductions, where savings could be used to fund new initiatives in various titles of the farm bill. The Administration and others contend that the private companies should be required to absorb more of the program losses, and that the reimbursement rate for their operating expenses needs to be reduced as a means of reducing federal costs. The insurance companies and many farm groups are concerned that significant reductions in federal support will negatively impact the financial health of the crop insurance industry and possibly jeopardize the delivery of crop insurance, particularly in high-risk areas.

Like the House- and Senate passed farm bills, the enacted 2008 farm bill contains several revisions to the crop insurance program, many of which are designed to reduce program costs. For the enacted crop insurance title, CBO has estimated net savings of $3.9 billion over five years (FY2008-2012), compared with estimated savings of $4.0 billion in the House bill and $3.7 billion in the Senate bill. Approximately $2.8 billion of the estimated savings in the enacted bill (as in the House and Senate bills) is achieved through changes in the timing of premium receipts from farmers, and payments to the companies, which has no effect on overall subsidies to participating farmers or insurance companies. A portion of the five-year savings is realized by requiring insurance companies and farmers to share more in program costs. The enacted bill increases the administrative fees paid by farmers for catastrophic crop insurance coverage (and for participation in the separate noninsured assistance program) to new levels that are higher than both the House and Senate bills. The bill also reduces reimbursement rates to private companies for their administrative and operating expenses by 2.3 percentage points. Conferees did not include a House provision that would have required the insurance companies to share more of their underwriting gains with the federal government.

Among its other provisions, the crop insurance title of the 2008 farm bill also (1) requires USDA to ensure that premiums are established at a level so that total premiums equal total indemnity payments over time; (2) allows USDA to periodically renegotiate its standard reinsurance agreement, which contains the obligations and financial terms of the relationship between the government and the participating private crop insurance companies; (3) reduces available mandatory funding for reimbursing private initiatives for the research and development of new crop insurance products, and revises the manner in which reimbursements are provided; and (4) provides $36 million in mandatory funding over 10 years for USDA to enhance its activities to reduce waste, fraud, and abuse within the crop insurance program.

For more background information on crop insurance, see CRS Report RL34207, *Crop Insurance and Disaster Assistance in the 2008 Farm Bill.*

Other Disaster Assistance Programs

Title XII includes other disaster assistance provisions, including the addition of the Small Business Disaster Response and Loan Improvements Act of 2008 (Subtitle B). This subtitle makes significant changes to the Small Business Administration's (SBA) response to disaster.

SBA has responsibility for making disaster loans to individuals and non-agricultural businesses (15 U.S.C. 636(b)). Hurricane Katrina, and the SBA response, which was widely viewed as slow and flawed, prompted may suggestions from Members and others on how to improve the disaster loan system. In general, this act requires SBA to implement new planning, management oversight, and reporting procedures and authorizes private lending to supplement the disaster loans that SBA makes directly.

SBA will continue to make six types of disaster loans for damages that were not covered by insurance:

- Personal property disaster loans (maximum $40,000) are made to individuals to replace or repair personal property such as cars and furniture. Home owners and renters may apply.
- Real property disaster loans (maximum $200,000) are made to home owners to repair or replace primary residences.
- Physical disaster loans (maximum $1.5 million) are made to businesses and nonprofits, regardless of size, to repair or replace real property, inventory, machinery, equipment, fixtures, etc. not covered by insurance.
- Economic injury disaster loans (maximum $1.5 million) are made to businesses to provide operating funds following a disaster. The act makes nonprofits eligible for these economic injury disaster loans and adds ice storms and blizzards to the list of disasters that can trigger these loans.
- Military reservist economic injury disaster loans (maximum $1.5 million) are made to businesses to help them overcome problems caused by a key employee being called to active military duty.
- GO Loans (Gulf Opportunity Pilot Loans, maximum $150,000), which were made to certain small businesses affected by Hurricane Katrina.

The act provides for SBA-guaranteed private disaster loans to supplement the current system of direct SBA disaster loans. Lenders in the SBA's Preferred Lender Program would be eligible to participate. The SBA would guarantee 85% of loans, subject to a maximum loan size of $2 million. Following a major disaster, the SBA can also hire private contractors to assist in processing disaster loan applications.

In the event of a major disaster with extraordinary levels of damage or disruption (such as Hurricane Katrina), SBA may make economic injury disaster loans to small businesses affected by the disaster regardless of location, but priority may be given to small businesses within the disaster area. Usually, only businesses within a disaster area are eligible.

The enacted farm bill also requires SBA to create an Immediate Disaster Assistance Program, providing loans to small businesses affected by a disaster with a maximum amount of $25,000 and an 85% SBA guarantee. SBA may defer payments on existing disaster loans for up to four years for those taking out new disaster loans. SBA is also required to assign at least 800 employees to the Office of Disaster Assistance and 1,000 employees to the Disaster Cadre. In FY2007, the SBA had 2,300 employees working on disaster loans; the Administration requested 921 employees for these functions for FY2009.

Title XIII: Commodity Futures

Title XIII of the enacted 2008 farm bill includes provisions that reauthorize appropriations for the Commodity Futures Trading Commission (CFTC) for FY2008-FY2013. The bill also makes several amendments to the Commodity Exchange Act to (1) clarify CFTC jurisdiction over retail financial contracts based on foreign currencies, (2) make the CFTC's anti-fraud authority applicable to certain off-exchange or over-the-counter derivatives contracts, (3) increase civil monetary and criminal penalties for violations, (4) permit cross-margining of accounts in security futures and options, and (5) establish CFTC regulation over certain exchange-like trading facilities that are currently exempt from most regulation.

The last section is the most controversial, and deals with an issue — sometimes referred to as the "Enron loophole" — that Congress has addressed several times since 2000.[28] These provisions of the bill would apply to electronic markets, other than regulated futures exchanges, where contracts based on energy commodities, metals, and other non-agricultural and non-financial commodities are traded. Under current law, such markets (or electronic trading facilities) are required to notify the CFTC of their operations, but are generally exempt from substantive regulation provided that small public investors are not permitted to trade there.

Under the enacted bill, if the CFTC determined (according to criteria set forth in the bill) that such an electronic market played a significant role in the price-setting process (that is, if market participants looked to prices generated there as a guide to their own transactions in the underlying commodities), the market would become subject to a set of regulatory "core principles." These principles amount to a regulatory regime roughly comparable to (but somewhat less extensive than) CFTC regulation of futures exchanges. Markets where "significant price discovery" contracts are traded will be required to prevent price manipulation, monitor trading, report daily figures on price and trade volume, guard against conflicts of interest, and disclose large positions held by individual traders. Failure to comply with these principles would give the CFTC grounds to suspend or revoke the market's registration.

Title XIV: Miscellaneous

The miscellaneous provisions in the 2008 farm bill cover various provisions that are discussed in other sections of this report, including in the research, energy, and rural development title sections. Below is a discussion of the first two subtitles, covering socially disadvantaged and limited resource producers (Subtitle A) and agricultural security (Subtitle B). The title also includes other miscellaneous provisions (Subtitle C), some of which are not separately detailed in this report.

Socially Disadvantaged and Limited Resource Producers

Several provisions in the enacted bill address outreach and assistance for socially disadvantaged farmers and ranchers and limited-resource farmers and ranchers, producers targeted by Section 2501 of the 1990 farm bill. Both the House- and Senate-passed farm bills contained this provision. Other farm bill titles in the bill contain similar provisions for the Section 2501 program, including conservation (Title II), farm credit (Title V), rural development (Title VI), agricultural research (Title VII), and crop insurance and disaster assistance (Title XII).

The enacted bill specifies that the Technical and Outreach Assistance Program is to be used to enhance the coordination, outreach, education, and assistance authorized under various USDA programs, and provides $75 million in mandatory funding through FY2012. The bill requires USDA to document the number, location, and economic contributions of socially disadvantaged and limited-resource farmers and ranchers. As part of the efforts to address the needs of socially disadvantaged and limited-resource farmers and ranchers, the bill also authorizes a new USDA Office of Advocacy and Outreach to carry out the Section 2501 program, and also to oversee the Minority Farmer Advisory Committee and carry out the functions of the Office of Outreach and Diversity previously handled by the Office of Assistant Secretary for Civil Rights. The bill also authorizes a new Office of Small Farms and Beginning Farmers and Ranchers, to be subsumed into the Office of Advocacy and Outreach.

The bill also addresses the so-called Pigford decision regarding the 1999 class action discrimination suit against USDA. Both the House- and Senate-passed farm bills contained this provision. The bill provides that Pigford claimants who have not had their cases determined on the merits may, in a civil action, obtain such a determination. The enacted bill further specifies steps USDA must take with regard to settling the claim and provides mandatory funding of $100 million for FY2008 to pay for successful claims.

Agricultural Security

The 2008 farm bill creates an Office of Homeland Security within USDA to coordinate the department's agroterrorism and agricultural disease efforts and to be a liaison with other federal agencies. It also creates an agricultural biosecurity communications center, and a competitive grant program for agricultural biosecurity and countermeasures development.

Regarding foreign animal diseases, the enacted bill adopts the Senate provision that would compel USDA to issue a permit to Department of Homeland Secutiry (DHS) to possess and work with live foot and mouth disease (FMD) virus at the proposed and yet-to-be-built National Bio- and Agro-Defense Facility, subject to compliance with USDA rules for handling "select agents." For more information, see CRS Report RL34 160, *The National Bio- and Agro-Defense Facility: Issues for Congress.*

Title XV: Trade and Tax Provisions

Supplemental Agricultural Disaster Assistance

During the congressional debate on the omnibus farm bill, some policymakers wanted to make permanent in the farm bill some level of disaster payments to supplement the crop insurance program. Consequently, Title XV authorizes a new $3.8 billion trust fund to cover the estimated cost of making agricultural disaster assistance available on an ongoing basis over the next four years (FY2008-FY2011) through five new programs.

The largest of the new farm disaster assistance programs authorized through the 2008 farm bill is a supplemental revenue assistance payment program for crop producers. The program is designed to compensate eligible producers for a portion of crop losses that are not eligible for an indemnity payment under the crop insurance program (i.e., the portion of losses that is part of the deductible on the policy). An eligible producer can receive a payment equal to 60% of the difference between a target level of revenue and the actual total farm revenue for the entire farm. The target level of revenue will be based on the level of crop insurance coverage selected by the farmer, thus increasing if a farmer opts for higher levels of coverage. To be eligible for a payment, a producer must be in or contiguous to a county that has been declared a disaster area by either the President or the Secretary of Agriculture. Payments are limited so that the disaster program guarantee level cannot exceed 90% of what income likely would have been in the absence of a natural disaster.[29]

The producer also must have at least the minimum level of crop insurance (CAT) coverage for insurable crops and participate in the NAP program for non-insurable crops. The statute makes an exception for the 2008 crop year by allowing producers who did not purchase crop insurance or NAP coverage in advance to be eligible for the program, as long as they pay the equivalent administrative fee for coverage within 90 days of enactment (September 16, 2008). Final payments for 2008 crop losses cannot be determined until late 2009, since a portion of the disaster payment formula is based on the national average market price of the commodity, which is determined at the end of the marketing year. For example, the 2008 marketing year for corn and soybeans ends September 30, 2009.

In addition to the supplemental crop revenue assistance payment program described above, the 2008 farm bill also authorizes and funds four smaller disaster programs also through FY20 11: (1) Livestock Indemnity Payments, which compensate ranchers at a rate of 75% of market value for livestock mortality caused by a disaster; (2) Livestock Forage Disaster Program, to assist ranchers who graze livestock on drought-affected pastureland or grazing land; (3) Emergency Assistance for Livestock, Honey Bess and Farm Raised Fish, which will provide up to $50 million to compensate these producers for disaster losses not covered under other disaster programs; and (4) the Tree Assistance Program, for orchardists and nursery growers who can receive a payment to cover 70% of the cost of replanting trees or nursery stock following a disaster (up to $100,000 per year per producer).

For more information on crop insurance and disaster assistance, see CRS Report RL34207, *Crop Insurance and Disaster Assistance in the 2008 Farm Bill*, and CRS Report RS2 1212, *Agricultural Disaster Assistance*.

Tax Provisions

The tax portions (Title XV) of the 2008 farm bill differ markedly from those in either the House- or the Senate-passed versions of the bill. The enacted bill's tax cuts consist of six groups, respectively containing provisions for revenue, an agriculture disaster reserve fund, conservation, energy, agriculture, and other provisions.

The single largest revenue-raising provision in the 2008 farm bill involves a change in the estimated tax payment of corporations. This provision increases the estimated tax payments of corporations due in the July through September 2012 by a factor of 7.75 percentage points, and raises approximately $4.5 billion in revenue through 2012. Other revenue-raising provisions limit the excess farming losses of certain taxpayers and modify the incentives related to alcohol fuels.

The single largest revenue-losing provision in the 2008 farm bill pertains to the agriculture disaster reserve fund, discussed above. Other revenue-losing provisions reduce the depreciable life of race horses 2 years or younger (from seven years to two years), increase the credit for cellulosic biofuel (to $1.01 per gallon less the amount of small-producer ethanol credit claimed and the alcohol mixture credit claimed for ethanol), and create a qualified

The enacted farm bill contains tax-related and revenue provisions related to conservation, energy, and agricultural provisions, among other revenue provisions. For example, among the conservation provisions, the bill authorizes a new type of tax-exempt private bond whose proceeds are used to finance $500 million in forest conservation; it also modifies income tax deductions for qualified timber gains.

The farm bill also includes several provisions which loosen the rules associated with the tax treatment of timber real estate investment trusts (REITs). For example, the allowable size of a REIT's taxable REIT subsidiary was expanded to 25% of REIT assets; the treatment of mineral royalty income as REIT qualified income; the reduction in the excise tax safe-harbor holding period from 4 years to 2 years for timber property sold to a qualified organization exclusively for conservation purposes.

Although the enacted farm bill contains many of the provisions from the Senate-passed bill and one of the two tax provisions in the House-passed bill, it does not include the largest single revenue-raising provision from either bill. The Senate bill included a provision designed to curtail the use of tax shelters, involving the codification of the judicial "economic substance" doctrine that has developed in court cases related to tax shelters.[30] In general terms, the doctrine denies the use of tax- reducing items, such as tax deductions and credits, generated by transactions that do not result in a meaningful change in the taxpayer's economic position. The House bill included a provision designed to curb what is sometimes termed "treaty shopping" — situations where a foreign firm with a U.S. subsidiary routes payments from its U.S. subsidiary through a subsidiary in another country so as to take advantage of tax-treaty benefits. These provisions were not included as part of the enacted 2008 farm bill.

Major Provisions of the Enacted 2008 Farm Bill (P.L. 110-246) Compared with Previous Law and the House- and Senate-Passed Bills (H.R. 2419)

PRIOR LAW/POLICY	HOUSE-PASSED BILL (H.R. 2419)	SENATE-PASSED SUBSTITUTE AMENDMENT (H.R. 2419)	ENACTED 2008 FARM BILL (P.L. 110-246)
"Farm Security and Rural Investment Act of 2002" *[7 U.S.C. 7901 note]*	"Farm, Nutrition, and Bioenergy Act of 2007" *[Sec. 1]*	"Food and Energy Security Act of 2007" *[Sec. 1]*	"Food, Conservation, and Energy Act of 2008" *[P.L. 110-246]*
TITLE I: COMMODITIES			
GRAINS, COTTON, PEANUTS, AND MINOR COMMODITIES (TITLE I)			
Definitions			
Agricultural Act of 1949: 7 U.S.C. 1421 et seq. as in effect before the suspensions under the 1996 farm bill (Federal Agricultural Improvement and Reform Act, P.L. 104-127). *[7 U.S.C. 7901(1)]*	**Agricultural Act of 1949:** same as prior law, with the addition of reference to suspensions under the 2002 farm bill and Sec. 1502(b) of this act. *[Sec. 1001(1)]*	No comparable definition.	No comparable definition.
No comparable definition.	No comparable definition.	**Average Crop Revenue Payment:** A payment made to producers under average crop revenue payment provisions. *[Sec. 1001(1)]*	**Average Crop Revenue Election Payment:** Adopts Senate definition, with change of name. *[Sec. 1001(1)]*
Base acres: the number of base acres established by the owner of the farm under base acre provisions. *[7 U.S.C. 7901 (2)]* Same definition for peanuts. *[7 U.S.C. 7951(1)]*	**Base acres:** the number of base acres of a covered commodity on a farm established under the 2002 farm bill (7 U.S.C. 7911, 7952), as in effect the day before enactment of this act, subject to adjustment. *[Sec. 1001(2)]*	**Base acres:** same as House definition, except covered commodity does not include peanuts. *[Sec. 1001(2)]* Same definition for peanuts. *[Sec. 1301(1)]*	**Base acres:** Adopts House provision, with special mention of peanuts. *[Sec. 1001(2)]* Same definition for peanuts. *[Sec. 1301(1)]*
No comparable definition.	**Comparable United States Quality:** upland cotton classified as Middling 1 3/32-inch cotton, micronaire of 3.7 to 4.2, strength 30 grams per tex, uniformity of 83. *[Sec. 1001(3)]*	No comparable definition.	No comparable definition.
Counter-cyclical payment: a payment to producers on a farm under counter-cyclical payment provisions. *[7 U.S.C. 7001(3)]* Same definition for peanuts. *[7 U.S.C. 7951(2)]*	**Counter-cyclical payment:** a payment to producers on a farm under traditional or revenue-based counter-cyclical payment provisions. *[Sec. 1001(4)]*	**Counter-cyclical payment:** a payment to producers on a farm under traditional counter-cyclical payment provisions. *[Sec. 1001(3)]* Same definition for peanuts. *[Sec. 1301(2)]*	**Counter-cyclical payment:** Adopts Senate provision. *[Sec. 1001(3)]* Same definition for peanuts. *[Sec. 1301(2)]*

(Continued)

PRIOR LAW/POLICY	HOUSE-PASSED BILL (H.R. 2419)	SENATE-PASSED SUBSTITUTE AMENDMENT (H.R. 2419)	ENACTED 2008 FARM BILL (P.L. 110-246)
Covered commodity: wheat, corn, grain sorghum, barley, oats, upland cotton, rice, soybeans, and other oilseeds. *[7 U.S.C. 7901(4)]*	**Covered commodity:** same as prior law, except adds peanuts. *[Sec. 1001(5)]*	**Covered commodity:** same as prior law, except differentiates between long and medium grain rice, and adds pulse crops. *[Sec. 1001(4)]*	**Covered commodity:** Adopts Senate definition. Includes wheat, corn, grain sorghum, barley, oats, upland cotton, long grain rice, medium grain rice, pulse crops, soybeans, and other oilseeds. *[Sec. 1001(4)]*
Direct payment: a payment made to producers on a farm under direct payment provisions. *[7 U.S.C. 7901(5)]*	**Direct payment:** same as prior law. *[Sec. 1001(6)]*	**Direct payment:** same as prior law. *[Sec. 1001(5)]* Same definition for peanuts. *[Sec. 1301(3)]*	**Direct payment:** Adopts Senate definition. *[Sec. 1001(5)]* Same definition for peanuts. *[Sec. 1301(3)]*
Effective price: for a covered commodity for a crop year, the price calculated by USDA under counter-cyclical payment provisions to determine whether payments are required for that crop year. *[7 U.S.C. 7901(6)]*	**Effective price:** same as prior law. *[Sec. 1001(7)]*	**Effective price:** same as prior law. *[Sec. 1001(6)]* Same definition for peanuts. *[Sec. 1301(4)]*	**Effective price:** Adopts Senate definition. *[Sec. 1001(6)]* Same definition for peanuts. *[Sec. 1301(4)]*
Extra long staple cotton: cotton that (A) is produced from pure strain varieties of the Barbadense species or any hybrid of the species, or other similar types of extra long staple cotton having characteristics needed for various end uses for which U.S. upland cotton is not suitable, and grown in irrigated or other designated U.S. cotton-growing regions; and (B) is ginned on a roller-type gin or, other authorized gin for experimental purposes. *[7 U.S.C. 7901(7)]*	**Extra long staple cotton:** same as prior law. *[Sec. 1001(8)]*	**Extra long staple cotton:** same as prior law. *[Sec. 1001(7)]*	**Extra long staple cotton:** Adopts House provision. *[Sec. 1001(7)]*
No comparable definition.	**Far East Price:** the Friday through Thursday average price quotation for the three lowest-priced growths of upland cotton, as quoted for Middling (M) 1 3/32-inch cotton, delivered C/F Far East. *[Sec. 1001(9)]*	No comparable definition.	No comparable definition.
Loan commodity: wheat, corn, grain sorghum, barley, oats, upland cotton, extra long staple cotton, rice, soybeans,	**Loan commodity:** same as prior law, except differentiates feed barley and malt barley; differentiates long, medium, and	**Loan commodity:** same as House definition, except does not differentiate types of barley; does not include small	**Loan commodity:** Adopts Senate definition, but differentiates graded wool and nongraded wool. Includes wheat,

(Continued)

PRIOR LAW/POLICY	HOUSE-PASSED BILL (H.R. 2419)	SENATE-PASSED SUBSTITUTE AMENDMENT (H.R. 2419)	ENACTED 2008 FARM BILL (P.L. 110-246)
other oilseeds, wool, mohair, honey, dry peas, lentils, and small chickpeas. *[7 U.S.C. 7901(8)]*	short grain rice; and includes peanuts. *[Sec. 1001(10)]*	grain rice directly (although included in definition of medium grain rice); excludes peanuts which are treated separately; and includes large chickpeas. *[Sec. 1001(8)]*	corn, grain sorghum, barley, oats, upland cotton, extra long staple cotton, long grain rice, medium grain rice, soybeans, other oilseeds, graded wool, nongraded wool, mohair, honey, dry peas, lentils, small chickpeas, and large chickpeas. *[Sec. 1001(8)]*
No comparable definition.	No comparable definition.	**Medium grain rice:** includes short grain rice. *[Sec. 1001(9)]*	**Medium grain rice:** Adopts Senate definition. *[Sec. 1001(9)]*
Other oilseed: sunflower seed, rapeseed, canola, safflower, flaxseed, mustard seed, or, if designated by the Secretary, another oilseed. Crambe and sesame seed were added by P.L. 108-7, Division A, Sec. 763. *[7 U.S.C. 7901(9)]*	**Other oilseed:** sunflower seed, rapeseed, canola, safflower, flaxseed, mustard seed, crambe, sesame seed, or, if designated by the Secretary, another oilseed. *[Sec. 1001(11)]*	**Other oilseed:** same as House definition, except adds camelina. *[Sec. 1001(10)]*	**Other oilseed:** Adopts House definition. *[Sec. 1001(10)]*
Payment acres: 85% of the base acres for the covered commodity on which direct payments and counter-cyclical payments are made. *[7 U.S.C. 7901(10)]* Same definition for peanuts. *[7 U.S.C. 7951(6)]*	**Payment acres:** same as prior law. *[Sec. 1001(12)]*	**Payment acres:** same as prior law. *[Sec. 1001(11)]* Same definition for peanuts. *[Sec. 1301(5)]*	**Payment acres:** Generally, 85% of base acres for the covered commodity for direct and counter-cyclical payments. Exception: 83.3% of base acres for direct payments only for crop years 2009-2011. *[Sec. 1001(11)]* Same definition for peanuts. *[Sec. 1301(5)]*
Payment yield: in general, the yield established under Sec. 1102 for a covered commodity. "Updated payment yield" means the yield established to calculate counter-cyclical payments. *[7 U.S.C. 7901(11)* Same definition for peanuts. *[7 U.S.C. 7951(7)]*	**Payment yield:** the yield established for direct payments and counter-cyclical payments for a farm for a covered commodity and peanuts under the 2002 farm bill as in effect on the day before the date of the enactment of this act. *[Sec. 1001(13)]*	**Payment yield:** same as House definition, except does not include peanuts. *[Sec. 1001(12)]* Same definition for peanuts. *[Sec. 1301(6)]*	**Payment yield:** Adopts Senate definition, and includes yields established for new commodities in Sec. 1102. *[Sec. 1001(12)]* Same definition for peanuts. *[Sec. 1301(6)]*
Producer: generally, an owner, operator, landlord, tenant, or sharecropper that shares in the risk of producing a crop and is entitled to share in the crop available for marketing from the farm, or would have shared had the crop been produced. For a grower of hybrid seed, the	**Producer:** same as prior law. *[Sec. 1001(14)]*	**Producer:** same as prior law. *[Sec. 1001(13)]* Same definition for peanuts. *[Sec. 1001(7)]*	**Producer:** Adopts Senate definition. *[Sec. 1001(13)]* Same definition for peanuts. *[Sec. 1301(7)]*

(Continued)

Prior Law/Policy	House-Passed Bill (H.R. 2419)	Senate-Passed Substitute Amendment (H.R. 2419)	Enacted 2008 Farm Bill (P.L. 110-246)
existence of a hybrid seed contract and other program rules shall not adversely affect the ability to receive a payment. *[7 U.S.C. 7901(12)]* Same definition for peanuts. *[7 U.S.C. 7951(8)]*			
No comparable definition.	No comparable definition.	**Pulse crop:** dry peas, lentils, small chickpeas, and large chickpeas. *[Sec. 1001(14)]*	**Pulse crop:** Adopts Senate definition. *[Sec. 1001(14)]*
Secretary: the Secretary of Agriculture. *[7 U.S.C. 7901(13)]* Same definition for peanuts. *[7 U.S.C. 7951(9)]*	**Secretary:** same as prior law. *[Sec. 1001(15)]*	No comparable definition.	No comparable definition.
State: each of the U.S. States, the District of Columbia, the Commonwealth of Puerto Rico, or U.S. territory/possession. *[7 U.S.C. 7901(14)]* Same definition for peanuts. *[7 U.S.C. 7951(10)]*	**State:** same as prior law. *[Sec. 1001(16)]*	**State:** same as prior law. *[Sec. 1001(15)]* Same definition for peanuts. *[Sec. 1301(8)]*	**State:** Adopts Senate definition. *[Sec. 1001(15)]* Same definition for peanuts. *[Sec. 1301(8)]*
Target price: the price per unit of a covered commodity used to determine the payment rate for counter-cyclical payments. *[7 U.S.C. 7901(15)]* Same definition for peanuts. *[7 U.S.C. 7951(11)]*	**Target price:** same as prior law. *[Sec. 1001(17)]*	**Target price:** same as prior law. *[Sec. 1001(16)]* Same definition for peanuts. *[Sec. 1301(9)]*	**Target price:** Adopts Senate definition. *[Sec. 1001(16)]* Same definition for peanuts. *[Sec. 1301(9)]*
United States: when used in a geographical sense, all of the States. *[7 U.S.C. 7901(16)]* Same definition for peanuts. *[7 U.S.C. 7951(12)]*	**United States:** same as prior law. *[Sec. 1001(18)]*	**United States:** same as prior law. *[Sec. 1001(17)]* Same definition for peanuts. *[Sec. 1301(10)]*	**United States:** Adopts Senate definition. *[Sec. 1001(17)]* Same definition for peanuts. *[Sec. 1301(10)]*
No comparable definition.	**United States Premium Factor:** the percentage by which the difference in the U.S. loan schedule premiums for Strict Middling (SM) 1 1/8-inch cotton and for M 1 3/32-inch exceeds the difference in the applicable premiums for comparable international qualities delivered C/F Far East. *[Sec. 1001(19)]*	No comparable definition.	**United States Premium Factor:** Adopts the House definition. *[Sec. 1001(18)]*

(Continued)

Prior Law/Policy	House-Passed Bill (H.R. 2419)	Senate-Passed Substitute Amendment (H.R. 2419)	Enacted 2008 Farm Bill (P.L. 110-246)
Base Acres and Program Yields			
Base acres: for each covered commodity on a farm, base acres are established by the owner's choice of (1) average of 1998-2001 plantings, or (2) the sum of production flexibility contract acreage under 1996 farm bill plus average oilseed acreage from 1998-2001. Accommodation for peanut acres, double cropping, and CRP acres. Base cannot exceed total crop land. **Payment acres** = 85% of base acres. ***[7 U.S.C. 7911]*** Base acres for peanuts also based on the 1998-2001 period. ***[7 U.S.C. 7952]***	**Base acres:** no choice of updating base acres or payment yields, except requires USDA to provide base acre adjustments when a CRP contract ends. ***[Sec. 1101]***	**Base acres:** same as House bill, except provides for adjustment to include pulse crops, camelina, or newly designated oilseed acreage. ***[Sec. 1101(a)-(c)(1)]*** Same provision for peanuts. ***[Sec. 1302(a)-(c)(1)]***	**Base acres:** Adopts Senate provision. ***[Sec. 1101(a)-(c)(1)]*** Same provision for peanuts. ***[Sec. 1302(a)-(c)(1)]***
No comparable provision.	No comparable provision.	**Required reduction of base acres:** suspend direct, counter-cyclical, and average crop revenue payments and reduce base acres for land that is no longer used for farming. Specifically, land that has been developed for "commercial or industrial use or has been subdivided and developed for multiple residential units or other nonfarming uses" unless producer demonstrates the land is devoted exclusively to agricultural production. ***[Sec. 1101(c)(2)]*** Same provision for peanuts. ***[Sec. 1302(c)(2)]***	Adopts Senate provision, with modification. Reduction in base acres is required for land that is "subdivided and developed for multiple residential units or other nonfarming uses" unless the producers "demonstrate that the land remains devoted to commercial agriculture production or is likely to be returned to agricultural use." ***[Sec. 1101(c)(2)]*** Same provision for peanuts. ***[Sec. 1302(c)(2)]***
No comparable provision.	No comparable provision.	Requires USDA to track reconstitutions of land and report to Congress to ensure that commercial or residential land is not eligible for payments. ***[Sec. 1101(c)(3)]*** Same provision for peanuts. ***[Sec. 1302(c)(3)]***	Adopts Senate provision. ***[Sec. 1101(c)(3)]*** Same provision for peanuts. ***[Sec. 1302(c)(3)]***
Direct payment yield: for each covered commodity on a farm, a direct payment yield is the yield established for the 1995 crop; the yield for oilseeds is the average	No provision to change payment yield; payment yields are continued from prior law by definition.	**Payment yields:** Establishes **payment yields** for designated oilseeds, camelina, or pulse crops using 1998-2001 farm yields, adjusted back to the national	**Payment yields:** Adopts Senate provision. ***[Sec. 1102]***

(Continued)

PRIOR LAW/POLICY	HOUSE-PASSED BILL (H.R. 2419)	SENATE-PASSED SUBSTITUTE AMENDMENT (H.R. 2419)	ENACTED 2008 FARM BILL (P.L. 110-246)
from 1998-2001, adjusted back to the national average from 1981-85. **Counter-cyclical payment yield** may be an updated yield using specified formulas. *[7 U.S.C. 7912]* Payment yields for peanuts are established using the 1998-2001 period. *[7 U.S.C. 7952]*		average from 1981-85. *[Sec. 1102]*	
No comparable provision.	No comparable provision.	Apportion base acres for long grain and medium grain rice based on average acreage planted to each type of rice in the applicable state during the 2003-2006 crop years to make counter-cyclical payments. Producers may elect to use farm-level planting history. Established totals of base acres, payment acres, and yields are maintained. *[Sec. 1107]*	Adopts the Senate provision. *[Sec. 1108]*
Prohibition on Small Payments			
No comparable provision.	**No payments under $25:** no payment will be made if the total direct payment to a producer on a farm for all covered commodities is less than $25. *[Sec. 1102(e)]* Same provision is made for counter-cyclical payments *[Sec. 1102(e)]*, and revenue-based counter-cyclical payments. *[Sec. 1104(i)]*	No comparable provision.	No direct, counter-cyclical, or average crop revenue election payments will be made on farms with less than a total of 10 bases acres, except for limited resource or socially disadvantaged farms. Requires USDA to report on the effect of the provision. *[Sec. 1101(d)]* Same provision for peanuts. *[Sec. 1302(d)]* Report language instructs USDA to allow for aggregation of farms when implementing the 10-acre requirement.
Producer Agreement			
Eligibility for payments requires producers to comply with conservation, wetland, and planting flexibility requirements; use base acres for agricultural or conserving use, and not for nonagricultural commercial or industrial use; control noxious weeds and	Same as prior law. *[Sec. 1105(a)]*	Same as prior law, except adds provision that land cannot be used for a residential use (including land subdivided and developed into residential units or other nonfarming uses, or that is otherwise no longer intended to be used in conjunction with farming (like Sec. 1101(c)(2)).	Adopts House provision, except adds a provision that land cannot be used for a residential use, and requires farmers in the ACRE program to report production. *[Sec. 1106(a)]* Same provision for peanuts. *[Sec. 1305(a)]*

(Continued)

PRIOR LAW/POLICY	HOUSE-PASSED BILL (H.R. 2419)	SENATE-PASSED SUBSTITUTE AMENDMENT (H.R. 2419)	ENACTED 2008 FARM BILL (P.L. 110-246)
maintain sound agricultural practices. *[7 U.S.C. 7915(a)]* Same provision for peanuts. *[7 U.S.C. 7955(a)]*		*[Sec. 1105(a)]* Same provision for peanuts. *[Sec. 1305(a)]* Same provision for Average Crop Revenue (ACR) program, except does not require compliance with planting flexibility provisions since ACR has its own planting flexibility rules. Requires USDA to certify entities receiving payments are producers. *[Sec. 1402(a)]*	
Sets requirements for transfer of interest in base acres. Requires acreage reports. Protects interests of tenants and sharecroppers and provides for sharing of payments on a farm on an equitable basis. *[7 U.S.C. 7915(b)-(e)]* Same provision for peanuts. *[7 U.S.C. 7955(b)-(e)]*	Same as prior law. *[Sec. 1105(b)-(e)]*	Same as prior law, except adds that no penalty shall be assessed for inaccurate acreage report unless producers knowingly and willfully falsified the report. *[Sec. 1105(b)-(e)]* Same provision for peanuts. *[Sec. 1305(b)-(e)]* Same provision for Average Crop Revenue program. *[Sec. 1402(b)-(e)]*	Adopts Senate provision, and applies to ACRE program. *[Sec. 1106(b)-(e)]* Same provision for peanuts. *[Sec. 1305(b)-(e)]*
Planting Flexibility			
Any crop may be planted on base acres, except restrictions are placed on planting of fruits, vegetables, and wild rice on base acres. Penalties apply if the fruit and vegetable restriction is violated. Provides an exception for lentils, mung beans, and dry peas. Exceptions provided for farms and producers with a history of double-cropping or history of growing fruits and vegetables (except that direct and counter-cyclical payments are reduced acre for acre for the year). *[7 U.S.C. 7916]* Same provision for peanuts. *[7 U.S.C. 7956]*	Same as prior law, and incorporates peanuts as a covered commodity. *[Sec. 1106(a)-(c)]*	Same as prior law, but the exception allows planting mung beans and pulse crops. *[Sec. 1106(a)-(c)]* Same provision for peanuts. *[Sec. 1306(a)-(c)]* Same provision for Average Crop Revenue program. *[Sec. 1403(a)-c)]*	Adopts Senate provision. *[Sec. 1107(a)-(c)]* Same provision for peanuts. *[Sec. 1306]*
No comparable provision.	Establishes a pilot Farm Flex project for planting tomatoes for processing on up to 10,000 base acres in Indiana during the 2008-2012 crop years. Base acres temporarily reduced for each acre of	Same as House provision for traditional direct and counter-cyclical program participants, except applies only to 2008 and 2009 crop years. *[Sec. 1106(d)]* No comparable provision for peanut base	Creates a pilot program beginning in 2009 in seven midwestern states to allow planting of fruits and vegetables for processing on base acres. Limited to cucumbers, green peas, lima beans,

(Continued)

Prior Law/Policy	House-Passed Bill (H.R. 2419)	Senate-Passed Substitute Amendment (H.R. 2419)	Enacted 2008 Farm Bill (P.L. 110-246)
	tomatoes, and are protected for future use. *[Sec. 1106(d)]*	acres in Sec. 1306. For participants in the ACR program, establishes pilot Farm Flex project for planting any fruit or vegetable for processing on up to 10,000 acres in certain states (IL, IN, IA, MI, MN, OH, WI). Available for the 2010-2012 crop years. Base acres temporarily reduced for each acre of fruit and vegetables, and protected for future use. *[Sec. 1403(d)]*	pumpkins, snap beans, sweet corn, and tomatoes grown for processing. States include Minnesota (34,000 acres), Wisconsin (9,000 acres), Michigan (9,000 acres), Illinois (9,000 acres), Indiana (9,000 acres), Ohio (4,000 acres), and Iowa (1,000 acres). Base acres are temporarily reduced for the year, but are restored for the next crop year and "considered planted" for any future base calculations. *[Sec. 1107(d)]*
Direct Payments			
Direct payments: available to producers on farms with payment yields and base acres. Covers 2002-2007 crop years. *[7 U.S.C. 7913(a)]* Direct payments for peanuts authorized separately. *[7 U.S.C. 7953(a)]*	**Direct payments:** continues prior law to cover 2008-2012 crop years. *[Sec. 1102(a)]*	**Direct payments:** continues prior law to cover 2008-2012 crop years. Excludes participants in the ACR program under Sec. 1401. *[Sec. 1103(a)]* Direct payments for peanuts authorized separately. *[Sec. 1303(a)]*	**Direct payments:** Continues prior law to cover 2008-2012 crop years. *[Sec.1103(a)]* Direct payment for peanuts continued separately *[Sec. 1303(a)]*
Direct payment rates: Wheat, bushel (bu.), $0.52 Corn, bu., $0.28 Grain sorghum, bu., $0.35 Barley, bu., $0.24 Oats, bu., $0.024 Upland cotton, lb., $0.0667 Rice, cwt., $2.35 Soybeans, bu., $0.44 Other oilseeds, lb., $0.0080 *[7 U.S.C. 7913(b)]* Peanuts, ton, $36 *[7 U.S.C. 7953(b)]*	Same as prior law; incorporates peanuts into same section. *[Sec. 1102(b)]*	Same as prior law, except differentiates between long grain rice and medium grain rice (both at $2.35 per cwt.). *[Sec. 1103(b)]* Peanuts, ton, $36 *[Sec. 1303(b)]*	Adopts Senate provision. *[Sec. 1103(b)]* Separate provision for peanuts. *[Sec. 1303(b)]*
Payment amount = Payment rate, times 85% of base acres, times direct payment yield. *[7 U.S.C. 7913(c)]* Same formula for peanuts. *[7 U.S.C. 7953(d)]*	Same as prior law. *[Sec. 1102(c)]*	Same as House provision. *[Sec. 1103(c)]* Separate provision for peanuts *[Sec. 1303(c)]*	Same as prior law, except a ratio of 83.3% of base acres is used for crop years 2009-2011. *[Sec. 1103(c)]* Separate provision for peanuts. *[Sec. 1303(c)]*
Timing: Generally paid after October 1 of the calendar year of the year of	**Timing:** Same as prior law, except (1) applies to peanuts in the same section,	Same as House provision. *[Sec. 1103(d)]* Separate provision for peanuts *[Sec.*	Adopts Senate provision. *[Sec. 1103(d)]* Separate provision for peanuts *[Sec.*

(Continued)

Prior Law/Policy	House-Passed Bill (H.R. 2419)	Senate-Passed Substitute Amendment (H.R. 2419)	Enacted 2008 Farm Bill (P.L. 110-246)
harvest. Advance payments up to 50% (later reduced to 22% by P.L. 109-171) beginning as early as December 1 of the calendar year before harvest, at the option of the producer. *[7 U.S.C. 7913(d)]* For peanuts: generally, before October 1 of the year of harvest. Similar advance payments. [*7 U.S.C. 7953(e)]*	and (2) the 22% advance payment option applies only to crop years 2008-2011. No advance payment for crop year 2012 and thereafter. *[Sec. 1102(d)]*	*1303(d)]*	*1303(d)]*
Counter-cyclical Payments			
Counter-cyclical payments: if the effective price for a covered commodity is less than the target price, a payment is available to producers on farms with payment yields and base acres. Covers 2002-2007 crop years. *[7 U.S.C. 7914(a)]* Counter-cyclical payments for peanuts authorized separately. *[7 U.S.C. 7954(a)]*	**Counter-cyclical payments:** same as prior law, except it covers 2008-2012 crop years. *[Sec. 1103(a)]*	**Counter-cyclical payments:** same as prior law, except it covers 2008-2012 crop years. Excludes participants in the ACR program. *[Sec. 1104(a)]* Counter-cyclical payments for peanuts authorized separately. *[Sec.1304(a)]*	**Counter-cyclical payments:** continues prior law to cover 2008-2012 crop years. *[Sec.1104(a)]* Counter-cyclical payments for peanuts continued separately *[Sec. 1304(a)]*
Effective price: the higher of (1) the national season average market price or (2) national average loan rate plus the direct payment rate. *[7 U.S.C. 7914(b)]* Same provision for peanuts. *[7 U.S.C. 7954(b)]*	Same as prior law, except applies to peanuts and clarifies that effective price for rice and barley are to be computed notwithstanding separate loan rates by type of rice or barley. *[Sec. 1103(b)]*	Same as prior law, except computed for rice using prices by type of rice. *[Sec. 1104(b)]* Same as prior law for peanuts. *[Sec. 1304(b)]*	Adopts Senate provision. *[Sec. 1104(b)]* Adopts Senate provision for peanuts. *[Sec. 1304(b)]*
Target prices for 2004-2007 crop years: Wheat, bu., $3.92 Corn, bu., $2.63 Grain sorghum, bu., $2.57 Barley, bu., $2.24 Oats, bu., $1.44 Upland cotton, lb., $0.7240 Rice, cwt., $10.50 Soybeans, bu., $5.80 Other oilseeds, lb., $0.1010	**Target prices:** Wheat, bu., $4.15 Corn, bu., $2.63 Grain sorghum, bu., $2.57 Barley, bu., $2.73 Oats, bu., $1.50 Upland cotton, lb., $0.70 Rice, cwt., $10.50 Soybeans, bu., $6.10 Other oilseeds, lb., $0.1150 Peanuts, ton, $495 *[Sec. 1103(c)]*	**Target prices:** Wheat, bu., $4.20 Corn, bu., $2.63 Grain sorghum, bu., $2.63 Barley, bu., $2.63 Oats, bu., $1.83 Upland cotton, lb., $0.7225 Long grain rice, cwt., $10.50 Medium grain rice, cwt., $10.50 Soybeans, bu., $6.00 Other oilseeds, cwt., $12.74 Dry peas, cwt., $8.33 Lentils, cwt., $12.82	**Target prices:** *2008 crop year* (same as prior law, except cotton lower) *[Sec. 1104(c)(1)]* Wheat, bu., $3.92 Corn, bu., $2.63 Grain sorghum, bu., $2.57 Barley, bu., $2.24 Oats, bu., $1.44 Upland cotton, lb., $0.7125 Long grain rice, cwt., $10.50 Medium grain rice, cwt., $10.50 Soybeans, bu., $5.80 Other oilseeds, cwt., $10.10

(Continued)

Prior Law/Policy	House-Passed Bill (H.R. 2419)	Senate-Passed Substitute Amendment (H.R. 2419)	Enacted 2008 Farm Bill (P.L. 110-246)
(Different target prices applied to 2002-2003 crop years). ***[7 U.S.C. 7914(c)]*** Peanuts, ton, $495 ***[7 U.S.C. 7954(c)]***		Small chickpeas, cwt., $10.36 Large chickpeas, cwt., $12.82 ***[Sec. 1104(c)]*** Peanuts, ton, $495 ***[Sec. 1304(c)]***	Peanuts, ton, $495 ***[Sec. 1304(c)]*** *2009 crop year* (same as 2008, except four new crops: dry peas, lentils, small and large chickpeas) ***[Sec. 1104(c)(2)]*** Wheat, bu., $3.92 Corn, bu., $2.63 Grain sorghum, bu., $2.57 Barley, bu., $2.24 Oats, bu., $1.44 Upland cotton, lb., $0.7125 Long grain rice, cwt., $10.50 Medium grain rice, cwt., $10.50 Soybeans, bu., $5.80 Other oilseeds, cwt., $10.10 Dry peas, cwt., $8.32 Lentils, cwt., $12.81 Small chickpeas, cwt., $10.36 Large chickpeas, cwt., $12.81 Peanuts, ton, $495 ***[Sec. 1304(c)]*** *2010-2012 crop years* (increases for wheat, sorghum, barley, oats, soybeans, and minor oilseeds) ***[Sec. 1104(c)(3)]*** Wheat, bu., $4.17 Corn, bu., $2.63 Grain sorghum, bu., $2.63 Barley, bu., $2.63 Oats, bu., $1.79 Upland cotton, lb., $0.7125 Long grain rice, cwt., $10.50 Medium grain rice, cwt., $10.50 Soybeans, bu., $6.00 Other oilseeds, cwt., $12.68 Dry peas, cwt., $8.32 Lentils, cwt., $12.81 Small chickpeas, cwt., $10.36 Large chickpeas, cwt., $12.81 Peanuts, ton, $495 ***[Sec. 1304(c)]***

(Continued)

PRIOR LAW/POLICY	HOUSE-PASSED BILL (H.R. 2419)	SENATE-PASSED SUBSTITUTE AMENDMENT (H.R. 2419)	ENACTED 2008 FARM BILL (P.L. 110-246)
Payment amount = target price minus effective price (if the difference is greater than 0), times 85% of base acres, times counter-cyclical payment yield. *[7 U.S.C. 7913(d)-(e)]* Same formula for peanuts. *[7 U.S.C. 7954(d)-(f)]*	Same as prior law. *[Sec. 1103(d)-(e)]*	Same as prior law. *[Sec. 1104(d)-(e)]* Same formula for peanuts. *[Sec. 1304(d)-(e)]*	Adopts Senate provision. *[Sec. 1104(d)-(e)]* Adopts Senate provision for peanuts. *[Sec. 1304(d)-(e)]*
Timing: generally, after the end of the 12-month marketing year. Advance payments are available; for the 2007 crop year, one advance payment of 40% of expected payment after first 6 months of marketing year (for 2002-2006, two advance payments, each 35% of the expected payment, in October of harvest year, and after February 1 of the next calendar year). *[7 U.S.C. 7914(f)]* Same provision for peanuts. *[7 U.S.C. 7954(g)]*	**Timing:** generally, the later of (1) the end of the 12-month marketing year, or (2) October 1 of the same calendar year as the end of the marketing year. Advance payments for 2008-2010 crop years: one advance payment of 40% after first 6 months of marketing year. No advance payments after 2010 crop year. *[Sec. 1103(f)]*	**Timing:** generally, beginning October 1 after the end of the marketing year. Advance payments same as in House bill. *[Sec. 1104(f)]* Same provision for peanuts. *[Sec. 1304(f)]*	Adopts Senate provision, with modification. *[Sec. 1104(f)]* Same provision for peanuts. *[Sec. 1304(f)]*
Revenue-based Counter-cyclical Payments			
No comparable provision.	**Revenue-based Counter-Cyclical Payments (RCCP):** an alternative to traditional counter-cyclical payments. Covers crop years 2008-2012. Producers have one opportunity to elect RCCP option soon after enactment. Traditional counter-cyclical payments remain the default if no election is made. *[Sec. 1104(a)]*	**Average Crop Revenue (ACR) program:** an alternative to traditional direct payments, counter-cyclical payments, and nonrecourse marketing loans for covered commodities and peanuts. Producers have one opportunity to elect ACR option: for 2010-12 crop years, 2011-12. or 2012. Traditional programs remain the default if no election is made. *[Sec. 1401(a)]*	**Average Crop Revenue Election (ACRE) program:** Adopts Senate approach, with significant modifications. An alternative to traditional counter-cyclical payments for covered commodities and peanuts, with a reduction in direct payments and marketing loan rates for participants. Producers can enter any year of the 2009-2012 crop years, but cannot return to the traditional counter-cyclical program. *[Sec. 1105(a)]*
	— No comparable provision. (Continue using traditional direct payments.)	— **Fixed payment component** = $15 per acre times 100% of base acres. *[Sec. 1401(b)(2)]*	— Continue traditional direct payments, but reduce them by 20% for ACRE participants. *[Sec. 1105(a)(1)]*
	— **Revenue-based payment** if national actual revenue per acre is less than the national target revenue per acre for the	— **Revenue-based component** if actual state revenue is less than a guaranteed level for the covered commodity. *[Sec.*	— **Revenue-based payment** based on a two-part trigger: (1) if actual state revenue is less than a guaranteed state

(Continued)

Prior Law/Policy	House-Passed Bill (H.R. 2419)	Senate-Passed Substitute Amendment (H.R. 2419)	Enacted 2008 Farm Bill (P.L. 110-246)
	covered commodity. ***[Sec. 1104(b)]***	***1401(b)(3)]***	level for the covered commodity, and (2) if actual farm revenue is less than a farm ACRE benchmark for the covered commodity. ***[Sec. 1105(b)]***
	— No comparable provision. (Continue using traditional marketing loan program.)	— **Recourse loans** available on actual production of a covered commodity. Loans must be repaid in full; traditional nonrecourse loans, loan deficiency payments, and marketing loan gains unavailable. ***[Sec. 1401(f)]***	— Continue using nonrecourse marketing loan program, but reduce loan rates by 30% for ACRE participants. ***[Sec. 1105(a)(1)]***
	National actual revenue per acre = national average yield for the year times the higher of (1) national season average market price, or (2) loan rate. An all-rice and all-barley loan rate will be used for those commodities. ***[Sec. 1104(c)]***	**Actual state revenue per acre** = actual state yield, times the ACR harvest price. Actual state yield is the actual quantity produced in the state during the crop year, divided by planted acres. ACR harvest price is the harvest price used to calculate revenue under Federal Crop Insurance program. ***[Sec. 1401(c)]***	**Actual state revenue** per acre = actual state yield, times the national average market price. **Actual state yield** is the actual quantity produced in the state during the crop year, divided by planted acres. **National average market price** is the greater of the national average price received during the 12-month marketing year, or the marketing loan rate after being reduced by 30%. ***[Sec. 1105(c)]***
	National target revenue per acre: Wheat, $149.92/acre Corn, $344.12/acre Grain Sorghum, $131.28/acre Barley, $153.30/acre Oats, $92.10/acre Upland cotton, $496.93/acre Rice, $548.06/acre Soybeans, $231.87/acre Other oilseeds, $129.18/acre Peanuts, $683.83/acre ***[Sec. 1104(d)]*** **National payment yield:** Wheat, 36.1 bu./acre Corn, 114.4 bu./acre Grain Sorghum, 58.2 bu./acre Barley, 48.6 bu./acre	**Average crop revenue guarantee per acre** = 90% times the expected state yield per planted acre, times the average of the pre-planting price for the crop year and the preceding 2 crop years. The expected state yield for a crop year is projected from a trend using 1980-2006 data. The pre-planting price is the price used to calculate revenue under the Federal Crop Insurance program, and cannot decrease or increase more than 15% from the preceding year. ***[Sec. 1401(d)]***	**ACRE program guarantee** per acre = 90% times the benchmark state yield, times the ACRE program guarantee price. The **benchmark state yield** is a 5-year Olympic average state yield. The **ACRE program guarantee price** is a 2-year average of the national average market price, as defined above. The ACRE program guarantee cannot change more than 10% from the previous year. If more than 25% of a state's acreage is irrigated and 25% is non-irrigated, separate guarantees shall apply. ***[Sec. 1105(d)]*** **Actual farm revenue** per acre = actual farm yield, times the greater of the national average price received during

(Continued)

PRIOR LAW/POLICY	HOUSE-PASSED BILL (H.R. 2419)	SENATE-PASSED SUBSTITUTE AMENDMENT (H.R. 2419)	ENACTED 2008 FARM BILL (P.L. 110-246)
	Oats, 49.8 bu./acre Upland cotton, 634 lb./acre Rice, 51.28 cwt./acre Soybeans, 34.1 bu./acre Other oilseeds, 1167.6 lb./acre Peanuts, 1.496 ton/acre *[Sec. 1104(e)]*		the 12-month marketing year or the marketing loan rate after being reduced by 30%. *[Sec. 1105(e)]*
	National payment rate = National target revenue per acre minus national actual revenue per acre (if difference greater than 0), divided by national payment yield. *[Sec. 1104(f)]*		**Farm ACRE benchmark revenue** per acre = the 5-year Olympic average farm yield, times the ACRE program guarantee price; plus the crop insurance premium per acre. *[Sec. 1105(f)]*
	Payment amount = National payment rate, times 85% of base acres, times payment yield. *[Sec. 1104(g)]*	**Revenue-based payment amount** = the average crop revenue program guarantee minus the actual state revenue (if the difference is greater than 0), times 85% of the base acres on the farm for the covered commodity, times the ratio of the actual production history (APH) on the farm divided by the expected state yield, times 90%. This formula multiplies a state-level payment rate per acre times 85% of base acres, then pro-rates the payment based on the farm's yield history compared to the expected state yield; the payment is then reduced by 10%. *[Sec. 1401(e)]*	**Payment amount** = the product of (1) the lesser of (a) the ACRE program guarantee minus actual state revenue or (b) 25% of the ACRE program guarantee, times (2) 83.3% (2009-2011) or 85% (2012) of the acreage planted of the covered commodity (not to exceed base acres of the commodity), times (3) the 5-year Olympic average farm yield divided by the 5-year Olympic average state yield. This formula multiplies a state-level payment rate per acre (up to a maximum of 25% of the guarantee level) times a percentage of planted acreage, then pro-rates the payment based on the farm's yield history compared to the state's yield history. *[Sec. 1105(g)]*
	Timing: Generally, later of (1) end of the 12-month marketing year, or (2) October 1 of the same calendar year as the end of the marketing year. Advance payments for 2008-2010 crop years: 40% of expected payment after the first 6 months of marketing year. No advance payments after 2010 crop year. *[Sec. 1104(h)]*	**Timing:** Beginning October 1 after the end of the marketing year for both the fixed payment and the revenue-based component. No advance payments. This delays the ACR direct payment component one year compared to traditional direct payments. *[Sec. 1401(b)(4)]*	**Timing:** Beginning October 1 after the end of the marketing year. No advance payments. *[Sec. 1105(b)(3)]*

(Continued)

Prior Law/Policy	House-Passed Bill (H.R. 2419)	Senate-Passed Substitute Amendment (H.R. 2419)	Enacted 2008 Farm Bill (P.L. 110-246)
Nonrecourse Marketing Loans and Other Recourse Loans			
Nonrecourse marketing loans: available for any amount of a loan commodity produced in crop years 2002-2007. Addresses commingled commodities, and requires compliance with conservation and wetlands requirements. ***[7 U.S.C. 7931]*** Nonrecourse marketing loans for peanuts authorized separately. ***[7 U.S.C. 7957(a)(1)-(3)]***	Same as prior law, except it covers 2008-2012 crop years, and includes peanuts. ***[Sec. 1201(a)-(d)]***	Same as prior law, except it covers 2008-2012 crop years, and excludes participants in the ACR program. ***[Sec. 1201]*** Nonrecourse marketing loans for peanuts authorized separately. ***[Sec. 1303(a)(1)-(4)]***	Generally continues prior law to cover 2008-2012 crop years. ***[Sec.1201]*** Nonrecourse marketing loans for peanuts continued separately ***[Sec. 1307(a)(1)-(3)]*** Deletes House and Senate provisions for commingled commodities and peanuts.
For peanuts, nonrecourse marketing loans available in crop years 2002-2007. May be obtained through marketing cooperative or association approved by USDA. Storage to be provided on a non-discriminatory basis and under any additional requirements. Payment of peanut storage costs authorized for 2002-2006 crops. ***[7 U.S.C. 7957(a)(4)-(7)]***	Same as prior law, except it covers 2008-2012 crop years, and payment for peanut storage costs is not authorized. ***[Sec. 1201(e)]***	For peanuts, same as House provision ***[Sec. 1307(a)(5)-(6), (8)]***, except it authorizes payment of storage, handling, and associated costs, and does so in such a way that handling and associated costs are not deducted from a producer's loan, but instead advanced when peanuts are placed under loan and repaid when peanuts are redeemed. ***[Sec. 1307(a)(7)]***	For peanuts, adopts Senate provisions, including payments for peanut storage, except storage payments begin with 2008 crop year. ***[Sec. 1307(a)(4)-(7)]***
Loan rates for 2004-2007 crop years: Wheat, bu., $2.75 Corn, bu., $1.95 Grain sorghum, bu., $1.95 Barley, bu., $1.85 Oats, bu., $1.33 Upland cotton, lb., $0.52 Extra long staple cotton, lb., $0.7977 Rice, cwt., $6.50 Soybeans, bu., $5.00 Other oilseeds, lb., $0.0930 Dry peas, cwt., $6.22 Lentils, cwt., $11.72 Small chickpeas, cwt., $7.43 Graded wool, lb., $1.00 Nongraded wool, lb., $0.40	**Loan rates:** Wheat, bu., $2.94 Corn, bu., $1.95 Grain sorghum, bu., $1.95 Malt barley, bu., $2.50 Feed barley, bu., $1.90 Oats, bu., $1.46 Base quality upland cotton, lb., $0.52 Extra long staple cotton, lb., $0.7977 Long grain rice, cwt., $6.50 Medium & short grain rice, cwt., $6.50 Soybeans, bu., $5.00 Other oilseeds, lb., $0.1070 Dry peas, cwt., $5.40 Lentils, cwt., $11.28 Small chickpeas, cwt., $8.54 Graded wool, lb., $1.10 Nongraded wool, lb., $0.40	**Loan rates:** Wheat, bu., $2.94 Corn, bu., $1.95 Grain sorghum, bu., $1.95 Barley, bu., $1.95 Oats, bu., $1.39 Base quality upland cotton, lb., $0.52 Extra long staple cotton, lb., $0.7977 Long grain rice, cwt., $6.50 Medium grain rice, cwt., $6.50 Soybeans, bu., $5.00 Other oilseeds, cwt., $10.09 Dry peas, cwt., $5.40 Lentils, cwt., $11.28 Small chickpeas, cwt., $7.43 Large chickpeas, cwt., $11.28 Graded wool, lb., $1.20 Nongraded wool, lb., $0.40	**Loan rates:** *2008 crop year* (same as prior law) Wheat, bu., $2.75 Corn, bu., $1.95 Grain sorghum, bu., $1.95 Barley, bu., $1.85 Oats, bu., $1.33 Base quality upland cotton, lb., $0.52 Extra long staple cotton, lb., $0.7977 Long grain rice, cwt., $6.50 Medium grain rice, cwt., $6.50 Soybeans, bu., $5.00 Other oilseeds, cwt., $9.30 Dry peas, cwt., $6.22 Lentils, cwt., $11.72 Small chickpeas, cwt., $7.43 Graded wool, lb., $1.00 Nongraded wool, lb., $0.40 Mohair, lb., $4.20

(Continued)

PRIOR LAW/POLICY	HOUSE-PASSED BILL (H.R. 2419)	SENATE-PASSED SUBSTITUTE AMENDMENT (H.R. 2419)	ENACTED 2008 FARM BILL (P.L. 110-246)
Mohair, lb., $4.20 Honey, lb., $0.60 (Different loan rates applied to 2002-2003 crop years.) *[7 U.S.C. 7932(b)]* Peanuts, ton, $355 *[7 U.S.C. 7957(b)]*	Mohair, lb., $4.20 Honey, lb., $0.60 Peanuts, ton, $355.00 *[Sec. 1202(a)]*	Mohair, lb., $4.20 Honey, lb., $0.72 *[Sec. 1202(a)]* Peanuts, ton, $355 *[Sec. 1307(b)]*	Honey, lb., $0.60 *[Sec. 1202(a)]* Peanuts, ton, $355 *[Sec. 1307(b)]* *2009 crop year* (same as 2008, except one new crop (large chickpeas) and decreases for dry peas and lentils. Wheat, bu., $2.75 Corn, bu., $1.95 Grain sorghum, bu., $1.95 Barley, bu., $1.85 Oats, bu., $1.33 Base quality upland cotton, lb., $0.52 Extra long staple cotton, lb., $0.7977 Long grain rice, cwt., $6.50 Medium grain rice, cwt., $6.50 Soybeans, bu., $5.00 Other oilseeds, cwt., $9.30 Dry peas, cwt., $5.40 Lentils, cwt., $11.28 Small chickpeas, cwt., $7.43 Large chickpeas, cwt., $11.28 Graded wool, lb., $1.00 Nongraded wool, lb., $0.40 Mohair, lb., $4.20 Honey, lb., $0.60 *[Sec. 1202(b)]* Peanuts, ton, $355 *[Sec. 1307(b)]* *2010-2012 crop years* (increases for wheat, barley, oats, minor oilseeds, graded wool, and honey) Wheat, bu., $2.94 Corn, bu., $1.95 Grain sorghum, bu., $1.95 Barley, bu., $1.95 Oats, bu., $1.39 Base quality upland cotton, lb., $0.52 Extra long staple cotton, lb., $0.7977 Long grain rice, cwt., $6.50 Medium grain rice, cwt., $6.50 Soybeans, bu., $5.00

(Continued)

PRIOR LAW/POLICY	HOUSE-PASSED BILL (H.R. 2419)	SENATE-PASSED SUBSTITUTE AMENDMENT (H.R. 2419)	ENACTED 2008 FARM BILL (P.L. 110-246)
			Other oilseeds, cwt., $10.09 Dry peas, cwt., $5.40 Lentils, cwt., $11.28 Small chickpeas, cwt., $7.43 Large chickpeas, cwt., $11.28 Graded wool, lb., $1.15 Nongraded wool, lb., $0.40 Mohair, lb., $4.20 Honey, lb., $0.69 *[Sec. 1202(c)]* Peanuts, ton, $355 *[Sec. 1307(b)]*
Adjustment of loans: establish a single loan rate in each county for each kind of "other oilseeds" *[7 U.S.C. 7932(c)]*	Same as prior law. *[Sec. 1202(b)]*	Same as prior law. *[Sec. 1202(b)]*	Same as prior law. *[Sec. 1202(d)]*
No comparable provision.	Establish a single county loan rate for corn and grain sorghum in each county; establish a single national average loan rate for corn and grain sorghum. *[Sec. 1202(c)(1)]*	Same as House provision *[Sec. 1202(d)]*	No comparable provision.
No comparable provision.	Administer the applicable loan, marketing loan, counter-cyclical and related programs using a single loan rate for corn and grain sorghum that is identical in each individual county. Any adjustment for location based on transportation shall be the same for corn and grain sorghum in each individual county. Allows adjustments for grade, type, and quality. *[Sec. 1202(c)(2)]*	Same as House provision, except does not specifically apply to counter-cyclical program. *[Sec. 1210(e)]*	No comparable provision.
Authorizes adjustments in the loan rates for any commodity based on differences in grade, type, quality, location, and other factors. Allows county loan rates as low as 95% of the U.S. average, if it does not increase outlays; prohibits adjustment of U.S. average loan rate. *[7 U.S.C. 7282]*	Amends prior law by excepting cotton and rice from the general provision for adjustment, with separate adjustment rules for cotton and rice. Encourages private sector consultation for cotton. For rice, prohibits adjustments except for grade and quality. *[Sec. 1505]*	Same as House provision, except the exception applies only to cotton, removes warehouse location differentials, and requires private sector consultation for cotton. *[Sec. 1210(a)-(d), (f)]* Same as prior law for peanuts. *[Sec. 1308]*	Adopts the Senate provision, with modifications to composition of private sector consultative committee. *[Sec. 1210]* Same basic provision for peanuts. *[Sec. 1308]*

(Continued)

PRIOR LAW/POLICY	HOUSE-PASSED BILL (H.R. 2419)	SENATE-PASSED SUBSTITUTE AMENDMENT (H.R. 2419)	ENACTED 2008 FARM BILL (P.L. 110-246)
Establish quality grades for dry peas as U.S. feed peas; for lentils as U.S. number 3 lentils; and for small chickpeas as U.S. number 3 small chickpeas that drop below a 20/64 screen. *[7 U.S.C. 7932(d)]*	No comparable provision.	Establishes grading basis for pulse crops based on a grade not less than grade number 2 or other factors, including fair and average crop quality (adjusted to reflect normal discounts for less than number 2 quality). *[Sec. 1202(c)]*	No comparable provision; however the statement of managers calls for USDA regulations that reflect number 2 quality.
Term of loans: 9 months after the day the loan is made; no extensions. *[7 U.S.C. 7933]* Same term for peanuts. *[7 U.S.C. 7957(c)]*	Same as prior law. *[Sec. 1203]*	Same as prior law. *[Sec. 1203]* Same provision for peanuts. *[Sec. 1307(c)]*	Adopts the Senate provision. *[Sec. 1203]* Same provision for peanuts. *[Sec. 1307(c)]*
Loan repayment: loans may be repaid at the lesser of (1) the loan rate plus interest, or (2) a rate determined by USDA that will minimize forfeitures, accumulation of stocks, storage costs, market impediments, and discrepancies in benefits across States and counties. Excludes upland cotton, rice, ELS cotton, confectionery and each other kind of sunflower seed (other than oil sunflower seed). *[7 U.S.C. 7934(a)]* Same provision for peanuts. *[7 U.S.C. 7957(d)]*	Same as prior law, except delineates long, medium, and short grain rice. *[Sec. 1204(a)]*	Same as prior law, except delineates long and medium grain rice. *[Sec. 1204(a)]* Same provision for peanuts. *[Sec. 1307(d)]*	Adopts Senate provision, except adds an option that the repayment rate is based on a 30-day average. *[Sec. 1204(a)]* Adopts Senate provision for peanuts. *[Sec. 1307(d)(1)]*
For upland cotton and rice, repayment may be at the lesser of the loan rate plus interest, or the prevailing world price for the commodity adjusted to U.S. quality and location. *[7 U.S.C. 7934(b)]*	Same as prior law, except delineates long, medium, and short grain rice. *[Sec. 1204(b)]*	Same as prior law, except delineates long and medium grain rice. *[Sec. 1204(b)]*	Adopts Senate provision. *[Sec. 1204(b)]*
For ELS cotton, repayment must be at the loan rate plus interest. *[7 U.S.C. 7934(c)]*	Same as prior law. *[Sec. 1204(c)]*	Same as prior law. *[Sec. 1204(c)]*	Same as prior law. *[Sec. 1204(c)]*
Prevailing world market prices for cotton and rice are determined and announced under USDA regulations, adjusted to U.S. quality and location. *[7 U.S.C. 7934(d)]*	Same as prior law, except specifies that the Far East price be used to determine the prevailing world market price. *[Sec.1204(d)]*	Same as prior law, except delineates long and medium grain rice. *[Sec. 1204(d)]*	Adopts the Senate provision. *[Sec. 1204(d)]*

(Continued)

PRIOR LAW/POLICY	HOUSE-PASSED BILL (H.R. 2419)	SENATE-PASSED SUBSTITUTE AMENDMENT (H.R. 2419)	ENACTED 2008 FARM BILL (P.L. 110-246)
Prevailing world market price for upland cotton adjusted if (a) it is less than 115% of the loan rate; and (b) the Friday through Thursday average price for the lowest priced U.S. growth for Middling (M) 1 3/32-inch cotton delivered C.I.F. Northern Europe is greater than the Friday through Thursday average price of five lowest-priced growths of upland cotton, as quoted for Middling (M) 1 3/32-inch cotton delivered C.I.F. Northern Europe. ***[7 U.S.C. 7934(e)(1)]***	Provides for adjustment to prevailing world market prices for rice and upland cotton. For rice, for U.S. quality and condition. For upland cotton, for U.S. quality and location (premiums for Comparable United States Quality and reduction to United States Premium Factor higher than Middling 1 3/32-inch; and costs to market the commodity. ***[Sec. 1204(e)]***	No comparable provision.	Adopts the House provision, without reference to Comparable United States Quality. ***[Sec. 1204(e)(1)-(2)(A)]***
Prevailing world market price for upland cotton further adjusted based on data including U.S. share of world exports, level of export sales and shipment, and other data USDA determines relevant. ***[7 U.S.C. 7934(e)(2)]***	Adjusts prevailing world market price for upland cotton further to minimize loan forfeitures and accumulation of stocks, improve marketing, and ensure competitiveness and transition between current-crop and future-crop price quotations. ***[Sec. 1204(f)]***	Same as House provision, except the further adjustment is to U.S. quality and location. With respect to transition, uses the term "insufficient" current-crop price quotations, rather than "less than three" current-crop price quotations in the House bill. ***[Sec. 1204(e)]***	Adopts the Senate provision. ***[Sec.1204(e)(2)(B)]***
For confectionary and other kinds of sunflower seeds (other than oil sunflower seed), loans may be repaid at the lesser of (1) the loan rate plus interest, or (2) the repayment rate for oil sunflower seed. ***[7 U.S.C. 7934(f)]***	Same as prior law. ***[Sec. 1204(g)]***	Same as prior law. ***[Sec.1204(f)]***	Same as prior law. ***[Sec. 1204(f)]***
For dry peas, lentils, and small chickpeas, loans shall be repaid at the quality grades for the applicable commodity specified in 7 U.S.C. 7932(d). ***[7 U.S.C. 7934(g)]***	Same as prior law. ***[Sec. 1204(h)]***	For pulse crops, loans shall be repaid at the quality grades for the applicable commodity as specified in Sec. 1202(c). ***[Sec. 1204(g)]***	No comparable provision.
Provide payment of storage for upland cotton, as allowed under general authorities of the CCC. ***[7 C.F.R. 1427.19(h)]***	Ends the practice of paying for upland cotton storage, handling and other costs starting with the 2011 crop. ***[Sec. 1510]***	Requires payment of cotton storage costs in same manner and at same rates as was provided for the 2006 crop, effective for 2008-12 crop years. ***[Sec. 1204(h)]***	Adopts Senate provision to pay cotton storage costs, except a 10% reduction applies to 2008-2011 crop years, and a 20% reduction in the 2012 crop year. ***[Sec. 1204(g)]***
No comparable provision.	No comparable provision.	No comparable provision.	Provides USDA authority to temporarily, an on a short-term basis only, adjust the

(Continued)

Prior Law/Policy	House-Passed Bill (H.R. 2419)	Senate-Passed Substitute Amendment (H.R. 2419)	Enacted 2008 Farm Bill (P.L. 110-246)
			repayment rates in the event of a severe disruption to marketing, transportation or related infrastructure. *[Sec. 1204(h)]* Same for peanuts. *[Sec. 1307(d)(2)]*
Loan deficiency payments (LDP): available to producers who agree to forego marketing loans. LDP computed by multiplying the payment rate (the amount that the loan rate exceeds the rate at which a marketing loan may be repaid) for the commodity times the quantity of the commodity produced. Loan deficiency payments available for unshorn pelts or hay and silage, even though they are not eligible for marketing loans. ELS cotton is not eligible. Payment rate determined using the rate in effect as of the date that producers request payment (producers do not need to lose beneficial interest). *[7 U.S.C. 7935]* Same provision for peanuts. *[7 U.S.C. 7957(e)]*	Same as prior law, except for 2008-2012 crop years. *[Sec. 1205]*	Same as House provision, except that for the 2008 crop year the payment rate is established as of the date that producers lose beneficial interest. *[Sec. 1205]* Same provision for peanuts. *[Sec. 1307(e)]*	Adopts House provision. *[Sec. 1205]* Same provision for peanuts. *[Sec. 1307(e)]*
Payments in lieu of LDP for grazed acreage of wheat, barley, oats, or triticale. *[7 U.S.C. 7936]*	Same as prior law. *[Sec. 1206]*	Same as prior law. *[Sec. 1206]*	Same as prior law. *[Sec. 1206]*
Special Marketing Loan Provisions for Upland Cotton: imposes a special import quota on upland cotton when U.S. prices exceed Northern European prices by more than 1.25¢ for 4 weeks. *[7 U.S.C. 7937(b)]* Limited global import quota is imposed on upland cotton when U.S. prices average 130% of the previous 3-year average of U.S. prices *[7 U.S.C. 7937(c)]*	Same as prior law, except uses Far East price. Special import quota defined. Limits imports under quota to 10 weeks of consumption by domestic mills. *[Sec. 1207(a)]* Same as prior law. *[Sec. 1207(b)]*	Same as House provision, except specifies the price of American cotton "delivered to a definable and significant international market." *[Sec. 1207(a)]* Same as prior law. *[Sec. 1207(b)]*	Adopts the Senate provision. *[Sec. 1207(a)]* Same as prior law. *[Sec. 1307(b)]*

(Continued)

Prior Law/Policy	House-Passed Bill (H.R. 2419)	Senate-Passed Substitute Amendment (H.R. 2419)	Enacted 2008 Farm Bill (P.L. 110-246)
No comparable provision.	Provides Economic Adjustment Assistance to Users of Upland Cotton via marketing certificates or cash payments of 4¢/lb. to domestic upland cotton users for all cotton uses regardless of origin for acquisition, construction, installation, modernization, development, conversion, or expansion of land, plant, buildings, equipment, facilities, or machinery. Effective through July 31, 2013. *[Sec. 1207(c)]*	Provides Economic Adjustment Assistance to Users of Upland Cotton via assistance of 4¢/lb. to domestic users of upland cotton for uses of all cotton regardless of origin for the same purposes as the House provision. Effective August 1, 2008, through June 30, 2013; payment rate drops to 0¢ on July 1, 2013, which terminates future funding. *[Sec. 1207(c)]*	Adopts the Senate provision, with modification. Effective August 1, 2008, through July 31, 2012 at 4¢/lb.; payment rate drops to 3¢ on August 1, 2012. *[Sec. 1207(c)]*
Special competitiveness program for ELS cotton provides marketing certificates or cash payments available to domestic users and exporters whenever the world market price for the lowest priced ELS cotton is below the prevailing U.S. price for a competing growth of ELS cotton for a 4-week period; and the lowest priced competing growth of ELS cotton is less than 134% of the loan rate for ELS cotton. Effective May 13, 2002, through July 31, 2008. *[7 U.S.C. 7938]*	Same as prior law. *[Sec. 1208]*	Same as prior law, except it does not specify form of payment (cash or certificates). *[Sec. 1208]*	Adopts the Senate provision. *[Sec. 1208]*
Recourse loans for high moisture feed grains and seed cotton: for farms that normally harvest corn or sorghum in a high moisture condition, recourse loans are available at rates set by the USDA. Recourse loans for seed cotton. Repayment at loan rate plus interest. *[7 U.S.C. 7939]*	Same as prior law. *[Sec. 1209]*	Same as prior law. *[Sec. 1209]*	Same as prior law. *[Sec. 1209]*
No comparable provision.	Requires a deadline for peanut loan repayment no later than June 30 of the year subsequent to the year in which the peanuts were harvested. Loan not redeemed by the deadline are deemed forfeited. *[Sec. 1210]*	No comparable provision.	No comparable provision.

(Continued)

Prior Law/Policy	House-Passed Bill (H.R. 2419)	Senate-Passed Substitute Amendment (H.R. 2419)	Enacted 2008 Farm Bill (P.L. 110-246)
No comparable provision.	Authorizes quality incentive payments for healthy oilseeds with special traits to enhance human health. Provides for discretionary appropriations of such sums as necessary. Crop years 2009-2013. USDA to solicit proposals; successful applicants enter contracts with producers and are reimbursed after premiums paid to producers. *[Sec. 1211]*	Similar to House provision, except it has fewer requirements for proposals, does not specify multi-year contracts, and provides for protection of proprietary information. Does not specify crop years, but authorizes discretionary appropriations of $400 million for the period FY2008-12. *[Sec. 1705]*	Adopts the Senate provision, with modifications, and adds compliance and penalty provisions. Authorization of such sums as necessary for FY2009-2012. *[Sec. 1605]*
Payment Limits			
Establishes payment limits on direct payments, counter-cyclical payments, and certain marketing loan benefits under the Food Security Act of 1985, as amended, to a "person" as broadly defined below *[7 U.S.C. 1308-1308-3(a)]*	Continues limits for direct payments and counter-cyclical payments, as amended below. Deletes payment limit for marketing loan program. Establishes direct attribution to natural person; eliminates 3 entity rule. *[Sec. 1503]*	Continues limits for direct payments and counter-cyclical payments, as amended below. Deletes payment limit for marketing loan program. Establishes direct attribution to natural person; eliminates 3 entity rule. *[Sec. 1703(a)]*	Continues prior law limits for direct payments and counter-cyclical payments. Deletes payment limit for marketing loan program. Establishes direct attribution to natural person; eliminates 3 entity rule. *[Sec. 1603(a)]*
Person: defined as an individual, partner in a general partnership or joint venture, trust, corporation, joint stock company, limited partnership, association, charitable organization, State agency, or political subdivision (except cooperative producer associations). *[7 U.S.C. 1308(e)]*	**Person:** "a natural person, and does not include a legal entity." *[Sec. 1503(b)(1)]*	**Person:** same as House definition. *[Sec. 1703(b)(2)]*	**Person:** adopts the House definition. *[Sec. 1603(b)(1)]*
No comparable definition.	**Legal entity:** an entity created under federal or state law that (1) owns land or an agricultural commodity, or (2) produces an agricultural commodity.	**Legal entity:** same as House definition. *[Sec. 1703(b)(2)]*	**Legal entity:** adopts the House definition. *[Sec. 1603(b)(1)]*
No comparable definition.	No comparable definition.	**Family member:** "an individual to whom a member in the farming operation is related as lineal ancestor, lineal descendant, sibling, or spouse." *[Sec. 1703(b)(1)]*	**Family member:** "a person to whom a member in the farming operation is related as lineal ancestor, lineal descendant, sibling, spouse, or otherwise by marriage." *[Sec. 1603(b)(1)]*
Maximum amount of payments per year to a person for the sum of all	**Maximum amount of payments** per year to a person or legal entity for the	**Maximum amount of payments** per year to a person or legal entity for the	**Maximum amount of payments** per year to a person or legal entity for the

(Continued)

PRIOR LAW/POLICY	HOUSE-PASSED BILL (H.R. 2419)	SENATE-PASSED SUBSTITUTE AMENDMENT (H.R. 2419)	ENACTED 2008 FARM BILL (P.L. 110-246)
covered commodities (except peanuts, wool, mohair, and honey): — Direct payments: $40,000 — Counter-cyclical payments: $65,000 — Marketing loan gains/LDP: $75,000 *[7 U.S.C. 1308(b)(1), (c)(1), (d)(1)]*	sum of all covered commodities, except peanuts: — Direct payments: $60,000 — Counter-cyclical payments: $65,000 *[Sec. 1503(a)(1)-(2)]* — Marketing loan gains/LDP: no limit. *[Sec. 1503(b)(2)]*	sum of all covered commodities, except peanuts: — Direct payments and fixed ACR payment: $40,000 — Counter-cyclical payments and revenue-based ACR payment: $60,000 — Marketing loan gains/LDP: no limit *[Sec. 1703(b)(2)]*	sum of all covered commodities, except peanuts: — Direct payments: $40,000 — Direct payments under ACRE: $40,000 minus the reduction required of an ACRE participant in Sec. 1105(a)(1). — Counter-cyclical payments: $65,000 — ACRE payments: $65,000 plus the reduction in the direct payment limit. — Marketing loan gains/LDP: no limit *[Sec. 1603(b)(2)]*
Maximum payment amount per year to a person for the sum of peanuts, wool, mohair, and honey: — Direct payments: $40,000 — Counter-cyclical payments: $65,000 — Marketing loan gains/LDP: $75,000 *[7 U.S.C. 1308(b)(2), (c)(2), (d)(2)]*	**Maximum payment amount** per year to a person or legal entity for peanuts: — Direct payments: $60,000 — Counter-cyclical payments: $65,000 *[Sec. 1503(a)(1)-(2)* — Marketing loan gains/LDP: no limit *[Sec. 1503(b)(2)]*	**Maximum payment amount** per year to a person or legal entity for peanuts: — Direct payments and fixed ACR payment: $40,000 — Counter-cyclical payments, and revenue-based ACR payment: $60,000 — Marketing loan gains/LDP: no limit *[Sec. 1703(b)(2)]*	**Maximum payment amount** per year to a person or legal entity for peanuts: — Direct payments: $40,000 — Direct payments under ACRE: $40,000 minus the reduction required of an ACRE participant in Sec. 1105(a)(1). — Counter-cyclical payments: $65,000 — ACRE payments: $65,000 plus the reduction in the direct payment limit. — Marketing loan gains/LDP: no limit *[Sec. 1603(b)(2)]*
No comparable provision.	**Direct attribution:** the total amount of direct and counter-cyclical payments are attributed to a person by taking into account direct and indirect ownership in a legal entity. Payments made directly to a person will be combined with the person's pro rata share of payments to a legal entity. Payments to a legal entity shall not exceed the limits above, and shall be attributed to persons with an	**Direct attribution:** same as House provision, except payments to a legal entity are reduced proportionately based on the ownership shares of a person or entity that exceeds the limit. *[Sec. 1703(b)(3)]*	**Direct attribution:** Adopts the Senate provision. *[Sec. 1603(b)(3)]*

(Continued)

Prior Law/Policy	House-Passed Bill (H.R. 2419)	Senate-Passed Substitute Amendment (H.R. 2419)	Enacted 2008 Farm Bill (P.L. 110-246)
	ownership interest. Attribution of payments to legal entities shall be traced through four levels of ownership (ownership of an entity by a person or another entity). If after four levels of ownership, the payment has not been allocated to a natural person, the payment to the first-level entity shall be reduced on a pro-rata basis. For joint ventures and general partnerships, payments shall not exceed the multiple of the limits for the number of persons and legal entities comprising the joint venture or general partnership. ***[Sec. 1503(b)(2)]***		
Payments to minor children generally are attributed to parents; marketing coops are not subject to the limits; trusts and estates qualify under certain rules; cash rent tenants that make a significant contribution of management but not of labor and equipment are ineligible; spouses are treated together except under certain conditions. ***[7 U.S.C. 1308(e)]***	Continue prior law rules for minor children, marketing cooperatives, trusts and estates, and cash rent tenants. ***[Sec. 1503(b)(2)]***	Same as House provision. ***[Sec. 1703(b)(3)]***	Adopts House provision. ***[Sec. 1603(b)(3)]***
States and political subdivisions are allowed to receive payments under the definition of person. Payment limits do not apply to land owned by a public entity to maintain a public school. ***[7 U.S.C. 1308(f)]***	Makes federal agencies, states and political subdivisions ineligible for payments, but tenants on such government-owned land may receive payments. An exception allows states and political subdivisions to receive payments to maintain a public school, but payment limits apply ***[Sec. 1503(b)(2)]***. However, existing law (7 U.S.C. 1308(f)) remains in effect, which exempts states and political subdivisions from payment limits to maintain a public school.	Same as House provision, except has no new exception for state and local governments to receive payments to maintain a public school. Such an exemption remains in prior law provision (7 U.S.C. 1308(f)), which is redesignated as subparagraph (g). ***[Sec. 1703(b)(3)]***	Same as House provision, except replaces 7 U.S.C. 1308(f) with new paragraph (g) that allows states and political subdivisions to receive payments to maintain a public school. A separate payment limit of $500,000 on total direct, counter cyclical, and ACRE payments applies to each state, except for states with less than 1.5 million population. ***[Sec. 1603(b)(3)]***
3-entity rule: no person may receive payments from more than three entities in which the person holds substantial	Repeals the 3-entity rule. Requires notification to USDA, including names and social security number or tax	Same as House provision. ***[Sec. 1703(c)]***	Adopts House provision. ***[Sec. 1603(c)]***

(Continued)

PRIOR LAW/POLICY	HOUSE-PASSED BILL (H.R. 2419)	SENATE-PASSED SUBSTITUTE AMENDMENT (H.R. 2419)	ENACTED 2008 FARM BILL (P.L. 110-246)
beneficial interest. *[7 U.S.C. 1308-1(a)]*	identification number. *[Sec. 1503(c)]*		
Requires a person or entity to be actively engaged in farming based on contributions of land, labor, equipment, and management, and requires profits be commensurate and at risk. *[7 U.S.C. 1308-1(b)]*	Continues prior law provisions and adds an exception that if one spouse is determined to be actively engaged, the other spouse shall be determined to meet requirements of personal labor or active personal management. *[Sec. 1503(d)]*	Same as House provision. *[Sec. 1703(d)]*	Adopts House provision. *[Sec. 1603(d)*
Disqualifies a person from payments in the crop year if the person adopted a scheme or device to evade payment limits. *[7 U.S.C. 1308-2]*	Disqualifies a person or entity for a 2-5 year period for evasion of payment limit rules. Benefits denied on a pro-rata basis according to ownership. *[Sec. 1503(b)(2)]*	Same as House provision, except adds joint and several liability for members of an entity regarding amounts payable to USDA, and authority for USDA to release a person from liability if they cooperate. *[Sec. 1703(e)]*	Adopts Senate provision. *[Sec. 1603(e)]*
No comparable provision.	Prior law shall apply to payments made for the 2007 crop year. *[Sec. 1503(e)]*	Same as House provision. *[Sec. 1703(g)]*	Prior law shall apply to payments made for the 2007 and 2008 crop years. *[Sec. 1603(h)]*
Adjusted Gross Income Limitation			
No firm cap (a cap without exceptions).	Sets a firm AGI cap of $1 million (no exceptions) to be eligible to receive direct and counter-cyclical payments, marketing loan gains or LDPs, and conservation benefits. Applies through the 2012 crop year. *[Sec. 1504(b)(1)]*	No firm cap.	Divides AGI into two parts: farm AGI and non-farm AGI; both are averages over a 3 year period. — Sets a firm cap of $500,000 non-farm AGI to receive any commodity program benefits, MILC, noninsured crop assistance, or disaster payments. — Sets a firm cap of $750,000 farm AGI to receive direct payments (but counter-cyclical, ACRE and marketing loan benefits may continue if farm AGI exceeds $750,000). *[Sec. 1604(a)]*
Sets a soft cap of $2.5 million Adjusted Gross Income Limitation (AGI) over a 3-year average for individuals or entities to be eligible to receive program payments.	Sets a soft AGI cap of $500,000, unless 66.66% of the 3-year average AGI is derived from farming, ranching, or forestry operations. Applies through the	Sets a gradually-declining soft AGI cap for direct payments, counter-cyclical payments, and marketing loan gains or LDPs, unless 66.66% of the 3-year	No comparable provision.

(Continued)

Prior Law/Policy	House-Passed Bill (H.R. 2419)	Senate-Passed Substitute Amendment (H.R. 2419)	Enacted 2008 Farm Bill (P.L. 110-246)
Limit may be exceeded if at least 75% of the AGI is derived from farming, ranching, or forestry operations. Applies to direct payments, counter-cyclical payments, marketing loan benefits, and conservation program payments for the 2003-07 crop years. *[7 U.S.C. 1308-3a]*	2012 crop year. *[Sec. 1504(b)(1)]*	average AGI is derived from farming, ranching, or forestry operations: — $2.5 million in crop year 2008, — $1 million in 2009, and — $750,000 in 2010-2012 *[Sec. 1704(c)]*	
AGI cap for conservation programs same as for commodity programs.	AGI cap for conservation programs same as for commodity programs.	For conservation programs, continues prior law level of $2.5 million AGI, with exception for 75% of AGI derived from farming, ranching or forestry. *[Sec. 1704(c)]*	For conservation programs, sets a soft cap of $1 million non-farm AGI, unless more than 66.66% of AGI is farm AGI. Provides USDA discretion to waive the limit for "environmentally sensitive land of special significance." *[Sec. 1604(a)]*
USDA determines types of income as derived from farming, ranching or forestry income. *[7 U.S.C. 1308-3a(b)(1)]*	Defines certain types of income as farming, ranching or forestry. Includes production of crops, livestock or raw forestry products; sale of land or rights; sale of equipment but not as a dealer; rental of land; supplying inputs and services to farmers; processing, storing and transporting agricultural products. *[Sec. 1503(b)(3)]*	Same as House provision, except does not limit sale of equipment to non-dealers and does not reference depreciable equipment; includes income from water or hunting rights; includes packing and shedding in processing and storing; and includes government payments from commodity and conservation programs. *[Sec. 1704(c)]*	Adopts Senate provision, with additional and expansion for the inclusion of livestock, insurance indemnities. Specifies that sale of inputs to farmers can be included if more than 66.66% of income is from farming. Generally, not to exclude anything reported on IRS Schedule F. *[Sec. 1604(a)]*
No comparable provision.	No comparable provision.	Allows the allocation of AGI among the individuals filing a joint tax return, under certain conditions. *[Sec. 1704(b)]*	Adopts the Senate provision. *[Sec. 1604(a)]*
Administrative Provisions			
Authorizes use of funds, facilities, and authorities of the Commodity Credit Corporation (CCC) to carry out Title I. Determinations by USDA shall be final. Allows promulgation of regulations, and adjusting expenditures if they will exceed allowable support levels under the Uruguay Round Agreements. *[7 U.S.C. 7991(a)-(c),(e)]*	Same as prior law. *[Sec. 1501]*	Same as prior law. *[Sec. 1701(a)-(d)]*	Same as prior law. *[Sec. 1601(a)-(d)]*

(Continued)

Prior Law/Policy	House-Passed Bill (H.R. 2419)	Senate-Passed Substitute Amendment (H.R. 2419)	Enacted 2008 Farm Bill (P.L. 110-246)
Advanced direct and counter-cyclical payments are taxable in the year received (rather than when producer has option to receive payment). *[7 U.S.C. 7991(d)]*	No comparable provision.	Same as prior law, applied to the 2008 farm bill. *[Sec. 1701(e)]*	Adopts Senate provision. *[Sec. 1601(e)]*
Suspends the permanent price support authority of the Agricultural Adjustment Act of 1938 and the Agricultural Adjustment Act of 1949 for the 2002-07 crops (covered commodities, peanuts, and sugar), and for milk through December 31, 2007. *[7 U.S.C. 7992]*	Same as prior law, except applies to 2008-2012 crop years, and milk through December 31, 2012. *[Sec. 1502]*	Same as House provision, except does not mention peanuts in paragraph (a). *[Sec. 1702]*	Adopts House provision. *[Sec. 1602]*
Exempts producers from liability for certain deficiencies in collateral to secure any nonrecourse loan. *[7 U.S.C. 7284]*	Same as prior law. *[Sec. 1506]*	Same as prior law. *[Sec. 1709]*	Same as prior law. *[Sec. 1606]*
Authorizes the use of commodity certificates, including to repay marketing loans. *[7 U.S.C. 7286]*	Same as prior law. *[Sec. 1507]*	Same as prior law. *[Sec. 1710]*	Same as prior law, except terminates authority to use commodity certificates to repay loans after the 2009 crop year. *[Sec. 1607]*
Requires that assignment of payments must be done in accordance with USDA regulations. *[7 U.S.C. 7995]*	Same as prior law. *[Sec. 1508]*	Same as prior law. *[Sec. 1711]*	Same as prior law. *[Sec. 1608]*
Requires tracking of program benefits under commodity and conservation titles that are made directly or indirectly to individuals and entities. *[7 U.S.C. 7997]*	Same as prior law. *[Sec. 1509]*	No comparable provision.	Same as prior law. *[Sec. 1609]*
Prohibits publication of cotton price forecasts in any governmental report, or bulletin. *[12 U.S.C. 1141j]*	Strikes the prior law prohibition on the publication of cotton price forecasts. *[Sec. 1511]*	Same as House provision. *[Sec. 1714]*	Adopts House and Senate provision. *[Sec. 1610]*
Allows payments to estates of deceased farmers *[7 U.S.C. 1308(e)(2)(B)(ii)]*, but without reference to a time period. USDA regulations establish a 2-year period for estates to qualify. *[7 C.F.R. 1400.206]*	Requires reports to Congress of deceased persons that received payments for more than two crop years following death. Establishes deadlines for notification of death, and denies payments and recoup losses for failure to comply. Reconcile	Prohibits any agricultural payment to any deceased individual or estate after two program years after the date of death. Require annual reports to Congress on the number and amount of payments to deceased individuals and the length of	Generally adopts the House approach. Requires regulations that describe the circumstances allowing payments to a deceased person to settle and estate, and to stop payments for those ineligible. Reconcile tax identification numbers with

(Continued)

Prior Law/Policy	House-Passed Bill (H.R. 2419)	Senate-Passed Substitute Amendment (H.R. 2419)	Enacted 2008 Farm Bill (P.L. 110-246)
	tax identification numbers with Internal Revenue Service (IRS) data twice a year to determine living status. *[Sec. 1512]*	time the estate has been open. *[Sec. 11073]*	IRS data twice a year to determine living status. *[Sec. 1611]*
Provide incentive payments to producers of hard white wheat on up to 2 million acres. Total mandatory funding of $20 million for the 2003-2005 crop years. *[7 U.S.C. 7999]*	No comparable provision.	Provides incentive payments to producers of hard white wheat of at least $20¢/bu. and at least $2/acre on up to 2.9 million acres. Mandatory funding of $35 million for the 2008-12 crop years. *[Sec. 1706]*	Adopts Senate provision, except authorizes discretionary appropriations rather mandatory funding. *[Sec. 1612]*
No comparable provision.	No comparable provision.	Authorizes compensation up to 50% of the cost of fungicides to control wheat scab in durum wheat. Authorize $10 million per year for FY2008-12, subject to appropriation. *[Sec. 1707]*	Adopts Senate provision. *[Sec. 1613]*
Provides farm storage facility loans under USDA regulations via the general authorities of the CCC. For commodities other than sugar, maximum term of loan is 7 years and $100,000 per borrower. *[7 C.F.R. 1436]*	No comparable provision.	Establishes a storage facility loan program for producers of grains, oilseeds, pulse crops, hay, renewable biomass, and other storable commodities (other than sugar) to construct or upgrade storage and handling facilities. Provides for 12-year terms and $500,000 maximum loans, as well as security and eligibility requirements. *[Sec. 1708]*	Adopts Senate provision, with modification to security and lien requirements. *[Sec. 1614]*
Authorizes cotton classification services be available to producers of cotton, and for the collection of fees and appropriations to pay for such services. *[7 U.S.C. 473a]*	Revises the authorization for cotton classification services through FY2012 to include leasing of property exceeding 5 years. *[Sec. 11302]*	Revises the authorization for cotton classification services for an indefinite time period, including consultation with the cotton industry, investment of funds, and long term lease of property. Provides authorization for appropriations. *[Sec. 1712]*	Adopts the Senate provision, with modification. *[Sec. 14201]*
Defines cotton-producing state, for purposes of a cotton research and promotion, using a historical measure of production. *[7 U.S.C. 2116(f)]*	Revises definition of cotton-producing state to explicitly include Kansas, Virginia, and Florida beginning with the 2008 crop. *[Sec. 11301]*	Same as House provision. *[Sec. 1713]*	Adopts the House provision. *[Sec. 14202]*
Instructs USDA to appoint committees of farmers in a fair and representative	No comparable provision.	For combined or consolidated area committees, requires 3-11 members that	Adopts the Senate provision, with modification for USDA to develop

(Continued)

PRIOR LAW/POLICY	HOUSE-PASSED BILL (H.R. 2419)	SENATE-PASSED SUBSTITUTE AMENDMENT (H.R. 2419)	ENACTED 2008 FARM BILL (P.L. 110-246)
manner. *[16 U.S.C. 590h(b)(5)(B)(ii)]*		are representative of the area and elected, and ensures representation of socially disadvantaged farmers. *[Sec. 1715]*	procedures to main representation of socially disadvantaged farmers. *[Sec. 1615]*
Authorizes USDA to collect commodity assessments from proceeds of marketing assistance loans, if assessment is required under state law. *[7 U.S.C. 7416a]*	No comparable provision.	Prohibits USDA from charging fees for the collection of commodity assessments in its agreement with the State. *[Sec. 1716]*	Adopts Senate provision. *[Sec. 1616]*
No comparable provision.	No comparable provision.	Requires that, if USDA approves a document containing signatures of applicants, it shall not subsequently determine the document to be inadequate or invalid. *[Sec. 1717]*	Adopts the Senate provision, with modification. *[Sec. 1617]*
No comparable provision.	No comparable provision.	Requires USDA to modernize the Farm Service Agency information technology systems to ensure timely and efficient program delivery. *[Sec. 1718]*	Requires a report by an outside party that describes USDA's technology problems and a plan to improve service. *[Sec. 1618]*
No comparable provision.	No comparable provision.	Requires USDA to consolidate geospatial database systems into a single system that is readily available to all agencies within two years of enactment. *[Sec. 1719]*	Adopts the Senate provision, with modification to limit disclosure of information. *[Sec. 1619]*
No comparable provision.	No comparable provision.	Allows the CCC to lease space for USDA agencies if the space is jointly occupied by the agencies. *[Sec. 1720]*	Requires a report on the cost of leasing procedures of the General Services Administration compared to USDA. *[Sec. 1620]*
No comparable provision.	No comparable provision.	Provides payments to "geographically disadvantaged farmers" in insular areas, Alaska, and Hawaii for transporting a commodity or input more than 30 miles. Reimbursement based on federal salary differentials defined elsewhere, with maximum of 25% transportation cost. Authorizes $15 million of discretionary appropriations annually. *[Sec. 6021]*	Adopts the Senate provision, with modification. *[Sec. 1621]*

(Continued)

Prior Law/Policy	House-Passed Bill (H.R. 2419)	Senate-Passed Substitute Amendment (H.R. 2419)	Enacted 2008 Farm Bill (P.L. 110-246)
No comparable provision.	No comparable provision.	No comparable provision.	Provides $50 million of mandatory funds from the CCC to implement Title I. *[Sec. 1622]*
Establishes a "Commission on the Application of Payment Limitations for Agriculture." *[7 U.S.C. 7993]*	No comparable provision.	Repeals the authorization for the payment limits commission. *[Sec. 1721(a)]*	Adopts the Senate amendment to repeal the commission. *[Sec. 1623(a)]*
Authorizes market loss assistance and other emergency assistance to persons that were eligible to receive assistance but did not receive assistance before a certain date. *[7 U.S.C. 8000]*	No comparable provision.	Repeals market loss assistance and other emergency assistance to persons that failed to receive assistance under earlier authorities. *[Sec. 1721(b)]*	Adopts the Senate amendment to repeal the continued assistance. *[Sec. 1623(b)]*
DAIRY (TITLE I)			
Dairy Price Support Program			
Mandatory support for farm price of milk at $9.90 per hundredweight (cwt.). Program authority expired on December 31, 2007, but was extended until March 15, 2008 by P.L. 110-161. *[7 U.S.C. 7981a-c]*	Mandates the direct support of cheese, nonfat dry milk, and butter at specified prices for five years (through December 31, 2012). This is a change from supporting the farm price of milk. *[Secs. 1401(a)-(b)]*	Similar to the House bill. *[Secs. 1601(a)-(b)]*	Adopts House provision. *[Sec. 1501(b)]*
Farm support price of $9.90 indirectly maintained by USDA offer to purchase butter, cheese, and nonfat dry milk from processors at prices determined by USDA that allow buyers to pay farmers at least the support price. *[7 U.S.C. 7981a-c]*	Specifies minimum purchase prices of: block cheese, $1.13/lb.; barrel cheese, $1.10/lb.; butter, $1.05/lb.; and nonfat dry milk, $0.80/lb (same levels currently used to support the farm price at $9.90 per cwt.) Allows USDA sale of acquired products when market prices rise to 110% of purchase price. *[Sec. 1401(b)]*	Similar to the House bill. *[Secs. 1601(b)-(c)]*	Adopts House provision. *[Sec. 1501(c)]*
No more than twice annually, USDA can adjust the purchase prices of butter and nonfat dry milk (reduce one and raise the other) in order to minimize acquisitions. *[7 U.S.C. 7981d]*	Allows reduction of mandated purchase prices when USDA acquisitions exceed specified levels. *[Sec. 1401(c)]*	No comparable provision.	Adopts House provision. *[Sec. 1501(d)]*

(Continued)

Prior Law/Policy	House-Passed Bill (H.R. 2419)	Senate-Passed Substitute Amendment (H.R. 2419)	Enacted 2008 Farm Bill (P.L. 110-246)
Milk Income Loss Contract Payments			
The 2002 farm bill mandated a new counter-cyclical payment program, the Milk Income Loss Contract (MILC) program. When the monthly fluid milk price falls below $16.94/cwt., all dairy farmers are paid an amount equal to 34% of the difference between $16.94 and the lower market price. Payments per farm are limited to 2.4 million lbs. of annual production. MILC authority expired Sept. 30, 2005, but several subsequent extensions continue it through March 15, 2008. ***[7 U.S.C. 7982]***	Extends the MILC program for five years, through September 30, 2012, at the current target price of $16.94/cwt. Payment rate remains at 34% of any deficiency between the market price and the target price, and eligible production continues to be capped at 2.4 mil. lbs. per farm per year. ***[Sec. 1406]***	Increases, through August 31, 2012, the payment rate to 45%, and raises the cap on eligible annual production to 4.15 mil. lbs. per farm. Payment rate and production cap would return to 34% and 2.4 mil. lbs. for the last month of program authority in September 2012. ***[Sec. 1602]***	For the period October 1, 2008 through August 31, 2012, increases the payment factor to 45%, and the annual eligible payment quantity to 2.985 million pounds. After that, payment factor and payment quantity revert to 34% and 2.4 million pounds, respectively. The $16.94 per cwt. payment rate must be adjusted to reflect feed cost increases above trigger levels, as specified in the final law. ***[Sec. 1506]***
Dairy Forward Pricing Program			
The FY2000 omnibus appropriations act authorized a pilot dairy forward pricing program implemented from mid-2000 until its required expiration date of December 31, 2004. It exempted handlers from having to pay farmers the federal order price when the forward contract price turns out to be lower. ***[7 U.S.C. 627]***	Authorizes a dairy forward pricing program similar to the pilot program of 2000-2004. Price paid by milk handlers under the contracts are deemed to satisfy the minimum price requirements of federal milk marketing orders. Applies only to milk purchased for manufactured products (Classes II, III, and IV), and excludes milk purchased for fluid consumption (Class I). Allows for new contracts until September 30, 2012, but no contract can extend beyond September 30, 2015. ***[Sec. 1402]***	Similar (but not identical) to the House bill. ***[Sec. 1606]***	Adopts House provision. ***[Sec. 1502]***
Dairy Export Incentive Program			
Provides cash bonus payments to U.S. dairy exporters, subject to World Trade Organization obligations to limit export subsidies. No DEIP bonuses have been awarded since FY2004. Legislative authority expires March 15, 2008. Intended to counter foreign (mostly EU)	Extends DEIP through December 31, 2012, with a reference to the Uruguay Round Agreements Act. ***[Sec. 1403]***	Extends DEIP through December 31, 2012. ***[Sec. 1603(a)]***	Adopts House provision. ***[Sec. 1503]***

(Continued)

PRIOR LAW/POLICY	HOUSE-PASSED BILL (H.R. 2419)	SENATE-PASSED SUBSTITUTE AMENDMENT (H.R. 2419)	ENACTED 2008 FARM BILL (P.L. 110-246)
dairy subsidies. *[15 U.S.C. 713a-14(a)]*			
Dairy Indemnity Program			
Authorizes payments to dairy farmers when a public regulatory agency directs removal of their raw milk from the market because of contamination by pesticides, nuclear radiation or fallout, or toxic substances and other chemical residues. Expires March 15, 2008. *[7 U.S.C. 450l]*	Extends the Dairy Indemnity Program fthrough December 31, 2012. *[Sec. 1405]*	Similar to the House bill. *[Sec. 1603(b)]*	Adopts House provision. *[Sec. 1505]*
Dairy Promotion and Research Program			
The Dairy Producer Stabilization Act of 1983 authorized a generic dairy product promotion, research, and nutrition education program, funded by a mandatory 15¢/cwt assessment on milk produced/marketed in the 48 contiguous states. Assessment extended to imports by Sec. 1505 of 2002 farm bill. Import assessment never collected because the exclusion of some states was considered inconsistent with WTO rules. Expires March 15, 2008. *[7 U.S.C. 4501-4514]*	Extends promotion and research program authority through Sep. 30, 2012. Amends the 1983 Act to require producers in all 50 states, the District of Columbia, and Puerto Rico to pay the 15¢/cwt. *[Sec. 1407]*	Extends program authority through Sep. 30, 2012. Does not address the issue involving the import assessment. *[Sec. 1604]*	Adopts House provision with changes to reduce the assessment rate on imported products to 7.5¢/cwt. Authorizes USDA to issue regulations on time and method of importer payments. *[Sec. 1507]*
Federal Milk Marketing Orders			
Federal milk marketing order rules issued by USDA place requirements on the first buyers or handlers of milk, including paying at least minimum prices for the milk depending on its end use. Perm-anent federal authority to regulate the handling of milk was first provided in the Agricultural Adjustment Act of 1933, and subsequently revised by the Agri-cultural Marketing Agreement Act of 1937, as amended. *[7 U.S.C. 601 et seq.]*	Creates a Federal Milk Marketing Order Review Commission to review and evaluate the current federal and similar state order systems. The 18-member Commission is to consider legislative and administrative options for: ensuring the competitiveness of farmers and processors, and simplifying and streamlining the federal order system. Report is due within two years of the first meeting. *[Sec. 1409]*	Creates a Federal Milk Marketing Order Review Commission, with same overall functions and purposes as the House bill, but with some differences in the appointment of members and issues to be studied. *[Sec. 1608]*	Creates a Federal Milk Marketing Order Review Commission with 14 members appointed by USDA; objectives of the commission are similar to but modified from the House version. *[Sec. 1509]*

(Continued)

PRIOR LAW/POLICY	HOUSE-PASSED BILL (H.R. 2419)	SENATE-PASSED SUBSTITUTE AMENDMENT (H.R. 2419)	ENACTED 2008 FARM BILL (P.L. 110-246)
When USDA amends federal orders, it must issue a notice of a hearing at least three days prior to the hearing. *[7 U.S.C. 608c(17)]*	Revises order amendment procedures by placing time constraints on USDA at various steps of the amendment process. *[Sec. 1404]*	Also revises amendment procedures by establishing a timetable for certain actions, but with some differences. *[Sec. 1605]*	Includes elements of both bills with regard to the time constraint provisions, avoidance of duplication, and use of feed and fuel costs for hearings involving adjustments to make allowances. *[Sec. 1504]*
In late April 2007, USDA announced an error in nonfat dry milk prices reported to them by manufacturers over the previous 12 months. The error contributed to lower farm milk prices than would otherwise have been the case.	Requires USDA, within 90 days of enactment, to submit a report to Congress on price reporting procedures for nonfat dry milk, and the effect these procedures have had on marketing order pricing since July 1, 2006. *[Sec. 1408]*	Similar to the House bill, except that the report is to be filed with the House and Senate Agriculture Committees. *[Sec. 1607)*	Adopts Senate provision. *[Sec. 1508]*
Mandatory Dairy Commodity Price Reporting			
Dairy Market Enhancement Act of 2000 requires manufacturers to report to USDA the price, quantity, and moisture content of dairy products sold. *[7 U.S.C. 1637b]*	No comparable provisions.	Requires manufacturers to report sales transactions daily. Requires USDA to publish the data each reporting day and compare it with other dairy market statistics on a quarterly basis. *[Secs. 1609 and 1610]*	Authorizes USDA to establish an electronic reporting system (subject to available funds), after which increased frequency in mandatory reporting of dairy product sales would be required. Provides for quarterly audits of submitted information and comparison with related dairy market statistics. *[Sec. 1510]*
SUGAR (TITLE I)			
No Net Cost Directive			
Requires USDA to the maximum extent practicable to operate the sugar nonrecourse loan program at no net cost by avoiding sugar forfeitures to the CCC. *[7 U.S.C. 7272 (g), 7 U.S.C. 1359bb (b), 7 U.S.C. 1359cc (b)(2)]*	Retains current no-net-cost requirement. *[Secs. 1301 and 1303(b)]*	Similar to the House bill. *[Secs. 1501 and 1504(b)]*	Continues no-cost requirement found in prior law. *[Secs. 1401, 1403]* Requires USDA to operate sugar-for-ethanol program (in Energy title) to ensure this no-cost directive is met. *[Sec. 9001]*
Price Support Levels, Loans and Payments			
Sets raw cane and refined beet sugar loan rates at 18.0¢ and 22.9¢/lb through FY2008. Expands loan eligibility to in-process sugars and syrups at 80% of the	Increases raw cane sugar and refined beet sugar loan rates to 18.5¢/lb. and 23.5¢/lb for FY2009 through FY2013. *[Sec. 1301]*	Increases raw cane sugar loan rate to 19.0¢/lb. by FY2013, in 1/4¢ increments beginning in FY2010. Increases beet sugar loan rate, to be set at 128.5% of the	Increases raw cane sugar loan rate to 18.75¢/lb. in FY2012 and FY2013, in 1/4¢ increments beginning in FY2010. Sets refined beet sugar loan rate at

(Continued)

PRIOR LAW/POLICY	HOUSE-PASSED BILL (H.R. 2419)	SENATE-PASSED SUBSTITUTE AMENDMENT (H.R. 2419)	ENACTED 2008 FARM BILL (P.L. 110-246)
applicable loan rates. Makes nonrecourse loans available to processors under certain conditions. Sets 9-month repayment term for such loans. *[7 U.S.C. 7272 (a, b, d, e, f)]*		raw cane rate in effect each year (e.g., reaching 24.42¢/lb. in FY2013). *[Sec. 1501]*	22.9¢/lb. in FY2009. Starting in FY2010, sets beet sugar rate equal to 128.5% of the raw cane rate in effect, (e.g., rising to 24.1¢/lb. in FY2012 and FY2013). Continues other provisions found in prior law. *[Sec. 1401]*
Authorizes CCC to accept bids from sugar processors to purchase USDA-owned sugar in conjunction with reduced production of new sugar crops. *[7 U.S.C. 7272 (g)]*	Continues in-kind authority. Stipulates that planted beets or cane diverted from production can only be used as bioenergy feedstock. *[Sec. 1301]*	Similar to the House bill. *[Sec. 1501]*	Continues in-kind authority and adds House/Senate provision. *[Sec. 1401]*
USDA now pays storage rates of 8¢/cwt. for raw cane and 10¢ per cwt. for refined beet sugar that has been forfeited under the nonrecourse loan program. *[15 U.S.C. 714b & 714c; 7 CFR Part 1423]*	No comparable provision.	Requires (only through crop year 2011) USDA minimum storage payment rates of 10¢/cwt. and 15¢/cwt. on forfeited raw cane and refined beet sugar. *[Sec. 1503]*	Adopts Senate provision. *[Sec. 1405]*
Authorizes CCC to provide financing to processors of domestic sugar to construct or upgrade storage and handling facilities. *[Sec. 1402]*	No comparable provision.	Retains authority, but stipulates that loans shall not require any prepayment penalty. *[Sec. 1502]*	Continues prior law and adds Senate provision. *[Sec. 1404]*
Marketing Allotments and Allocations			
To avert loan forfeitures, USDA limits the amount of sugar processors can sell each year (according to a national "overall allotment quantity" (OAQ) divided between cane and beet sectors, then allocated to individual processors). The OAQ must accommodate WTO and NAFTA import commitments (1.532 million short tons). If imports are larger, USDA's authority to implement allotments is suspended. *[7 U.S.C. 1359aa, 1359bb, 1359cc, and 1359dd]*	Continues purpose and structure of marketing allotments and allocations, but changes some key provisions. Changes formula to require USDA to set OAQ at not less than 85% of estimated human food and beverage sugar use. Eliminates allotment suspension provision. *[Sec. 1303(a)-(d)]*	Similar to the House bill. *[Sec. 1504(a)-(d)]*	Continues marketing allotment authority and adopts House/Senate provisions requiring USDA to set OAQ at not less than 85% of estimated U.S. human consumption, and eliminating allotment suspension trigger. *[Sec. 1403(a)-(d)]*
Directs USDA to reassign unused raw cane and beet sugar marketing allocations	Requires that any reassignment of unused cane and beet allocations to imports [in	Similar to the House bill. *[Sec. 1504(e)]*	Adopts House/Senate change to prior law. *[Sec. 1403(e)]*

(Continued)

PRIOR LAW/POLICY	HOUSE-PASSED BILL (H.R. 2419)	SENATE-PASSED SUBSTITUTE AMENDMENT (H.R. 2419)	ENACTED 2008 FARM BILL (P.L. 110-246)
first to other cane states and beet processors, respectively; *second* to cane processors within each state; *third* to sales of sugar in CCC's inventory; and *fourth* to imports. *[7 U.S.C. 1359ee]*	the fourth step] must be met by imports "of raw cane sugar." *[Sec. 1303(e)]*		
Sugar Provision Related to Bioenergy Programs — see section on Energy Programs (below)			
Trade-Related Provisions			
In accord with a 1994 trade commitment, USDA sets an annual global sugar import quota of not less than 1.256 mil. short tons. USTR allocates the quota among eligible countries, and also administers preferential sugar import quotas for free trade agreement partner countries. Effective January 1, 2008, Mexico can ship duty free an unlimited amount of sugar to the U.S. market.	Makes no changes to import quota commitments found in various trade agreements and laws.	Makes no changes to import quota commitments.	Makes no change to current U.S. trade commitments.
Requires USTR in 2002-07 to reallocate unused country quota allocations to other quota-holding countries with sugar to sell. *[7 U.S.C. 1359kk]*	Repeals requirement for reallocating sugar import quota shortfalls. *[Sec. 1303(i)]*	Similar to the House bill. *[Sec. 1504(i)]*	Adopts House/Senate repeal provision. *[Sec. 1403(i)]*
USDA has discretion to increase the size of global raw cane and refined sugar import quotas when domestic sugar supplies are inadequate to meet U.S. demand at reasonable prices. ***[Chapter 17, additional note 5, of the U.S. Harmonized Tariff Schedule; 19 CFR Part 2001, Subpart A]***	Requires USDA to set quotas for raw cane and refined sugar at the minimum level necessary to comply with U.S. trade agreement obligations. In cases of sugar shortages, supplies are to be increased first by reassigning allotment deficits to imports of raw cane sugar, second by increasing the refined sugar quota, and third by increasing raw cane sugar quota. *[Sec. 1303(i)]*	Similar to the House bill. *[Sec. 1504(j)]*	Adopts House/Senate provision on setting initial import quotas at minimum levels and laying out steps to be followed to increase imports in the event of a sugar shortage. *[Sec. 1403(j)]*
To protect domestic sugar prices, USDA regulated the flow of sugar imports from large quota holders (through 2005).	Requires USDA to establish "orderly shipping patterns" for major suppliers of sugar to the U.S. market. *[Sec. 1303(i)]*	No comparable provision.	Deletes House "shipping patterns" provision.

(Continued)

PRIOR LAW/POLICY	HOUSE-PASSED BILL (H.R. 2419)	SENATE-PASSED SUBSTITUTE AMENDMENT (H.R. 2419)	ENACTED 2008 FARM BILL (P.L. 110-246)
The U.S.-Mexican agreement on bilateral market access for sugar and high-fructose corn syrup created an industry and government task force to address problems that might arise after the elimination of tariffs on sweeteners on January 1, 2008. ***[Exchange of Letters between USTR and Mexico's Secretariat of Economy, July 27, 2006]***	No comparable provision.	Expresses sense of Senate that U.S. & Mexican governments should coordinate their sugar policies to be consistent with U.S. international commitments, to avoid disruptions of each country's sweetener markets (sugar and high-fructose corn syrup). ***[Sec. 1505]***	Deletes Senate provision.
The U.S. withdrew from the International Sugar Organization (ISO) in 1992 because of opposition to the allocation of country contributions to ISO's budget.	Requires the Secretary of Agriculture to work with the Secretary of State to restore U.S. membership in the ISO within one year. ***[Sec. 1302]***	Similar to the House bill. ***[Sec. 1504]***	Adopts House provision. ***[Sec. 1402]***
TITLE II: CONSERVATION			
Program Definitions and Funding			
Sec. 1201 of the Food Security Act of 1985 (FSA) (P.L. 99-198, or the 1985 farm bill), as amended, defines 18 terms. ***[16 U.S.C. 3801]***	No provisions. (*Note*: some terms added by the Senate bill in this section are defined for specific conservation programs, as noted below.)	Adds definitions of beginning farmer or rancher, Indian tribe, nonindustrial private forest land, socially disadvantaged farmer or rancher, and technical assistance. Authorizes USDA to employ a test of net worth or other measure to qualify. ***[Sec. 2001]***	Adopts Senate provision with changes. Removes the test of net worth. Adopts the 1990 farm bill definition of a socially disadvantaged farmer or rancher, with changes to define farm, integrated pest management, person and legal entity, and livestock. ***[Sec. 2001]***
Sec. 1241(a) of the FSA, as amended, authorizes mandatory funding through FY2007 to carry out various conservation programs. ***[16 U.S.C. 3841]*** *Note*: Authorized funding levels for various programs is provided in individual program sections below.	Extends reauthorization through FY2012 with funding specified for CSP, FPP, EQIP, and WHIP. ***[Sec. 2401(a)]***	Extends reauthorization through FY2012 with funding specified for CSP, FPP, EQIP, WHIP, GRP, and the Voluntary Public Access and Habitat Incentives Program. ***[Sec. 2401(a)]***	Extends reauthorization through FY2012 with the following in additional new budget authority: CSP ($1.1 billion); EQIP ($3,393 million); and FPP ($773 million). ***[Sec. 2701]***
Highly Erodible and Wetland Conservation			
Secs. 1211-1212 of the FSA, as amended, makes violators of the conservation compliance program ineligible for certain program benefits, with some exceptions	No comparable provision.	Adds a second level of review by the state or district director, with technical concurrence from USDA's Natural Resources Conservation Service (NRCS)	Adopts Senate provision, providing for review of good faith determinations related to highly erodible land conservation. ***[Sec. 2002]***

(Continued)

PRIOR LAW/POLICY	HOUSE-PASSED BILL (H.R. 2419)	SENATE-PASSED SUBSTITUTE AMENDMENT (H.R. 2419)	ENACTED 2008 FARM BILL (P.L. 110-246)
from full loss of eligibility. ***[16 U.S.C. 3811-3812a and 3812f]***		if the Secretary has determined that this exception should apply. ***[Sec. 2101]***	
Secs. 1221-1222 of the FSA, as amended, makes swampbuster ineligible for certain program benefits, with some exceptions from full loss of eligibility. ***[16 U.S.C. 3821-3824, and 3822h]***	No comparable provision.	Add a second level of review by the state or district director, with technical concurrence from NRCS if the Secretary has determined that this exemption should apply. ***[Sec. 2201]***	Adopts Senate provision, providing for review of good faith determinations related to wetland conservation. ***[Sec. 2003]***
Comprehensive Conservation Enhancement Program			
The 1990 farm bill amended Sec. 1230 of the FSA to establish a program later renamed the Comprehensive Conservation Enhancement Program (CCEP). The CCEP, which includes the Conservation Reserve Program (CRP), Wetlands Reserve Program (WRP), and the Environmental Quality Incentives Program (EQIP), promotes long-term protection for environmentally sensitive lands through easements and technical/ financial assistance. ***[16 U.S.C. 3830]*** *Note:* Administration of CCEP, the subject of Sec. 1243, is described below.	No comparable provision. (*Note:* Amendments to Sec. 1243 described below in the "other conservation programs" subsection.)	Deletes Section 1243 in prior law, and moves some provisions, amended, into this (and other) sections. Extends CCEP through FY2012. Makes changes that reduce administrative burdens, streamline the application process, and promote partnerships. Deletes EQIP from CCEP and adds the Healthy Forests Reserve Program. Adds a new exception whereby USDA may exceed the enrollment limitation when a state or local regulation prohibits agricultural water use, requiring USDA to enroll the land within 180 days of receiving a request and pay a rental rate that reflects the rate prior to implementing the regulation. ***[Sec. 2301]***	Does not reauthorize the CCEP. The Healthy Forest Reserve Program is retained in the forestry title ***[Sec. 8205]***; the county acreage cap is addressed elsewhere ***[Sec. 2708]***. The agreement adopts a provision to exclude CREP acreage and continuous CRP acreage from the 25% cap if the county government concurs, and further specifies this provision is separate and distinct from the existing waiver authority. ***[Sec. 2106]*** Additional guidance is provided in the Managers statement.
The 1990 farm bill amended Sec. 1243 of the FSA to authorize administration of CCEP. Provisions include avoiding duplication of required conservation plans, limiting enrollment under CRP and WRP to 25% of the cropland in a county, protecting the interests of share croppers and tenants, allowing approved sources to provide technical assistance, and using up to 5% of the funds from the mandatory funded conservation programs to foster cooperation through partnerships. ***[16 U.S.C. 3844]***	Amends administration provisions by moving sections on acreage enrollment limits, tenant protection, and obtaining technical assistance. Establishes a new Cooperative Conservation Partnership Initiative to carry out projects/initiatives using competitive (2-5 years) grants. Specifies 14 criteria to be used in reviewing applications and 9 project priorities. Specifies duties of participant and USDA. Specifies program will be funded with 10% of funds for CSP, EQIP, and WHIP. The federal share for	Amends to streamline application process, add new endangered species provisions, and establish new partnerships and cooperation projects for special projects (up to 5 years) with multiple producers and eligible partners to address state conservation recommendations. Specifies five project purposes, lists application contents, and identifies USDA's duties and priorities when selecting projects (including 14 priority water project areas); also requires monitoring and evaluation. Specifies	Adopts House provision with changes and names the initiative the Cooperative Conservation Partnership Initiative (CCPI). ***[Sec. 2707]*** Allows USDA to make consider local circumstances, goals, and objectives, and provides for adjustments to provide producers preferential enrollment in the applicable program as part of the special project. Applies to all USDA conservation programs except CRP, WRP, FPP and GRP. The stated intent is to provide for applications that include innovative

(Continued)

Prior Law/Policy	House-Passed Bill (H.R. 2419)	Senate-Passed Substitute Amendment (H.R. 2419)	Enacted 2008 Farm Bill (P.L. 110-246)
	each project will be at least 75% of costs; 90% of the funds will be allocated at the state level (incentives and bonus payments may be used for specified purposes). Limits administrative costs to 5% of any grant. *[Sec. 2403]*	funding of 10% of the mandatory funds allocated to each state (except CRP, CSP, WRP, and the new Conservation Stewardship Program), with 75% of funds for intra-state and 25% for multi-state projects. *[Sec. 2405]*	combinations of covered initiative programs, and applications that might work in tandem with the enhancement programs under CRP or WRP. Additional guidance is provided in the Managers statement.
Conservation Reserve Program			
Sec. 1231(a-d) of the 1985 farm bill (FSA) authorizes the CRP; the program is currently authorized through CY2007 at 39.2 million acres. *[16 U.S.C. 3831(a-d)]*	Extends authorization through CY2012; retains current acreage enrollment limit (39.2 million acres). *[Sec. 2101(a-b)]*	Extends authorization through CY2012. Retains current acreage enrollment limit. Adds pollinator habitat to the general purposes. Expands eligible land to include some types of marginal pasture-land and land enrolled in a new flooded farmland program. *[Sec. 2311(a-c)]*	Extends authorization through CY2012, and allows USDA to address issues under State, regional, and national conservation initiatives. Caps enrollment at 32 million acres *[Sec. 2103]*. Clarifies that alfalfa grown as part of a rotation practice is a commodity for crop history purposes. *[Secs. 2101-2102, 2105]* Provides for pollinator habitats. *[Secs. 2706, 2708]*
Sec. 1231(f) of the FSA, as amended, lists priority areas as the Chesapeake Bay region (PA, MD, VA) , the Great Lakes Region, and Long Island Sound. *[16 U.S.C. 3831f]*	Deletes states but retains Chesapeake Bay region. *[Sec. 2101(b)]*	Similar to the House bill; also adds to the list the prairie pothole region, the Grand Lake St Mary's Watershed, and the Eastern Snake Plain Aquifer. *[Sec. 2311(d)]*	Adopts House provision to include all States that make up the Chesapeake Bay Region as the Conservation Priority Area. *[Sec. 2104]*
Sec. 1231(h) of the FSA, as amended, authorizes a one million acre pilot program within the CRP for wetlands and buffer areas. *[16 U.S.C. 3831h]*	Extends program through CY2012. *[Sec. 2101(e)]*	Extends program through CY2012; expands eligibility to include certain shallow water areas and certain agricultural drainage water treatment collection areas, and expands the eligible buffer acreage. Directs USDA to establish the maximum size of the buffer acreage to be enrolled along with eligible lands. Increases the maximum wetland size to 40 contiguous acres and makes all acres eligible for payment. *[Sec. 2311(e)]*	Adopts Senate provision with changes. amendment. Enrollment is capped at 100,000 acres in any State and 1 million acres total. Adds conforming changes to the Emergency Forestry Conservation Reserve Program. Expands enrollment of wetland and buffer acreage to include land that had been cropped during 3 of 10 crop years prior to 2002 and after 1990 and is subject to a natural overflow of a prairie wetland. *[Sec. 2106]*
Sec. 1232(a)(7) of the FSA, as amended, specifies a duty of participants is limiting commercial uses, including haying and grazing on enrolled lands; allows	Allows managed haying and grazing to control invasive species, and adds detail on allowed uses, enrolled lands, and adjustments to annual contract payments.	Allows managed haying and grazing to control invasive species and permits managed haying and grazing that is a part of a conservation plan. *[Sec. 2311(h)]*	Adopts House provision with changes, allowing for routine grazing, including grazing to control invasive species. Additional guidance is provided in the

(Continued)

Prior Law/Policy	House-Passed Bill (H.R. 2419)	Senate-Passed Substitute Amendment (H.R. 2419)	Enacted 2008 Farm Bill (P.L. 110-246)
managed haying/grazing under certain circumstances. *[16 U.S.C. 3832a(7)]*	*[Sec. 2101(f)]*		Managers statement. *[Sec. 2108]* Grants "management on land" should not result in a reduced payment, if done in accordance with the contract. *[Sec. 2107]*
Sec. 1234(c) of the FSA, as amended, establishes a framework for calculating annual rental payments. *[16 U.S.C. 3834c]*	Requires USDA to conduct and make available an annual survey of dryland and irrigated cropland cash rental rates in all counties with more than 20,000 acres of crop and pasture land. *[Sec. 2101(g)]*	Similar to the House bill; also requires USDA to give preference to local owners or operators when considering competing offers providing equivalent benefits. *[Sec. 2311(j)]*	Requires USDA to survey annually the per acre estimates of county cash rents paid to contract holders, and requires that USDA give priority to offers from local residents if conservation benefits are equivalent among offers. *[Sec. 2110]*
Sec. 1235(e) of the FSA, as amended, allows USDA to terminate CRP contracts after 5 years if contract was in effect before 1/1/95. *[16 U.S.C. 3835e]*	Allows USDA to terminate any contract after 5 years, but prohibits terminating contracts for land enrolled under a continuous signup. *[Sec. 2101(i) and (j)]*	Allow USDA to terminate a contract if a retired or disabled producer has endured financial hardship because of taxes on rental payments. *[Sec. 2311(k)]*	Adopts House provision. *[Sec. 2111]*
No comparable provision.	No comparable provision.	No comparable provision.	Specifies a 50% federal share of cost sharing payments relating to trees, windbreaks, shelterbelts, and wildlife corridors. *[Sec. 2109]*
The 2002 farm bill amended Sec. 1244(a) of the FSA to authorize USDA to provide incentives to beginning farmers/ranchers and Indian tribes to participate in conservation programs. *[16 U.S.C. 3844(a)]*	Facilitates the transfer of CRP land from a retiring owner to a beginning / socially-disadvantaged producer to return land to production, and allows new owner to begin land improvements or start organic certification process one year before CRP contract expires. *[Sec. 2101(h)]*	No comparable provision. (*Note*: Support for socially disadvantaged and limited resource farmers/ranchers are in other bill sections.)	Adopts House provision with modifications. *[Sec. 2111]*
No comparable provision.	No comparable provision.	Creates new Flooded Farmland Program for the Prairie Pothole region within the CRP. Allows continuous enrollment. Eligible land parcels must exceed 5 acres, been incapable of production preceding 3 crop years, have a cropping history, and have no natural outlet. *[Sec. 2312]*	Deletes this section and modifies CRP and WRP to accomplish the intent of the Senate amendment, including expanding eligible lands under the CRP pilot program for wetlands and buffer areas *[Sec. 2106]* and expands eligible lands under WRP. *[Sec. 2201]*
No comparable provision.	No comparable provision.	Creates new Wildlife Habitat Program for CRP participants with established softwood pine stands using management	Deletes this section and modifies CRP to accommodate the intent of the Senate amendment. Additional guidance is

(Continued)

PRIOR LAW/POLICY	HOUSE-PASSED BILL (H.R. 2419)	SENATE-PASSED SUBSTITUTE AMENDMENT (H.R. 2419)	ENACTED 2008 FARM BILL (P.L. 110-246)
		practices that benefit wildlife (contracts up to 5 years). Program ends September 30, 2011. *[Sec. 2313]*	provided in the Managers statement.
Wetlands Reserve Program			
The 1996 farm bill amended Sec. 1237(a) of the FSA to authorize WRP, stating its purpose to restore and protect wetlands. *[16 U.S.C. 3837a]*	Adds to the purposes to create and to enhance wetlands, and to purchase floodplain easements. *[Sec. 2102(a)]*	No comparable provision.	Amends purposes to restore, protect, or enhance wetlands on private or tribal lands. *[Sec. 2201]*
Sec. 1237(b) of the FSA, as amended, sets maximum enrollment at 2.275 million acres, with an annual calendar year enrollment goal of 250,000 acres. *[16 U.S.C. 3837b]*	Sets maximum enrollment at 3.605 million acres. Sets an annual fiscal year enrollment goal of 250,000 acres, of which not more than 10,000 acres may be flood plain easements. *[Sec. 2102(b)]* New section adds language authorizing WRP from FY2008-12. *[Sec. 2402(h)]*	Sets annual fiscal year enrollment goal of 250,000 acres, with no enrollment after FY2012. *[Sec. 2321]*	Sets maximum enrollment at 3.041 million acres. Sets annual fiscal year enrollment goal of 250,000 acres through FY2012. *[Sec. 2202-2203]*
Sec. 1237(c) of the FSA, as amended, establishes requirements for eligible lands through 2007. *[16 U.S.C. 3837c]*	Adds riparian areas to eligible wetlands, and makes eligible floodplain land flooded in the past calendar year or at least twice in the past 10 years, and land that contributes to flood water storage, flow, or erosion control. *[Sec. 2102(c)]*		Expands eligible lands under WRP to include cropland or grassland that was used for agricultural production prior to flooding from the natural overflow of a closed basin lake or pothole. *[Sec. 2203]* Adds terms to "meet habitat needs of specific wildlife species." *[Sec. 2204]*
Sec. 1237(e) of the FSA, as amended, identifies ineligible land to include lands already planted to timber in the CRP. *[16 U.S.C. 3837e]*	Expands ineligible lands to include floodplains where restoration practices would not be productive or the land is already protected. *[Sec. 2102(d)]*	No comparable provision.	Expands ineligible lands to include farmed wetland or converted wetland, together with the adjacent land that is functionally dependent on the wetlands, except wetlands converted before December 23, 1985. *[Sec. 2203]*
Sec. 1237A(f) of the FSA, as amended, states compensation to be paid in cash (in 5 to 30 payments) and not to exceed the fair market value, as reduced by the easement. *[16 U.S.C. 3837a(f)]*	Limits compensation to lowest of 4 options: percentage of the fair market value; percentage of market value determined by a survey; a geographic cap; or a landowner's offer. Allows USDA to use non-federal contributions to administer program *[Sec. 2102(e)]*	Limits compensation to lowest of 3 options: an amount necessary to encourage enrollment; a limit for a geographic area; or a landowner's offer. Compensation may be in 1 to 30 payments. *[Sec. 2322(b)(3-4)]*	Adopts House provision with changes, revising the process for determining the value of easements and contracts by requiring USDA to provide the lowest amount of compensation based on a comparison of the fair market value of the land, a geographic cap, or an offer

(Continued)

PRIOR LAW/POLICY	HOUSE-PASSED BILL (H.R. 2419)	SENATE-PASSED SUBSTITUTE AMENDMENT (H.R. 2419)	ENACTED 2008 FARM BILL (P.L. 110-246)
			made by the landowner. Provides that easements with values less than $500,000 be paid out over 1-30 years; easements with values greater than $500,000 are to be paid out over 5-30 years. *[Sec. 2208]* Provides for the repeal of payment limitations (exception for State agreements for new Wetlands Reserve Enhancement Program. *[Sec. 2209]*
Sec.1237C(c) of the FSA, as amended, lists three considerations USDA is to use when considering offers for WRP contracts. *[16 U.S.C. 3837c(c)]*	Adds new additional criteria for ranking offers (conservation benefits; cost-effectiveness; and offer of a financial contribution) and conservation benefits of floodplain lands. *[Sec. 2102(f)]*	No comparable provision.	Adopts House provision. *[Sec. 2207]*
Sec. 1237D(c)(4) of the FSA, as amended, waives limits for public entities receiving payments through the wetland and environmental enhancement programs. *[16 U.S.C. d(c)(4)]*	Replaces provision with a new language on Wetland Reserve Enhancement program, where states contribute funds so as to increase payments. *[Sec. 2102(g)]*	Authorizes a Wetlands Reserve Enhancement Program (WREP). Makes a conforming change to allow payments for 30-year contracts. *[Sec. 2322(c)]*	Adopts Senate provision authorizing WREP for agreements with States similar to CREP. Authorizes a Reserved Rights Pilot program. *[Sec. 2205-2206]*
No comparable provision.	No comparable provision.	Requires a report to House and Senate Agriculture Committees by 1/1/2010 on the implications of long-term easements on USDA resources. *[Sec. 2322(d)]*	Adopts Senate provision. *[Sec. 2210]*
Conservation Security Program			
The 2002 farm bill amended the FSA to establish the Conservation Security Program (CSP) for FY2003-11. Defines eligible producers and eligible lands and excluded lands (land enrolled in multiyear land retirement programs and land not in crop production at least 4 of the preceding 6 years). Specifies terms for 3 tiers of conservation contracts. Identifies topics that may be addressed in contracts. Contracts are 5 years under tier 1, and 5 to 10 years under tiers 2 and 3.	Establishes a new CSP for FY2012-2017. Eligible producers must submit an offer addressing "at least one priority resource of concern to a minimum level of management intensity." Eligible land excludes incidental forest land. Limits program to one type of contract of 5 years; describes five elements to be in all contracts, but eliminates list of topics to be addressed. Prohibits termination of contracts, without penalty, by a producer who is required to modify a contract.	Authorizes through FY2012 a new CSP as a conforming amendment, and reauthorizes current CSP for existing contracts only. Future CSP contracts would be replaced by a Comprehensive Stewardship Incentives Program consisting of a Conservation Stewardship Program with similar provisions of the existing CSP and EQIP. Eligible land must have been planted to crops 4 of preceding 6 years. Specifies contract are for 5 years, with renewal under certain	Defines program terms for the new CSP. Adopts elements of the Senate provision. Expands eligible lands to include nonindustrial private forest land (limited to not more than 10% of total annual acres under the program). Permits 5-year extension of contracts. Excludes under the program, land used for cropland that had not been planted, considered to be planted, or devoted to crop production for 4 of the 6 years prior to the date of enactment of the act (unless the land had

(Continued)

PRIOR LAW/POLICY	HOUSE-PASSED BILL (H.R. 2419)	SENATE-PASSED SUBSTITUTE AMENDMENT (H.R. 2419)	ENACTED 2008 FARM BILL (P.L. 110-246)
Specifies circumstances and requirements for modifying, terminating, and renewing contracts. Contracts may be renewed for 5 to 10 years. Defines 15 terms pertaining to the program. *[16 U.S.C. 3838]*	Allows contracts to be renewed for one additional 5 year period. Adds new provisions on evaluating offers from organic producers. Defines twelve terms that are new terms or differ from prior law. *[Sec. 2103(a)]*	conditions. Allows for terminating and changing contracts. Specifies how to evaluate contract offers, producer duties, enhancement terms, and supple-mental payments. Defines 15 terms. *[Secs. 2391 and 2341]*	previously been enrolled in CRP; had been maintained in a long term crop rotation; or was incidental land needed for efficient operation). *[Sec. 2301]*
Secs 1238A(d-3) of the FSA, as amended, specifies terms for 3 tiers of conservation contracts. Identifies topics that may be addressed in contracts. Contracts are 5 years under tier 1, and 5 to 10 years under tiers 2 and 3. Specifies circumstances and requirements for modifying, terminating, and renewing contracts. Contracts may be renewed for 5 to 10 years *[16 U.S.C. 3838a(d-e)]*	Limits program to 1 type of contract of 5 years and describes 5 elements to be in all contracts, but eliminates list of topics to be addressed. Contracts could no longer be terminated, without penalty, by a producer who is required to modify a contract. Contracts may be renewed for 1 additional 5 year period. New provisions on evaluation of offers and coordination with organic certification are added. *[Sec. 2103(a)]*	Defines eligible land and eligible producers; land must have been planted to crops 4 of preceding 6 years. Contents of contracts are specified and are for 5 years, with renewal if certain conditions are met. Specifies considerations in evaluating contract offers, producer duties, enhancement terms, and supplemental payments. Adds provisions on terminating and changing contracts. *[Sec. 2341]*	Adopts Senate provision with modifications. Authorizes supplemental payments for producers who adopt a beneficial crop rotation that provide significant conservation benefits and are not limited to a particular crop, cropping system, or region of the country. Allows for on-farm conservation research and demonstration activities and pilot-testing as part of contract offers. Allows for contract modification. *[Sec 2301]*
Sec. 1238C of the FSA, as amended, specifies that duties of the Secretary include making payments early in each fiscal year, the components of payments for each tier, annual payment limits for each tier ($20,000 for tier 1, $35,000 for tier 2, and $45,000 for tier 3), minimum requirements for practices, and requirements for implementing regulations *[16 U.S.C. 3838c]*	Alters duties of the Secretary to include identification of priority resources of concern at the state level (limited to 5 concerns in any geographic area of a state). Limits total payments under a contract to $150,000 (5 years); allows for the environmental needs associated with agriculture to be considered in state allocations; requires USDA to compile data of specified program contract and payment topics. *[Sec. 2103 (a)]*	Alters duties of the Secretary to allow for continuous enrollment (allowing a producer to apply at any time during the year), providing assistance to producers, and maintaining contract and payment information that will support program monitoring and evaluation, and enabling specialty crop producers to participate. Specifies an acreage allocation, limiting payments to $240,000 (6-year). *[Secs. 2391 and 2341]*	Provides that state acreage allocations be based on each state's proportion of eligible acres to the total eligible acres nationwide (available to all producers, not only specific watersheds/geographic regions), allowing for input from USDA. Directs USDA to adopt continuous enrollment, but allows for USDA to determine when to rank applications. Directs USDA to provide technical assistance to specialty crop and organic producers. Directs USDA to encourage producers who are transitioning from land retirement programs to enroll in CSP and other working lands programs. Limits payments to $200,000 in any 5-year period. *[Sec. 2301]*
No comparable provision.	No comparable provision.	Provides for enrollment of up to 79.628 million acres and attempted annual enrollment of 13.273 million acres, nationwide and at a average annual cost	Deletes Senate provision. Enrolls an additional 12.8 million acres annually FY2008-2017. *[Sec. 2301]*

(Continued)

PRIOR LAW/POLICY	HOUSE-PASSED BILL (H.R. 2419)	SENATE-PASSED SUBSTITUTE AMENDMENT (H.R. 2419)	ENACTED 2008 FARM BILL (P.L. 110-246)
		of $19 per acre. Provides for small farm participation, and allocates to each state each year the lesser of 20,000 acres or 2.2% of the eligible land. *[Sec. 2341]*	
No comparable provision.	No comparable provision.	Requires regulations to be issued within 180 days of enactment. *[Sec. 2341]*	Deletes Senate provision. Directs USDA to promulgate regulations. *[Sec. 2301]*
Sec. 1241(a)(3) of the FSA, as amended, authorizes mandatory funding for the CSP at $1.954 billion for FY2006-10 and $5.65 billion from FY2006-15. ***[16 U.S.C. 3841(a)(3)]***	Prohibits any new contracts under the terms of the 2002 CSP, such that no new CSP contracts may be entered into after October 1, 2007 (although payments may be continued until contracts expire). For contracts signed before 10/1/07, provides a total of $1.5 billion in mandatory funding for FY2007-12, and $1.9 billion for FY2012-17. For contract signed after 10/1/11, provides $0.5 billion for FY2012 and $4.6 billion for FY2012-17. ***[Sec. 2401(b)]***	Prohibits any new contracts under the terms of the 2002 CSP, such that no new CSP contracts may be entered into after October 1, 2007 (although payments may be continued until contracts expire). Authorizes $2.3 billion in mandatory funding for contracts entered into before the date of enactment, (available until spent) and an unspecified amount for the new CSP (enrollment in the new program is measured in acres rather than dollars). ***[Sec. 2401(a)(3-4)]***	Adopts Senate provision. The Manager's report states that the bill provides for a $1.1 billion increase in budget authority above current baseline (FY2008-2017). Provides such sums as necessary to carry out existing CSP contracts. *[Sec. 2701]*
Environmental Quality Incentives Program			
The 1996 farm bill amended Sec. 1240 of the FSA to authorize EQIP, stating its purpose as promoting production and environmental quality as compatible goals, and optimizing environmental benefits by working in 5 specified areas. ***[16 U.S.C. 3839aa]*** Defines 6 terms: beginning farmer or rancher, eligible land, land management practice, livestock, practice, and structural practice.	Adds forest management and organic transition to the program purposes. Revises the descriptions of 2 of the 5 purposes to recognize energy conservation and conservation on forest lands. ***[Sec. 2105(a)]*** Adds forestry, forest management practices, and coordinated implementation to the "land management practice" definition. Adds alpacas and bison to the "livestock" definition. Adds definitions of "integrated pest management" and "socially disadvantaged farmer or rancher." ***[Sec. 2105(b)]***	Adds forest management to the statement of program purposes, and recognizes pollinators and fuels management in the amplifying statements. ***[Secs. 2351, 2352, and 2354]*** Adds aquaculture to the "eligible land" definition; forestry is added to the "land management practice" definition; adds conservation planning practices to "practices;" defines "producer" to include custom feeding businesses and contract growers; and adds firebreaks to "structural practice." ***[Sec. 2352]***	Adds forest management to purposes, and adds language regarding forest lands on EQIP program plan and duplication. ***[Secs. 2505-2506]*** Adopts the Senate provisions with an amendment to modify eligible land. ***[Sec. 2502]*** Allows for technical assistance to farmers that promote pollinator habitats, and farmers transitioning to organic farming, among activities. ***[Sec. 2501]*** Further clarifies duties of the Secretary. ***[Sec. 2507]***
Sec. 1240B(a-c) of the FSA, as amended, authorizes EQIP through FY2010. Eligible practices are defined. Contracts	Reauthorizes through FY2012. Expands types of eligible practices to include organic certification using technical	Reauthorizes through FY2012. Expands permitted practices to include conservation planning. Limits contracts	Reauthorizes through FY2012. Limits payments to 75% of costs. *[Sec. 2503]* Lowers individual payment limits from

(Continued)

PRIOR LAW/POLICY	HOUSE-PASSED BILL (H.R. 2419)	SENATE-PASSED SUBSTITUTE AMENDMENT (H.R. 2419)	ENACTED 2008 FARM BILL (P.L. 110-246)
are 1 to 10 years. *[16 U.S.C. 3839aa-2(a-c)]*	services from approved providers, and improved energy efficiency, renewable energy systems. *[Sec. 2105(c)]*	to a maximum of 5 years. Removes prohibitions on bidding down. *[Sec. 2353(a-c)]*	$450,000 to $300,000 during any 6-year period , except in cases of special environmental significance (including projects involving methane digesters), allowing USDA to raise the limit to not more than $450,000. *[Sec. 2508]* Directs USDA to develop criteria for evaluating applications to address national, State, and local conservation priorities, allowing for prioritization and grouping of applications based on cost-effectiveness, how address the designated resource concern(s), how best fulfills the purpose of EQIP, and if improves conservation practices or systems. *[Sec. 2504]*
Sec. 1240B(d)(2) of the FSA, as amended, allows limited resource and beginning producers to receive not more than a 90% federal cost share. *[16 U.S.C. 3839aa-2(d)]*	Sets the federal cost share at 90% for socially disadvantaged producers. Provides increased federal cost-share of 90% for using gasifier technology for certain purposes. *[Sec. 2105(d-e)]*	Sets the federal cost share at 90% for socially disadvantaged producers. Allows for advanced payments to purchase materials and contracting. *[Sec. 2353(c)]*	Adopts Senate proposal. *[Sec. 2503]*
Sec. 1240B(e) of the FSA, as amended, provides incentive payments to perform land management practices, with special emphasis given to practices that promote specified goals. *[16 U.S.C. 3839aa-2(e)]*	Expands purposes for incentive payments: (1) receiving technical services from approved third party providers, (2) developing a comprehensive nutrient management plan, and (3) implementing energy efficiency and renewable energy projects. Pollinator habitats will receive special emphasis. *[Sec. 2105(f)]*	Expands purposes receiving special emphasis to include predator species protected under the Endangered Species Act, gray wolves, grizzly bears, and black bears. *[Sec. 2353(c)(3)]*	Adopts House provision. *[Secs. 2503. 2706 and 2708]* Includes special rule that USDA may accord significance to practices promoting residue, nutrient, air quality, pest, and invasive species management; pollinator habitat; and animal carcass management technology. The conference report recognizes as consistent with the purposes of EQIP options to deter predators protected by the Endangered Species Act, and also delisted populations.
Sec. 1240B(g) of the FSA, as amended, requires that 60% of payments go to practices related to livestock production requirement. *[16 U.S.C. 3839aa-2(g)]*	Extends through FY2012 the 60% of payments to livestock production requirement. *[Sec. 2105(g)(2)]*	Similar to the House bill. *[Sec. 2353]*	Extends through FY2012 the 60% of payments to livestock production requirement. *[Sec. 2503]*

Prior Law/Policy	House-Passed Bill (H.R. 2419)	Senate-Passed Substitute Amendment (H.R. 2419)	Enacted 2008 Farm Bill (P.L. 110-246)
No comparable provision.	Requires USDA to reserve at least 5% of program funds for beginning and socially disadvantaged producers for at least 90 days after the program funds have been made available. *[Sec. 2105(g)]*	Amends the cost-share rate exception for beginning and socially disadvantaged farmers or ranchers to allow variable payment, not to exceed 90%, and authority to provide advance payments up to 30% for the purchase of materials or contracting. *[Sec. 2353]*	Adopts Senate provision for advance payments for beginning, socially disadvantaged and limited resource farmers or ranchers and deletes Senate provision for guaranteed loan eligibility. Adopts Senate provision with an amendment for cost-share rates and advance payments for beginning, socially disadvantaged, and limited resource farmers or ranchers. *[Sec. 2503]*
No comparable provision.	No comparable provision.	Gives priority to improving water conservation and air quality, under certain conditions. Requires participants to have/expect at least $15,000 in gross sales from farming. *[Sec. 2353(c)(6)]*	Deletes Senate provision.
No comparable provision.	Expands eligibility to market agencies and custom feeders. *[Sec. 2105(h)]*	Expands 'producer' eligibility to include a custom feeding business and a contract grower or finisher. *[Sec. 2352]*	Adopts Senate provision with an amendment to modify eligible land. *[Sec. 2502]*
Sec. 1240C of the FSA, as amended, gives higher priority for participation in EQIP to producers using cost-effective conservation practices and practices that address national conservation priorities. *[16 U.S.C. 3839aa-3]*	Identifies 5 priorities for program applications. Specifies a streamlined evaluation process for operations with substantial and sound environmental management systems involving a limited number of practices. *[Sec. 2105(i)]*	Adds a higher priority for improving conservation practices or systems in place at the time of the contract offer. *[Sec. 2354]*	Adopts House provision with changes to prioritize State, regional, or local resource concerns, and to allow for the grouping of applications of similar agriculture operations. *[Sec. 2502]*
Sec. 1240E of the FSA, as amended, defines the general contents of a producer's EQIP plan, and calls on USDA to avoid duplication with other conservation plans. *[16 U.S.C. 3839aa-5]*	Adds to the planning requirements the need to be consistent with forest plans, and allows as an acceptable plan consideration of an air or water quality permit that meet regulatory requirements as an acceptable plan. *[Sec. 2105(k)]*	Includes forestry language similar to House bill, but also allows a producer organization to act on behalf of its membership in submitting applications or conducting similar activities to facilitate program participation. *[Sec. 2356]* Also establishes a Chesapeake Bay Watershed Conservation Program under EQIP to assist producers in applying conservation practices on agricultural/nonindustrial private forestland in the Bay watershed to address natural resource concerns related to agriculture, funded at $165 million for	Adopts House forestry provision *[Sec. 2502]*, but strikes Senate provision on producer organizations.

Prior Law/Policy	House-Passed Bill (H.R. 2419)	Senate-Passed Substitute Amendment (H.R. 2419)	Enacted 2008 Farm Bill (P.L. 110-246)
		FY2008-12. *[Sec. 2361]*	
Sec. 1240F of the FSA, as amended, provides for USDA funding, information, and training to develop and implement conservation plans. ***[16 U.S.C. 3839aa-6]***	Lists 3 criteria that must be met before USDA can provide assistance for practices with a primary purpose of water conservation. ***[Sec. 2105 (l)]***	No comparable provision.	Deletes House provision.
Sec. 1240H of the FSA, as amended by the 2002 farm bill, provides for a competitive grants program within EQIP, on a matching basis, to implement innovative conservation practices; examples listed are using market systems in pollution reduction, and using innovative practices, such as storing carbon in soil; no funding is specified. ***[16 U.S.C. 3839aa-8]***	Adds to the Conservation Innovation Grants provisions under EQIP. Adds detail on qualities of eligible projects, establishes a pilot program for conservation planning in the Chesapeake Bay watershed, and adds a new subsection to assist producers who are meeting state and local regulatory air quality requirements. Provides funding from EQIP: $30 million (FY2008) rising to $75 million (FY2012), with specified funds for air quality and for organic and specialty crop producers. ***[Sec. 2105(m)]***	Adds to the Conservation Innovation Grants provisions under EQIP. Adds nonindustrial private forest lands to the list of potential recipients of innovative technologies. Adds two items to the list of examples: (1) transfers of innovative technologies to nonindustrial private forest land in production, and (2) assistance for specialty crop production. ***[Sec. 2358]***	Adopts House provisions related to forest resource management and air quality. Provides $37.5 million annually (FY2009-2012) to implement air quality plans. ***[Sec. 2509]*** The conference report states conservation programs should recognize the use of innovative technology such as enhanced efficiency fertilizers.
Sec. 1240I of the FSA, as amended by the 2002 farm bill, creates a Ground and Surface Water Conservation Program (GSWCP) within EQIP for activities that will result in a net savings of ground or surface water; lists 6 types of eligible activities (improve irrigation systems, for example), and provides mandatory funding of $25 million in FY2002, growing to $60 million annually in FY2004-07. ***[16 U.S.C. 3839-aa-9]***	Replaces GSWCP with a Regional Water Enhancement Program to address water quality, make eligible governmental entities (including irrigation and water districts) and Indian tribes, and to implement program on a regional scale through cooperative agreements. Expands the list of eligible activities and requires the Secretary to identify priority areas. Lists 5 priority areas, which together may receive no more than 50% of the available funds. Establishes a process for soliciting/selecting proposals and developing implementation agreements. Provides mandatory funds of $60 million annually through FY2012 (limits administrative expenses to no more than 3% of the total). ***[Sec. 2106]***	Maintains GSWCP and provides an increase in funding from $60million to $65 million annually. Provides funding for each state that received funding under the program in previous years (simple average of funds provided for FY2002-2007 or the amount provided in 2007, whichever is greater), except for states over the Ogallala Aquifer, which will receive not less than the greater of $3 million or the average of funds provided for FY2002-2007. Provides at least $20 million for the Eastern Snake Plain Aquifer. ***[Sec. 2359]***	Replaces GSWCP with the Agricultural Water Enhancement Program (AWEP) under EQIP. Provides mandatory funding: $73 million annually (FY2009-2010), $74 million (FY2011), and $60 million annually (FY2012 and each year thereafter). Recognizes the purpose as addressing water quality/quantity concerns on agricultural land, with the role of AWEP partners as leveraging federal funds and encouraging producers to address these concerns. The Managers report emphasizes the importance of addressing groundwater management in the Ogallala region, promoting water use efficiency projects, converting irrigated farming operations to a dryland farming; and providing assistance to construct onfarm reservoirs/irrigation ponds in drought-stricken areas. Identifies six

(Continued)

PRIOR LAW/POLICY	HOUSE-PASSED BILL (H.R. 2419)	SENATE-PASSED SUBSTITUTE AMENDMENT (H.R. 2419)	ENACTED 2008 FARM BILL (P.L. 110-246)
			priority areas: The Eastern Snake Plain Aquifer region, Puget Sound, the Ogallala Aquifer, the Sacramento River watershed, Upper Mississippi River Basin, the Red River of the North Basin, and the Everglades. *[Sec. 2510]* Follows same payment limits as under EQIP. *[Sec. 2508]* Provides for a transition period for the existing GSWCP through September 30, 2008. *[Sec. 2903]*
Sec. 1240I(c)(2) of the FSA, as amended by the 2002 farm bill, provides $50 million to carry out water conservation activities in the Klamath River basin (OR, CA) *[16 U.S.C. 3839aa-9(c)(2)]*	Lists the Klamath River basin as one of the 5 listed priority areas under the Regional Water Enhancement Program. *[Sec. 2106 (b)(2)]*	No comparable provision. (*Note*: The Klamath Basin is listed as 1 of 14 priority areas in the Partnerships and Cooperation Program; see above.)	Deletes House provision.
No comparable provision.	No comparable provision.	Adds program at end of EQIP to assist farmers who are converting to organic production (with contracts of 3-4 years). Payments are limited to $20,000 per year. *[Sec. 2360]*	Adopts Senate provision. Includes organic practices as eligible management systems and limits payment to an aggregate of $80,000 in any 6-year period. *[Secs. 2501 and 2503]*
Sec. 1241(a)(6) of the FSA, as amended, authorizes EQIP funding, rising from $0.4 billion in FY2002 to $1.3 billion in FY2010. *[16 U.S.C. 3841(a)(6)]*	Authorizes EQIP funding: $1.25 billion (FY2008), $1.6 billion (FY2009), $1.7 billion (FY2010), $1.8 billion (FY2011), and $2 billion (FY2012). *[Sec. 2401(d)]*	Authorizes EQIP funding: $1.27 billion annually (FY2008-09), $1.3 billion each (FY2010-FY2012). *[Sec. 2401(a)(7)]*	Provides additional budget authority. Authorizes EQIP funding: $1.2 billion (FY2008); $1.337 billion (FY2009); $1.45 billion (FY2012); $1.588 billion (FY2011); and $1.75 billion (FY2012). *[Sec. 2701]*
Farmland Protection Program			
The 1996 farm bill amended Sec. 1238H of the FSA to authorize the Farmland Protection Program (FPP), defining eligible entity, land, Indian tribes, and programs. *[16 U.S.C. 3838h-i]* The program, as amended, provides for the purchase of conservation easements to protect topsoil by limiting the land's nonagricultural uses subject to a pending	Reauthorizes program, and renames to Farm and Ranchland Protection Program (FRPP). Expands eligible land definition to include historic and archaeological resources. States will be certified (reviewed every 3 years) to participate and receive program funds based on 4 listed requirements. States may spend up to 10% of funds for administrative costs.	Reauthorizes the program. Modifies definition of eligible forest land, and makes eligible other land that is needed for efficient administration of an easement. Changes purpose of program from protecting topsoil to "protecting agricultural use and related conservation values." Adds new requirements for cooperative agreements with participants,	Reauthorizes the program through 2012, but does not rename program. Changes administrative requirements, appraisal methodology, and terms and conditions of cooperative agreements. Adopts terms/conditions for cooperative agreements similar to Senate provision. Clarifies the purpose of the program as protecting land for agricultural use by

(Continued)

PRIOR LAW/POLICY	HOUSE-PASSED BILL (H.R. 2419)	SENATE-PASSED SUBSTITUTE AMENDMENT (H.R. 2419)	ENACTED 2008 FARM BILL (P.L. 110-246)
offer. The federal cost may not exceed 50% of the value of the easement; the value of a charitable donation by the seller may not exceed 25% of the value of the easement. If multiple applications are comparable, USDA may not use cost alone to determine which ones will be funded.	Lists terms and conditions for agreements with eligible entities (reviewed every 3 years). Provides that USDA may require a contingent right to enforce easement, and requires the use of a conservation plan for highly erodible cropland. Retains a federal contingent right of enforcement or executory limitation in an easement to ensure its enforcement. Provides cost-share assistance for purchasing an easement, but assistance may not exceed 50% of the appraised fair market value of the easement. The fair market value is determined by an appraisal using an industry-approved method. ***[Sec. 2110]***	cost-sharing, and protection of the federal investment. Allows USDA to enter into cooperative agreements with eligible entities under certain circumstances, and requires the protection of federal investment through an executory limitation. Limits the amount USDA can share in the costs of purchasing the easement to 50% of the appraised fair market value and establishes minimum amounts entities pay based on the amount of landowner contributions. Requires appraisals based on uniform standards of professional appraisal practice or any other industry- approved standard. ***[Sec. 2371]***	limiting nonagri-cultural uses. Adopts Senate provision to modify the definition of eligible land to include forestland and other land that contributes to the economic viability of an operation. Establishes a certification process similar to the House bill for all eligible entities. To become certified, entities must have the authority and resources to enforce easements, polices in place that are consistent with the purposes of the program, and clear procedures to protect the integrity of the program. Includes a limit on impervious surfaces consistent with agricultural activities, and clarifies agreement terms for certified and non-certified entities. ***[Sec. 2401-2402]***
Sec. 1241(a)(4) of the FSA, as amended, authorizes mandatory funding for the FPP at; $50 million in FY2002, $100 million in FY2003, $125 million in FY2004 and FY2005, $100 million in FY2006, and $97 million in FY2007. ***[16 U.S.C. 3841(a)(4)]***	Mandatory funding for the renamed Farm and Ranchland Protection Program is authorized at; $125 million in FY2008, $150 million in FY2009, $200 million in FY2010, $240 million in FY2011, and $280 million in FY2012. ***[Sec. 2401(c)]***	Mandatory funding for the FPP is authorized at $97 million annually from FY2008 through FY2012. ***[Sec. 2401(a)(5)]***	Provides additional budget authority. Authorizes FPP funding: $97 million (FY2008); $121 million (FY2009); $150 million (FY2010); $175 million (FY2011); and $200 million (FY2012). ***[Sec. 2701]***
Grassland Reserve Program			
The 2002 farm bill amended Sec. 1238N of the FSA to authorize the Grasslands Reserve Program (GRP), setting maximum enrollment for at 2.0 million acres (all enrolled parcels in at least 40 contiguous acres). Requires 40% of land enrolled in 10-20 year, and 60% in 30 year agreements. ***[16 U.S.C. 3838n]***	Sets the GRP enrollment ceiling at an additional 1.34 million acres, with at least 60% of these acres to be enrolled using 30 year rental agreements and easements. ***[Sec. 2104(a) and (b)]***	Adds definitions: eligible entity, eligible land, and permanent conservation easement. Eligible entity and authority would allow for USDA to enter cooperative agreements with entities to purchase easements. Provides for GRP enrollment options through a 30 year contract, 30 year easement, and permanent easement. ***[Sec. 2381]***	Adopts an acreage enrollment goal of an additional 1.22 million acres by 2012. Includes 10-, 15-, and 20-year rental contracts and permanent easements. Deletes House priority for 60% of acreage in long term contracts; retains law that 60% of funds would be dedicated to easements, while 40% of funds would be dedicated to short term contracts. Adopts Senate definition of eligible entity, authority, and eligible land (with technical corrections). Adopts a priority for enrollment of CRP land

(Continued)

Prior Law/Policy	House-Passed Bill (H.R. 2419)	Senate-Passed Substitute Amendment (H.R. 2419)	Enacted 2008 Farm Bill (P.L. 110-246)
			with a modification to clarify that the priority applies upon expiration of the CRP contract. *[Sec. 2403]*
No comparable provision.	Allows USDA to transfer certain land currently in the CRP into the GRP, but limits the total in any calendar year to no more than 10% of GRP acres enrolled. Requires USDA pay the lowest of four specified ways to calculate fair market value. *[Sec. 2104(c)]*	Allows USDA to transfer certain land currently in the CRP to be transferred to a permanent easement under GRP, but limits the total transferred in any calendar year to 10% of the total funding available for the GRP in that year. *[Sec. 2381]*	Adopts House provision regarding the method for determining fair market value with a technical correction. *[Sec. 2403]*
Sec. 1238O of the FSA, as amended, specifies the duties and requirements of landowners in the GRP, terms of easements and agreements, and how applications are to be evaluated. *[16 U.S.C. 3838o]*	No comparable provision.	Specifies landowner duties and USDA considerations in evaluating offers. Specifies how to determine compensation levels and technical assistance. Specifies terms/conditions that apply to GRP contracts/easements, such as permitted and prohibited uses, minimum requirements for cooperative agreements, and other considerations. *[Sec. 2381]*	Adopts Senate provisions with changes. *[Sec. 2403]*
No comparable provision.	Authorizes a Grasslands Reserve Enhancement Program. *[Sec. 2104 (d)]*	No comparable provision.	Deletes House provision.
Sec. 1238Q(a) of the FSA, as amended, allows USDA to transfer the title of an easement in the GRP to a state or private organization. *[16 U.S.C. 3838q(a)]*	Requires USDA to transfer the title of an easement to a private organizations or a state. *[Sec. 2104 (e)]*	Provides authority for USDA to enter into cooperative agreements with eligible entities for those entities to purchase, own, enforce, and monitor easements. *[Sec. 2381]*	Adopts the Senate amendment provision for cooperative agreements between USDA and eligible entities with a modification to the language specifying that eligible entities shall assume costs of administering and enforcing easements. Adopts a requirement for a contingent right of enforcement. *[Sec. 2403]*
Sec. 1241(a)(5) of the FSA, as amended, limits funding for the GRP to a total of $254 million from FY2003-07. *[16 U.S.C. 3841(a)(5)]*	No comparable provision. (*Note:* Sets acreage enrollment limit in GRP provisions, but no funding limit.)	Total GRP funding limited to $240 million for FY2008-12, with no acreage enrollment limit. *[Sec. 2401(a)(6]*	Deletes Senate provision.

(Continued)

PRIOR LAW/POLICY	HOUSE-PASSED BILL (H.R. 2419)	SENATE-PASSED SUBSTITUTE AMENDMENT (H.R. 2419)	ENACTED 2008 FARM BILL (P.L. 110-246)
Wildlife Habitat Incentives Program			
The 1996 farm bill amended Sec. 1240N of the FSA to authorize Wildlife Habitat Incentives Program (WHIP), providing cost-sharing to landowners who improve habitat, with up to 15% of the total made available in any years for agreements that are longer than 15 years. ***[16 U.S.C. 3839bb-1]***	Reauthorizes WHIP through FY2012; allows additional funds to be used to meet regulatory requirements that "reduces the economic scope of the producer's operation;" increases portion of funds for long-term agreements from 15% to 25% of funding. ***[Sec. 2112]***	Reauthorizes WHIP through FY2012; increases portion of funds for long-term agreements from 15% to 25% of funding; requires USDA to give priority to projects that foster the goals of state, regional, and national fish and wildlife conservation plans. ***[Sec. 2393]***	Limits program eligibility to focus on lands "for the development of wildlife habitat on private agricultural land, nonindustrial private forest land, and tribal lands." Raises limit on cost-share payments for long-term projects to 25% and limits total payments to $50,000 per year. Allows USDA to provide priority to projects that address issues raised by State, regional, and national conservation initiatives. ***[Sec. 2602]***
Sec. 1241(a)(7) of the FSA, as amended, authorizes mandatory funding raising from $15 million to $60 million between FY2002-04, and $85 million annually (FY2005-07). ***[16 U.S.C. 3841(a)(7)]***	Mandatory funding for WHIP is authorized at $85 million annually through FY2012. ***[Sec. 2401(e)]***	Similar to the House bill. ***[Sec. 2401(a)(8)]***	Reauthorizes program through FY2012 at current levels. ***[Sec. 2701]***
Other Conservation Programs			
The 2002 farm bill amended the FSA to provide grants to implement a Farm Viability Program. Authorizes appropriations "such sums as are necessary" through FY2007. ***[16 U.S.C. 3838j]***	Reauthorizes discretionary funding for program through FY2012. ***[Sec. 2111]***	Reauthorizes discretionary funding for program through FY2012. ***[Sec. 2396]***	Reauthorizes the program through 2012. ***[Sec. 2402]***
The 1996 farm bill amended Sec. 1240M(e) of the FSA to authorize the Conservation of Private Grazing Land Program. Authorizes appropriations of $60 million annually through FY2007. ***[16 U.S.C, 3839bb(e)]***	Extends authorization of appropriations through FY2012. ***[Sec. 2108]***	Extends authorization of appropriations through FY2012. ***[Sec. 2392]***	Extends authorization of appropriations through FY2012. ***[Sec. 2601]***
The 2002 farm bill amended Sec. 1240O of the FSA to authorize a Grassroots Source Water Protection Program to assist state rural water associations that operate wellhead and groundwater	Authorizes $20 million annually in discretionary funds (FY2008-12) and one-time funding of $10 million in mandatory funding to remain available until spent. ***[Sec. 2107]***	Authorizes $20 million annually in discretionary funding (FY2008-12). ***[Sec. 2394]***	Adopts Senate provision. ***[Sec. 2603]***

(Continued)

PRIOR LAW/POLICY	HOUSE-PASSED BILL (H.R. 2419)	SENATE-PASSED SUBSTITUTE AMENDMENT (H.R. 2419)	ENACTED 2008 FARM BILL (P.L. 110-246)
protection programs. Authorizes appropriations of $5 million annually through FY2007. *[16 U.S.C. 3839bb-2]*			
The 2002 farm bill amended Sec. 1240P of the FSA to authorize a Great Lakes Program for Soil Erosion and Sediment Control, and provides $5 million annually through FY2007. *[16 U.S.C. 3839bb-3]*	Extends authorization of appropriations through FY2012. *[Sec. 2109]*	Extends authorization of appropriations through FY2012; specifies program will help implement recommendations of a collaborative restoration strategy, giving priority to certain projects. *[Sec. 2395]*.	Reauthorizes program, and authorizes $5 million annually FY2008-2012. *[Sec. 2604]*
Sec. 524(b)(1) of the Federal Crop Insurance Act authorizes an Agricultural Management Assistance (AMA) program for listed states that have historic low participation rates in the Federal Crop Insurance Program. *[7 U.S.C. 1524(b)]*	Adds Hawaii and Virginia to the list of eligible states. Allocates 50% of funds to USDA's NRCS; 10% to Agricultural Marketing Service (for organic certification assistance); and 40% to the Risk Management Agency. *[Sec. 2201]*	Reauthorizes the program through FY2012; adds Idaho to the list of eligible states. *[Sec. 2601]*	Adopts House provision with changes to include Hawaii as an eligible State. Provides an additional $25million in mandatory funding (FY2008-2012). *[Sec. 2801]*
Secs. 1528-1537 of the 1981 farm bill (Agriculture and Food Act of 1981, P.L. 108-7) authorizes the Resource Conservation and Development Program (RC&D) to develop and implement a regional plan to address conservation, water/land management, or community development. *[16 U.S.C. 1528-1527]*	Amends RC&D program to provide a designated coordinator to assist each approved area. Eliminates requirement to submit a program evaluation to the House and Senate Agriculture Committees before June 30, 2005. *[Sec. 2202]*	Similar to House provision, clarifying that an area plan must be developed through a locally led process, and that the planning process, and that the planning process must be conducted by a local council. Provides for a coordinator to improve technical assistance to councils, as designated by USDA. *[Sec. 2605]*	Adopts Senate provision. *[Sec. 2805]*
Sec. 14(h) of the Watershed Protection and Flood Prevention Act (P.L. 106-472) authorizes discretionary and mandatory funding for a Small Watershed Rehabilitation Program. *[16 U.S.C. 1012]*	Authorizes $50 million annually in mandatory funding (FY2009-12); extends FY2007 discretionary funding level through FY2012. *[Sec. 2203]*	Authorizes such sums as necessary in discretionary funding annually (FY2008-12). *[Sec. 2604]*	Adopts House provision and provides $100 million in mandatory funding for FY2009 to remain available until expended. *[Sec. 2803]*
The 2002 farm bill amended Sec. 1241(d) of the FSA to authorize a program to promote regional equity, giving each state a total of at least $12 million annually from certain mandatory programs. *[16 U.S.C. 3841d]*	Annual funding for regional equity is raised to at least $15 million *[Sec. 2404]*	Annual funding for regional equity is raised to at least $15 million, and crop insurance payments are added to this calculation. Directs USDA to update state allocation formulas. *[Sec. 2402]*	Adopts Senate provision with changes. *[Sec. 2703]*

(Continued)

PRIOR LAW/POLICY	HOUSE-PASSED BILL (H.R. 2419)	SENATE-PASSED SUBSTITUTE AMENDMENT (H.R. 2419)	ENACTED 2008 FARM BILL (P.L. 110-246)
The 2002 farm bill amended Sec. 1242 of the FSA to authorize delivery of technical assistance directly or using a third party provider and specifies how providers are to be approved by USDA. Authorizes cooperative agreements with non-federal entities to provide technical assistance. *[16 U.S.C. 3842]*	Expands use of third party providers using contracts. Specifies providers should get at least prevailing market rates, calls for a review/update of all technical assistance specifications, including the needs of specialty crop producers. *[Sec. 2402]*	Expands use of third party providers using contracts. Directs USDA to develop national certification criteria and approve established state standards. Provides funding through each conservation program, specifies minimum and maximum contract terms, among other considerations. Includes similar provisions for specialty crop producers as the House bill. *[Sec. 2404]*	Adopts Senate provision with changes regarding the delivery of technical assistance. *[Sec. 2706]*
The 2002 farm bill amended Sec. 1244(a) of the FSA to authorize USDA to provide incentives to beginning farmers/ranchers and Indian tribes to participate in conservation programs. *[16 U.S.C. 3844(a)]*	Expands access to program incentives to include socially disadvantaged and limited resource farmers and ranchers. Requires USDA to develop a streamlined application process. *[Sec. 2405]*	Requires USDA to develop a streamlined application process for conservation programs. Provides for Safe Harbor assurances to the landowner under the Endangered Species Act. Allows producers to apply for programs through a producer organization. *[Sec. 2405]*	Adopts House provision with changes to include certain acreage limitations and exemptions, and also a pollinator protection provision. Requires USDA report to Congress on the completion of the requirements not later than 1 year after enactment. *[Sec. 2708]*
The 1990 farm bill amended Sec. 1261 of the FSA to authorize state technical committees (STC), including members and interests to be represented, outlining duties, and specifying that committees are advisory with no implementation or enforcement authority. *[16 U.S.C. 3861-3862]*	Specifies STC have at least 12 producers representing agriculture. Removes requirement for persons knowledgeable about conservation. Adds new provisions creating subcommittees issue areas. Describes responsibilities in more general terms. *[Sec. 2408]*	Adds non-industrial private forest land owners to the list of groups represented on the STC. Requires USDA to develop standard operating procedures to be used by the State technical committee in the development of technical guidelines for the implementation of the conservation provisions of this title. Makes local work groups subcommittees of the State technical committee. *[Sec. 2501]*	Adopts House provision with changes to require USDA to develop standard committee operating procedures, updates the names of participating agencies, and deletes the requirement for establishing specific issue-area subcommittees. Requires that public notice be given for meetings of the State technical committee and adds local working groups as subcommittees. Deletes the requirement for establishing specific issue-area subcommittees. *[Sec. 2711]*
The 1996 farm bill amended Sec. 351 of the FSA to authorize a National Natural Resources Conservation Foundation to raise private funds that will be used to promote conservation. Program has never been implemented. *[16 U.S.C. 5801-5809]*	No comparable provision.	Amends numerous provisions authorizing the Foundation. *[Sec. 2606]*	Deletes Senate provision.

(Continued)

Prior Law/Policy	House-Passed Bill (H.R. 2419)	Senate-Passed Substitute Amendment (H.R. 2419)	Enacted 2008 Farm Bill (P.L. 110-246)
The 2002 farm bill amended Sec. 2507 of the FSA to authorize USDA to transfer $200 million of CCC funds to the Bureau of Reclamation for water to at-risk natural desert terminal lakes. *[43 U.S.C. 2211note]*	No comparable provision.	Amends the desert terminal lakes provision to allow funds to be used to lease water or to purchase land and related interests in the Walker River Basin. *[2607]*	Adopts Senate provision with changes. Provides $175 million in mandatory funding. *[Sec. 2807]*
New Conservation Programs			
No comparable provision.	Authorizes a new Chesapeake Bay Program for Nutrient Reduction and Sediment Control to carry out restoration, enhancement, and preservation projects. Identifies four specified watersheds. The non-federal cost share for each project will be at least 35%, but will not exceed $5 million. Sets mandatory funding at $10 million (FY2008), rising to $55 million (FY2012). *[Sec. 2301]*	No comparable provision. (*Note*: The Chesapeake Bay program is authorized as a part of EQIP.)	Adopts House provision with changes. Renames program Chesapeake Bay Watershed Program. Applies to all tributaries, backwaters, and side channels, including watersheds, draining into the Chesapeake Bay, but gives priority to the Susquehanna, Shenandoah, Potomac, and Patuxent Rivers. Provides mandatory funds of $23 million (FY2009); $43 million (FY2010); $72 million (FY2011); and $50 million (FY2012). *[Sec. 2605]*
No comparable provision.	The so-called "Open Fields" provision authorizes state grants through a new Voluntary Public Access and Habitat Incentive Program to encourage land-owners to provide public access for wildlife-dependent recreation. Sets application contents and award priorities. Authorizes discretionary funding of $20 million annually through FY2012. *[Sec. 2303]*	Similar to the House bill, and includes a priority to States where the location of participating lands would be available to the public and provides $20 million per year in mandatory funding annually (FY2008-2012). *[Sec. 2399 and Sec. 2401(a)(9)]*	Adopts Senate provision with an amendment, providing $50 million in mandatory funds for the period FY2009-2012. Includes a 25% reduction for the total grant amount if the opening dates for migratory bird hunting in the State are not consistent for residents and non-residents. *[Sec. 2606]*
No comparable provision.	No comparable provision.	Creates a new Conservation Access program, requiring 10% of the funds (or acres in the cases of WRP and CRP) be used to assist beginning and socially disadvantaged farmers and ranchers with annual gross sales of $15,000 or more. *[Sec. 2403]*	Adopts Senate provision with changes. Provides that 5% of CSP acres and 5% of EQIP funds be used to assist beginning farmers or ranchers, and an additional 5% of each to assist socially disadvantaged farmers or ranchers. *[Sec. 2704]*

PRIOR LAW/POLICY	HOUSE-PASSED BILL (H.R. 2419)	SENATE-PASSED SUBSTITUTE AMENDMENT (H.R. 2419)	ENACTED 2008 FARM BILL (P.L. 110-246)
No comparable provision.	Authorizes a new Muck Soils Conservation Program for eligible land, defined by five characteristics. Authorizes appropriations of $50 million annually through FY2012, with payments between $300-$500 per acre. ***[Sec. 2303]***	No comparable provision.	Deletes provision.
No comparable provision. (*Note:* The 2002 farm bill limited CRP payments to $50,000 per year, set payment limits for each of 3 tiers in the CSP, and limited EQIP payments to $450,000 for all contracts in any 6-year period. ***[16 U.S.C. 3834, 3838c, and 3839aa-7]***	Authorizes new payment limits, deleting existing conservation payment limit language. Limits annual payments to $60,000 for any single program; limits total payments to $125,000 under all conservation (except WRP, FRPP, GRP). Defines how payments should be attributed to individuals. ***[Sec. 2409]***	No comparable provision.	Deletes provision. Places payment limits within each section of the bill and deletes this section.
No comparable provision.	Requires USDA to submit an annual report on specialty crop producer participation in conservation programs, including how to improve producer program access. ***[Sec. 2406]***	No comparable provision.	Deletes provision. The managers report states it has modified the compliance and performance provisions of Section 1244 of FSA to accommodate the intent of the House bill.
No comparable provision.	Authorizes a new provision to develop agriculture and forestry based environmental services to promote market-based conservation. Specifies use of USDA-funded research, contracts, and award grants. Establishes a USDA-led Environmental Services Standards Board of senior federal officials to facilitate the development of credit markets and disseminate performance standards to federal agencies. Authorizes $50 million in discretionary funding, with appropriated amounts to remain available until spent. ***[Sec. 2407]***	Authorizes a new provision to develop agriculture and forestry based environmental service markets, giving priority to developing carbon storage. Directs USDA to use a collaborative process with specified government and non-government interests to develop a framework and identifies relevant framework components (quantification, accounting, and verification). Requires three reports to Congress. Authorizes discretionary funding of "such sums as are necessary." ***[Sec. 2406]***	Adopts House provision with changes. Directs USDA to work in consultation with other federal and state government agencies, nongovernmental interests and other interested persons, as determined by USDA, to establish technical guidelines for measuring environmental services and to establish a verification process (allowing for consideration of third party verifiers). Directs USDA to focus initially on carbon markets. Does not authorize funds, expecting USDA to use available resources. ***[Sec. 2709]***
No comparable provision.	Adds income from affiliated packing and handling operations to definition of farm income when calculating adjusted gross	No comparable provision.	Deletes provision.

(Continued)

Prior Law/Policy	House-Passed Bill (H.R. 2419)	Senate-Passed Substitute Amendment (H.R. 2419)	Enacted 2008 Farm Bill (P.L. 110-246)
	income limitation to determine eligibility for conservation programs. *[Sec. 2501]*		
No comparable provision.	Allows USDA to encourage development of voluntary sustainable practices for specialty crops. *[Sec. 2502]*	No comparable provision.	Deletes provision.
No comparable provision.	Requires USDA to develop information on the importance of productive farmland and designate at least one farmland information center to distribute this and related information. Specifies federal matching funds of at least $400,000 (not exceeding 0.5% of the amount provided to implement the FRPP). *[Sec. 2503]*	No comparable provision.	Deletes provision.
No comparable provision.	Requires USDA to contract with a peanut producer for a 4 year crop rotation; authorizes appropriations up to $10 million annually (FY2008-FY2012). *[Sec. 2504]*	Similar to House provision, but provided within CSP, directing USDA to provide additional payments to producers who agree to adopt resource-conserving crop rotations to achieve optimal crop rotations. *[Sec. 2341]*	Adopts Senate provision. *[Sec. 2301]*
No comparable provision. (*Note*: See the Conservation Security Program, above, in existing programs.)	See Section 2103 on the Conservation Security Program (described above), for some related changes. For example, the House bill defines "priority resources of concern;" however, the House bill does not create a new program.	Authorizes a new Comprehensive Stewardship Incentives Program to coordinate administration of a new Conservation Stewardship Program (see above) and EQIP. Addresses defined resource concerns, meets regulatory demands, encourages conservation, and promotes conservation and production as compatible goals. *[Sec. 2341]*	Deletes Senate provision. Renames CSP the Conservation Stewardship Program. *[Sec. 2301]*
No comparable provision.	No comparable provision.	Authorizes a Discovery Watershed Demonstration Program to reduce loss of nutrients into surface waters in 30 small watersheds in the Upper Mississippi River basin. Authorizes discretionary funds as are necessary. *[Sec. 2397]*	Deletes provision.

(Continued)

PRIOR LAW/POLICY	HOUSE-PASSED BILL (H.R. 2419)	SENATE-PASSED SUBSTITUTE AMENDMENT (H.R. 2419)	ENACTED 2008 FARM BILL (P.L. 110-246)
No comparable provision.	No comparable provision.	Authorizes an Emergency Landscape Restoration Program to repair landscapes damaged by natural events. Replaces two others emergency conservation and watershed programs. *[Sec. 2398]*	Deletes provision.
No comparable provision.	No comparable provision.	Directs USDA to assist producers who apply for programs indirectly through certain organizations, if this will increase participation and program benefits; payment limits apply to each producer, not the organization. *[Sec. 2405]*	Deletes provision. Places payment limits within each section of the bill and deletes this section.
No comparable provision.	No comparable provision.	Authorizes a new Agriculture Conservation Experienced Service Program, such that USDA can enter into agreements with organizations to provide technical assistance using qualified individuals 55 years or older. *[Sec. 2602]*	Adopts Senate provision, but limits individuals employed under this authority to providing only technical assistance (excluding administrative tasks). *[Sec. 2710]*
No comparable provision.	No comparable provision.	Amends the Soil Conservation and Domestic Allotment Act of 1935 by providing definitions and creates new technical assistance provisions. Reauthorizes the Soil and Water Resources Conservation Act of 1977 through 2028; requires a national appraisal of soil, water and related resources to be issued every 10 years. *[Sec. 2603]*	Adopts Senate provision. *[Sec. 2802]*
No comparable provision.	The "sodsaver" provision noncropland (including native grassland and pastureland) planted to an insurable crop ineligible for crop insurance for the first 4 years of planting. *[Sec. 11007]*	Makes native sods planted to an insurable crop (over 5 acres) ineligible for crop insurance and the noninsured crop disaster assistance program. Directs USDA to report within 180 days of enactment, and annually thereafter, on changes in cropland acreage, by county, since 1995. *[Sec. 2608]*	Makes native sods planted to an insurable crop (over 5 acres) ineligible for crop insurance and the noninsured crop disaster assistance program for the first 5 years of planting. May apply to virgin prairie converted to cropland in the Prairie Pothole National Priority Area, if elected by the state. *[Sec. 12020]*

(Continued)

Prior Law/Policy	House-Passed Bill (H.R. 2419)	Senate-Passed Substitute Amendment (H.R. 2419)	Enacted 2008 Farm Bill (P.L. 110-246)
No comparable provision.	No comparable provision.	Requires that no producers in Texas lose program benefits as a result of participating in a study of the Ogallala Aquifer's recharge potential. *[Sec. 2609]*	Adopts Senate provision. *[Sec. 2901]*
No comparable provision.	No comparable provision.	Amends the Federal Insecticide, Fungicide, and Rodenticide Act (FIFRA) *[7 U.S.C. 136o(d)]* to require the State Department to pay expenses incurred by EPA employees associated with certain international activities. *[Sec. 2610]* Amends Sec. 33 of FIFRA *[7 U.S.C. 136w-8]* to allow the EPA Administrator to waive a portion of the pesticide registration service fee under certain circumstances. *[Sec. 2612]*	Adopts the Senate provision on payment of expenses. *[Sec. 14209]* Deletes Senate provision making technical corrections to pesticide registration, but includes a container recycling provision. *[Sec. 14109]*
Sec. 202(a) of the Colorado River Salinity Control Act of 1974 authorizes DOI to construct, operate, and maintain the specific salinity control units as the initial stage of the Colorado River Basin salinity control program. *[43 U.S.C. 1592(a)]*	No comparable provision.	Amends the act to create a basin states program implementing specified salinity control activities. Requires DOI to consult with the Colorado River Basin Salinity Control Advisory Council related to assistance in the form of grants, grant commitments, or the advancement of funds to federal or non-federal entities. Requires a planning report to Congress describing the proposed program implementation; stipulates that no funds may be expended until 30 days after the report is submitted. *[Sec. 2611]*	Adopts Senate provision. The Mangers report states this provision to be fiscally neutral both as to appropriations and as to draws on the basin funds. States there are no changes to the cost share ratios already established in the act; the percentage split between the two funds; or the 15% cap requirement on the basin states cost share derived from the Upper Colorado River Basin Fund. *[Sec. 2806]*
No comparable provision.	No comparable provision.	No comparable provision.	Authorizes USDA to accept contributions to support conservation programs to establish a sub-account for each USDA conservation program to accept contri-butions of non-Federal funds. Provides that contributions of non- Federal funds received for a conservation program be deposited and shall be available to USDA, without further appropriation and until expended. *[Sec. 2702]*

(Continued)

Prior Law/Policy	House-Passed Bill (H.R. 2419)	Senate-Passed Substitute Amendment (H.R. 2419)	Enacted 2008 Farm Bill (P.L. 110-246)
No comparable provision.	No comparable provision.	No comparable provision.	Direct USDA to submit to Congress an annual report regarding enrollments and assistance under conservation programs, including (1) WRP, FPP, and GRP payments valued at $250,000 or greater; (2) EQIP payments for land determined to have special environmental significance; and (3) AWEP payments subject to the waiver of adjusted gross income limitations. Allows for waivers granted by USDA to protect environmentally sensitive land of special significance. *[Sec. 2705]*
No comparable provision.	No comparable provision.	No comparable provision.	Amends the Soil and Water Resources Conservation Act of 1977 to require USDA to conduct two comprehensive appraisals and inventory of soil, water, and related natural resource conservation (completed by year-end 2010 and 2015). Requires a report in early 2012 on the types of improvements to appraisals and programs. *[Sec. 2804]*
No comparable provision.	No comparable provision.	No comparable provision.	Other miscellaneous provisions: — Names the National Plant Materials Center at Beltsville, MD, in honor of Norman A. Berg. *[Sec. 2902]* — Directs USDA, in consultation with the CCC, to promulgate regulations not later than 90 days after the date of enactment, necessary to implement title II. *[Sec. 2904]*
Title III: Agricultural Trade and Aid			
P.L. 480 Food Aid			
The Agricultural Trade Development and Assistance Act of 1954 *[7 U.S.C. 1691a]*, as amended, authorizes the use of U.S.	No comparable provision.	Renames the law the "Food for Peace Act." Deletes language making expansion of U.S. agricultural exports an	Adopts Senate provisions. *[Secs. 3001-3003]*

(Continued)

PRIOR LAW/POLICY	HOUSE-PASSED BILL (H.R. 2419)	SENATE-PASSED SUBSTITUTE AMENDMENT (H.R. 2419)	ENACTED 2008 FARM BILL (P.L. 110-246)
agricultural commodities and local currencies to combat world hunger and malnutrition; promote sustainable development; expand international trade; develop and expand markets for U.S. agricultural exports; foster development of private enterprise and democratic participation in developing countries; and prevent conflicts. Declares the Sense of Congress that the President should initiate consultations with other food aid donors to consider appropriate food aid levels to meet needs of developing countries, and that the U.S. should increase its food aid contribution.		objective of the program. Replaces the Sense of Congress with language stating that, in negotiations, the President shall seek higher levels of food aid to meet legitimate needs of developing countries; ensure that nongovernmental organizations, recipient governments, and international organizations continue to be eligible to receive food aid resources and to implement programs; and options for providing food aid shall not be subject to limitation with respect to in-kind commodities, funds for commodity procurement, and monetization, under certain conditions. ***[Secs. 3001-3003]***	
Trade and Development Assistance (P.L. 480 Title I) provides for concessional financing , i.e., long-term, low-interest loans to developing countries to purchase U.S. agricultural commodities. ***[7 U.S.C. 1701]***	Makes no changes in P.L. 480 Title I.	Renames Title I "Economic Assistance and Food Security." ***[Sec. 3004]*** Makes improving trade capacity of the recipient country an activity that can be supported by local currency payments for P.L. 480 Title I loans.	Adopts Senate provision. ***[Sec. 3004]***
Emergency and Private Assistance Programs (P.L. 480 Title II) provide U.S. agricultural commodities for emergency and nonemergency assistance. Provides that private voluntary organizations and cooperatives that carry out Title II programs may receive not less than 5% nor more than 10% of available funds (for establishing new programs or meeting other administrative directives). ***[7 U.S.C. 1722(e)(1)]*** No comparable provision.	Reauthorizes through FY2012, with changes as outlined below. Provides that not less than 7% and not more than 12% of funds available for Title II for support of eligible organizations. Funds may also be used for "developing, implementing and improving monitoring systems of program receiving funds" under Title II. ***[Secs. 3001(b)(1)and (b)(4)]*** No comparable provision.	Reauthorizes through FY2012, with changes as outlined below. Provides not less than 7.5% of total funds be available from all sources to support eligible organizations. Inserts language on "meeting specific administrative, management, personnel, programmatic, and operational activities, and internal transportation and distribution costs" for new and existing programs in foreign countries. ***[Sec. 3008(2)(A)(ii)]*** Authorizes the USAID Administrator to use funds to assess the types and quality of agricultural commodities used for food	Reauthorizes through FY2012, with changes as outlined below. Provides that the share of Title funds that can be used for administrative and distribution expenses will be from 7.5% to 13%. ***[Sec. 3008]*** Authorizes $4.5 million for fiscal years 2009-2011 to be used to study and improve food aid quality. ***[Sec. 3008]***

(Continued)

PRIOR LAW/POLICY	HOUSE-PASSED BILL (H.R. 2419)	SENATE-PASSED SUBSTITUTE AMENDMENT (H.R. 2419)	ENACTED 2008 FARM BILL (P.L. 110-246)
		aid, and to adust products and formulations to meet nutrient needs of target populations. *[Sec. 3008 (3)(h)]*	
Provisions regarding Private Voluntary Organizations (PVOs) and cooperatives allow for the sale or barter of Title II commodities by PVOs and coops. Sec. 203 allows for the monetization (sale) of not less than 15% of the total of all commodities distributed each fiscal year under non-emergency programs. Proceeds can be used for certain specified purposes. *[7 U.S.C. 1723]*	No comparable provision.	Provides for the inclusion of activities involving micro-enterprise and village banking as an authorized use of Title II monetization (sales) proceeds. *[Sec. 3009]*	Maintains monetization uses as prescribed in 7 U.S.C.1723. *[Sec. 3009]*
USDA make 2.5 mmt of commodities available annually through FY2007. Of that amount, a minimum of 1.875 mmt of commodities is available for non-emergency programs annually through FY2007 (and may be waived under certain circumstances only). *[7 U.S.C. 1724]*	Extends authorized levels of assistance through FY2012. *[Sec. 3001(d)]*	Similar to the House bill. *[Sec. 3010]*	Adopts House bill. *[Sec. 3010]*
The Food Aid Consultative Group (FAC) is composed of the Administrator of USAID, the Secretary of Agriculture, and representatives of PVOs, coops, indigenous NGOs in recipient countries, and U.S. producer groups, who review overall program effectiveness. *[7 U.S.C. 1725]*	Extends FAC through FY2012 and requires USAID, in close consultation with the FAC, to submit a report to certain congressional committees on efforts to achieve an integrated and effective food assistance program. *[Sec. 3001(f)]*	Extends FAC through FY2012. Adds representatives of the maritime transportation sector involved in overseas shipping of commodities to the member list. Requires biannual consultations between USAID and the FAC, and requires FAC consultations on issuing draft regulations. *[Sec. 3011]*	Extends the FAC through FY2012 and requires that a representative of the maritime transportation sector be included in the Group. *[Sec. 3011]*
Title II program administration provides for various administrative procedures including identification of developing country recipients, deadlines for USAID acceptance (or rejection) of a proposal, specifying reasons for denial of proposals, issuance of regulations, and deadlines for submission of commodity	Provides for program oversight, monitoring, and evaluation, and requires that systems be established to accomplish these tasks. Requires an implementation report be prepared, to be reviewed by GAO, along with annual reporting. Authorizes appropriations up to $15 million of funds be made available	Adds language allowing for payment to the World Food Program for indirect support costs of donated commodities, along with a report to the relevant congressional committees on such payments. Clarifies the authority to pay indirect costs associated with funds received or generated for programs to	Adopts the House language to require specific oversight, monitoring and assessment activities and provides up to $22 million of Title II funds for monitoring and assessment activities for non-emergency programs. No more than $8 million of these funds may be used for the Famine Early Warning System

(Continued)

Prior Law/Policy	House-Passed Bill (H.R. 2419)	Senate-Passed Substitute Amendment (H.R. 2419)	Enacted 2008 Farm Bill (P.L. 110-246)
orders. Requires the Administrator to report by December 1 each year to relevant committees on program, countries, and commodities approved to date and on the total amount of funds approved for transportation and administrative costs. *[7 U.S.C. 1726a]*	annually (FY2008-12). Requires procedures be developed for providing commodities overseas in a timely manner and according to delivery schedules. Changes the date of the required program reports, among other changes. *[Sec. 3001 (g)]*	PVOs and coops, and requires that project reports be submitted in a form that can be readily displayed for public use on the USAID website. *[Sec. 3012]*	Network. Provides $2.5 million (of the $22 million) to upgrade information technology systems in FY2009 to enhance monitoring of Title II non-emergency programs. *[Sec. 3012]*
Provisions on assistance for stockpiling and rapid transportation, delivery, and distribution of shelf-stable prepackaged foods. Authorizes appropriated grants at $3 million annually. *[7 U.S.C. 1726b]*	Reauthorizes program and increases appropriations authority from $3 million to $7 million annually for FY2008-12. *[Sec. 3001(h)]*	Reauthorizes program and increases appropriations authority from $3 million to $8 million annually (FY2008-12). *[Sec. 3013]*	Adopts Senate provision, increasing the level that can be appropriated to assist in developing shelf-stable, pre-packaged foods for food aid programs to $8 million annually. *[Sec. 3013]*
No comparable provision.	Authorizes to be appropriated not less than $40 million to carry out Sec. 491 of the Foreign Assistance Act of 1961 (22 U.S.C.2292) to provide for famine prevention and relief. *[Sec. 3016]*	Establishes a pilot program for local and regional purchase of commodities for emergency food aid. Authorizes appropriations of $25 million annually for FY2009-12. *[Sec. 3014]*	Establishes a pilot program for local and regional purchase of commodities for famine prevention to be conducted by USDA with $60 million in mandatory funding (FY2009-2012). *[Sec. 3206]*
Provisions on commodity determinations *[7 U.S.C. 1731]* establish criteria for eligibility of commodities for disposition.	No comparable provision.	Strikes the requirement that a USDA determination of domestic supply needs be made before a commodity is available for food aid. *[Sec. 3015]*	Adopts Senate provision. *[Sec. 3014]*
Provisions on the use of the CCC lists the expenses that may be covered by the CCC in carrying out food aid programs. *[7 U.S.C. 1736]*	No comparable provision.	Adds costs incurred to improve food aid quality to the list of expenses that can be covered by CCC through advanced appropriations acts. *[Sec. 3016]*	Adopts Senate provision. *[Sec. 3016]*
Provisions regarding prepositioning authorize USAID to use Title II (and Title III) funds to procure transport, and store commodities for prepositioning (limited to $2 million). *[7 U.S.C. 18736(c)]*	Extends authorization for prepositioning through FY2012. Increases the limit that may be used to preposition commodities from $2 million to $8 million. Authorizes assessment and possible establishment of additional prepositioning sites. *[Sec. 3001 (I)]*	Extends authorization for prepositioning through FY2012. Increases cap on funds available for prepositioning from $2 million to $4 million, among other requirements for non-emergency assistance agreements. *[Sec. 3017(1)]*	Adopts Senate provision, but increases funding to $10 million and adds House language on studying additional prepositioning sites. *[Sec. 3017 and Sec. 3018(a)]*
Annual reporting requirements specify that an annual report on P.L. 480 food aid programs be submitted by January 15 of	Changes the annual report submission date from January 15 to March 1. Requires an assessment on the progress	Changes the date of report submission from January 15 to April 1 each year, and requires its distribution to the public by	The conference report consolidates a number of reporting requirements and date changes from both House and

(Continued)

PRIOR LAW/POLICY	HOUSE-PASSED BILL (H.R. 2419)	SENATE-PASSED SUBSTITUTE AMENDMENT (H.R. 2419)	ENACTED 2008 FARM BILL (P.L. 110-246)
each year to the relevant congressional committees.	to reduce food insecurity in countries receiving U.S. assistance. *[Sec. 3001(j)]*	electronic and other means. *[Sec. 3017(2)]*	Senate bills. *[Sec. 3018]*
Authority to enter into new P.L. 480 agreements provided through December 31, 2007. *[Sec. 408, 7 U.S.C. 1736b]*	Extends the authority to enter into P.L. 480 agreements to December 31, 2012. *[Sec. 3001(k)]*	Similar to the House bill. *[Sec. 3018]*	Adopts House provision. *[Sec. 3020]*
Authorizes to be appropriated such sums as may be necessary to carry out P.L. 480 food aid programs. *[7 U.S.C. 1736f]*	Authorizes appropriations of such sums as may be necessary for Title I and III programs, and appropriations of $2.5 billion for Title II emergency and non-emergency programs. *[Sec. 3001(l)]*	Reauthorizes discretionary appropriations for P.L. 480 programs and strikes authority to transfer funds between P.L. 480 Titles. *[Sec. 3019]*	Adopts House provision. *[Sec. 3020]*
No comparable provision.	Requires that non-emergency food assistance of not less than $450 million be available for non-emergency programs. *[Sec. 3013]*	Provides that not less than $600 million be available annually for FY2008-12 to for Title II non-emergency programs. *[Sec. 3019(b)]*	Specifies funds for non-emergency food aid: $375 million (FY2009), $400 million (FY2010), $425 million (FY2011), and $450 million (FY2012). Authority can be waived only if the President determines that an extraordinary food emergency exists, that resources from the Bill Emerson Trust have been exhausted, and the President has submitted a request for additional appropriations to Congress needed. *[Sec. 3022]*
The Micronutrient Fortification Program establishes a micro-nutrient fortification program in food aid recipient countries through year-end 2007. *[7 U.S.C. 1736g-2]*	Extends authorization for the program through December 31, 2012, and amends purposes. *[Sec. 3001(m)]*	Reauthorizes the program through December 31, 2012. Adds new authority to improve food quality, safety, and other aspects. Eliminates limitation to five countries. *[Sec. 3020]*	Adopts Senate provision. *[Sec. 3023]*
The John Ogonowski and Doug Bereuter Farmer-to-Farmer Program authorizes voluntary technical assistance to raise farm production/incomes in developing and middle income countries, emerging markets, and in Sub-Saharan Africa and the Caribbean Basin. *[7 U.S.C. 1737]*	Provides minimum program funding of $10 million. Authorizes appropriations of $10 million to carry out the program in Sub-Saharan African and Caribbean Basin countries, and $5 million for all other countries. *[Sec. 3001(n)]*	Extends program through 2012. *[Sec. 3022]*Adopts House provision. *[Sec. 3024]*	

(Continued)

PRIOR LAW/POLICY	HOUSE-PASSED BILL (H.R. 2419)	SENATE-PASSED SUBSTITUTE AMENDMENT (H.R. 2419)	ENACTED 2008 FARM BILL (P.L. 110-246)
Other Food Aid Programs			
The Food for Progress Act provides commodities to support countries that have made commitments to expand free enterprise in their agricultural economies. ***[7 U.S.C. 1736o]***	Extends program through 2012. ***[Sec. 3004]***	Extends program through 2012. Raises the cap on coverage of costs of transporting commodities to $48 million annually (FY2009-10). ***[Sec. 3106]***	Adopts House provision and requires the USDA to establish a project in Malawi under this program. ***[Sec. 3105]***
The McGovern-Dole International Food for Education and Child Nutrition Program makes available U.S. agricultural commodities, financial and technical assistance to carry out food for education and child nutrition programs in foreign countries. Authorizes CCC funding of $100 million in FY2003 and such sums as may be necessary in FY2004-07. ***[7 U.S.C. 1736o-1]***	Extends program through 2012. Authorizes USDA to carry out the program, providing mandatory (CCC) funding as follows: $0 (FY2008); $140 million (FY2009); $170 million (FY2010); $230 million (FY2011); $300 million (FY2012); and $0 (FY2013). ***[Sec. 3005]***	Extends program through 2012. ***[Sec. 3107]***	Adopts Senate provision, except that it provides $84 million in mandatory funds for this program for FY2009, to be available until expended. ***[Sec. 3106]***
The Bill Emerson Humanitarian Trust establishes a reserve of up to 4 million metric tons (mmt) of wheat, rice, corn and sorghum to meet emergency food needs in developing countries when there are unanticipated needs or when U.S. domestic supplies are short. The Trust can also hold cash. ***[7 U.S.C. 1736f-1 note]***	Reauthorizes through 2012. ***[Sec. 3006]***	Reauthorizes through 2012 ***[Sec. 3201(6)]***, with some changes including (1) provides that the Trust can be held as a combination of cash and commodities (not to exceed 4 mmt.) ***[Sec. 3201(1)]***; (2) allows the commodities in the Trust to be exchanged for funds available under Title II or the McGovern-Dole Program, or for sale in the market (in some cases) ***[Sec. 3201(2)]***; (3) allows the funds in the Trust to be invested in low-risk short-term securities or instruments; and (4) lists the rules for the release of commodities or funds from the Trust; defines emergency in cases of release ***[Sec. 3201(3), (4)]***.	Adopts Senate provision with the following modifications: removes the 4 million metric ton cap; does not allow for the exchange of funds available under Title II or the McGovern-Dole program; and does not require the transfer of foregone storage charges into the Trust. ***[Sec. 3201]***
Export Programs			
Export Credit Guarantees provide for the repayment of credit made available to finance commercial sales of U.S.	Repeals the supplier credit guarantee program and the intermediate credit guarantee program. Lifts the 1%	Similar to the House bill; also clarifies how USDA should evaluate the creditworthiness of participating	Adopts Senate provision, but in lieu of reducing the term for short-term (GSM-102) guarantees beginning in FY2012, it

PRIOR LAW/POLICY	HOUSE-PASSED BILL (H.R. 2419)	SENATE-PASSED SUBSTITUTE AMENDMENT (H.R. 2419)	ENACTED 2008 FARM BILL (P.L. 110-246)
agricultural commodities. *[7 U.S.C. 5622]*	origination fee cap. Reduces the term of short term credit guarantees to six months starting in FY2008. *[Sec. 3002]*	countries. Reduces the minimum volume of guarantees annually from $5.5 billion to $5 billion. *[Sec. 3101]*	includes a $40 million annual cap on the credit subsidy for the program, and requires the CCC to make available each year guarantees not more than the lesser of $5.5 billion or the sum of guarantees supported by $40 million in budget authority plus the amount of guarantees that the CCC can make available from unobligated prior fiscal year balances. *[Sec. 3101]*
The Market Access Program (MAP) provides for CCC funding of export market development for U.S. agricultural commodities by eligible trade organizations. Authorizes CCC funding of $200 million annually. *[7 U.S.C. 5623]*	Makes organic agricultural commodities eligible for MAP promotions. Increases annual MAP funding from $200 million (FY2007) to $225 million (FY2008-12). *[Sec. 3003]*	Makes organic agricultural products eligible for CCC funds. Increases funding above baseline level of $200 million by raising it $10 million annually until FY2011, with $200 million for FY2012. *[Sec. 3102]*.	Adopts Senate provision, but maintains funding at the level of $200 million annually. *[Sec. 3102]*
Technical Assistance for Specialty Crops (TASC) authorizes USDA to address barriers prohibiting or threatening exports of U.S. specialty crops. Authorizes CCC funds of $2 million annually (FY2002-07). *[7 U.S.C. 5680]*	Extends authority for CCC funding of TASC through FY2012. Increases funding from $2 million annually to $4 million in FY2008, and rising to $10 million in FY2011-12. *[Sec. 3007]*	Extends authority to use CCC funds for TASC through FY20012 and increases funding by $19 million over the baseline level. *[Sec. 1835]*	Adopts House provision with annual funding ramped up to $9 million in FY2011 and FY2012. *[Sec. 3203]*
No comparable provision.	Authorizes USDA to enhance U.S. participation in international standard-setting bodies (e.g., Codex Alimentarius Commission; International Plant Protection Convention; and the World Animal Health Organization). Authorizes appropriations as necessary for FY2007-12. *[Sec. 3009]*	No comparable provision.	Deletes House provision.
Foreign Market Development Cooperator Program (FMDP) requires USDA to establish and carry out a program to maintain and develop foreign markets for U.S. agricultural commodities and products. *[7 U.S.C. 5721]*	Extends through FY2012. *[Sec. 3010]*Increases mandatory funding for FMDP from its current annual level of $34.5 million for FY2007 by $5 million annually for FY2008-09, by $10 million in FY2010. Reverts to baseline levels in 2011 and thereafter.	Adopts House provision. *[Sec. 3104]*	

PRIOR LAW/POLICY	HOUSE-PASSED BILL (H.R. 2419)	SENATE-PASSED SUBSTITUTE AMENDMENT (H.R. 2419)	ENACTED 2008 FARM BILL (P.L. 110-246)
The Emerging Markets Program (EMP) *[7 U.S.C. 5622 note]* promotes U.S. agricultural exports in emerging markets. Authorizes direct credits or export credit guarantees of not less that $1 billion for exports to emerging markets. Requires a portion of export credit guarantees be made available to establish or improve facilities and services for U.S. products.	Reauthorizes through FY2012. *[Sec. 3011]*	Reauthorizes through FY2012. *[Sec. 3202 (1)]* Permits waiving the requirement that U.S. goods be used in the construction of a facility, if such goods are not available or their use is not practicable. Provides for a guarantee that the term of the depreciation schedule for the facility will not exceed 20 years. *[Sec. 3202(2)]*	Adopts Senate provision. *[Sec. 3201]*
The Export Enhancement Program (EEP) (Section 301, Agricultural Trade Act of 1978) authorizes the CCC to encourage the sale of U.S. agricultural exports in world markets at competitive prices. The CCC may pay a bonus to exporters in markets where the competition is subsidized. *[7 U.S.C. 5651]*	Extends through FY2012. *[Sec. 3012]*	Repeals authority for EEP. *[Sec. 3103]*	Adopts Senate provision. *[Sec. 3103]*
Other Trade Provisions			
No comparable provision.	Requires USAID to contribute to the Global Crop Diversity Trust for germ plasm conservation (up to $60 million over 5 years, but may not exceed one-fourth of the total of funds contributed to the Trust from all sources). *[Sec. 3014]*	Similar to the House bill. *[Sec. 3021]*	Adopts House provision. *[Sec. 3202]*
No comparable provision.	Directs USAID and USDA to submit a report on efforts to improve planning for food and transportation procurement, including efforts to eliminate bunching of food purchases. *[Sec. 3015]*	No comparable provision.	Adopts House provision. *[Sec. 3022]*
No comparable provision.	No comparable provision.	Requires USDA, in cooperation with the Depart. of Labor, to develop standards to certify that U.S. agricultural importers were not produced with the use of abusive forms of child labor. *[Sec. 3104]*	Adopts Senate provision, modified to establish a consultative group of interested stakeholders changed with developing recommendations and guidelines for monitoring and verifying whether food products were made with child labor. *[Sec. 3104]*

(Continued)

Prior Law/Policy	House-Passed Bill (H.R. 2419)	Senate-Passed Substitute Amendment (H.R. 2419)	Enacted 2008 Farm Bill (P.L. 110-246)
The Biotechnology and Agricultural Trade Program ***[7 U.S.C. 5679]*** provides for a biotechnology and agricultural trade program to remove, resolve, or mitigate nontariff barriers on U.S. agricultural commodities produced through biotechnology. Authorizes appropriations of $6 million annually (FY2002-07).	No comparable provision	Extends authority through FY2012. ***[Sec. 3203]***	Deletes provision.
No comparable provision.	Authorizes USDA to provide technical assistance to limited resource persons that are involved in trade disputes. Authorizes appropriations as necessary. ***[Sec. 3008]***	Similar to the House bill. ***[Sec. 3204]***	Deletes House and Senate provision.
No comparable provision.	No comparable provision	Requires the Secretary of Health and Human Services to report to Congress on the importation and use of high protein food ingredients. ***[Sec. 3206]***	Deletes Senate provision.
Provision Regarding U.S.-Canada Softwood Lumber Agreement — see section on Forestry Programs (below)			
Title IV: Nutrition			
Reauthorization			
Authorities in the Food Stamp Act and other laws covered by the nutrition title generally expired in FY2007. ***[7 U.S.C. 612c note, 1431e(2), 2020(t), 2025(h), 2025(k), 2026(b), 2028(a), 2034, 2036, 3007(a), 3171 note, 7508(a)]***	Extends all expired authorities in the Food Stamp Act and other laws covered by the nutrition title through FY2012. ***[Sec. 4016, 4019-4021, 4025, 4027, 4028, 4201, 4202, 4203]*** Policy amendments made in the House bill generally are part of permanent law.	Extends all significant expired authorities in the (renamed) Food Stamp Act and other laws covered by the nutrition title indefinitely, with the exception of funding for community food projects (extended through FY2012). ***[Sec. 4110, 4701, 4801-4803]*** Policy amendments generally terminate with FY2012.	With some exceptions, extends expired authorities in the (renamed) Food Stamp Act and other laws covered by the nutrition title through FY2012. Includes those governing required funding or appropriations authorizations for the Food Stamp program (including grants for simplified application projects), nutrition assistance block grants for Puerto Rico and American Samoa, The Emergency Food Assistance Program (TEFAP), the Food Distribution Program on Indian Reservations (FDPIR), the Commodity Supplemental Food Program (CSFP), and the Senior Farmers' Market Nutrition Program (SFMNP). Extends

(Continued)

PRIOR LAW/POLICY	HOUSE-PASSED BILL (H.R. 2419)	SENATE-PASSED SUBSTITUTE AMENDMENT (H.R. 2419)	ENACTED 2008 FARM BILL (P.L. 110-246)
			other expired authorities indefinitely: community food project grants, authority to reduce federal payments for state food stamp administrative costs, cash-payment food stamp pilot projects, and grants for employment/training programs for food stamp recipients. *[Sec. 4406]* Policy amendments generally are part of permanent law, as in the House bill.
Food Stamp Program			
The Food Stamp program is established under the Food Stamp Act. *[7 U.S.C. 2011 et seq]*	Renames the Food Stamp program: the "Secure Supplemental Nutrition Assistance Program" (SSNAP). *[Sec. 4001]*	Renames the Food Stamp program: the "Food and Nutrition Program." Renames the Food Stamp Act: "The Food and Nutrition Act." *[Sec. 4001]*	Renames the Food Stamp program the Supplemental Nutrition Assistance Program (SNAP). Renames the Food Stamp Act "The Food and Nutrition Act." *[Sec. 4001]*
Sec. 5(e)(1) of the Food Stamp Act specifies that when determining benefits and, in some cases, judging income eligibility, a standard portion of households' monthly income be disregarded (deducted), as a "standard deduction." The minimum standard deduction is $134 per month per household (with differing amounts for Alaska, Hawaii, the Virgin Islands, and Guam). *[7 U.S.C. 2014(e)(1)]*	Increases the minimum standard deduction to $145 per month, annually indexed to inflation beginning with FY2009. Provides comparable increases for Alaska, Hawaii, the Virgin Islands, and Guam. *[Sec. 4006]*	Increases the minimum standard deduction to $140 per month, annually indexed to inflation beginning with FY2009. Provides comparable increases for Alaska, Hawaii, the Virgin Islands, and Guam. *[Sec. 4102]*	Increases the minimum standard deduction to $144 per month in FY2009, annually indexed to inflation beginning with FY2010. Provides comparable increases for Alaska, Hawaii, the Virgin Islands, and Guam. *[Sec. 4102]*
Sec. 5(e)(3) of the Food Stamp Act specifies that when determining benefits and, in some cases, judging income eligibility, dependent care expenses related to employment, training, or education be disregarded (deducted) from household monthly income. This deduction is capped at $200 per month for each dependent under age 2 and $175 per month for all other dependents. *[7 U.S.C. 2014(e)(3)]*	Removes the caps on dependent care deductions. *[Sec. 4007]*	Same as the House bill. *[Sec. 4103]*	Removes the caps on dependent care deductions, as in the House bill and Senate amendment. *[Sec. 4103]*

PRIOR LAW/POLICY	HOUSE-PASSED BILL (H.R. 2419)	SENATE-PASSED SUBSTITUTE AMENDMENT (H.R. 2419)	ENACTED 2008 FARM BILL (P.L. 110-246)
Sec. 8(a) of the Food Stamp Act sets the minimum benefit for households of one and two persons at $10 per month. *[7 U.S.C. 2017(a)]*	Increases the minimum benefit to 10% of the inflation-indexed cost of the "Thrifty Food Plan" (the maximum benefit) for a one — person household. *[Sec. 4013]*	Same as the House bill. *[Sec. 4109]*	Increases the minimum benefit to 8% of the inflation-indexed cost of the "Thrifty Food Plan" (the maximum benefit) for a one-person household. *[Sec. 4107]*
Under terms stipulated in USDA appropriations acts since FY2005, combat-related military pay is excluded from household income when determining eligibility and benefits. (Most recently, the Consolidated Appropriations Act, 2008, P.L. 110-161.)	Places into law an exclusion for combat-related military pay. *[Sec. 4005]*	Same as the House bill. *[Sec. 4101]*	Places into law an exclusion for combat-related military pay, as in the House bill and Senate amendment. *[Sec. 4101]*
Sec. 5(g) of the Food Stamp Act specifies that eligible households' liquid assets may not be more than $2,000 (or $3,000 for a household with an elderly or disabled member). *[7 U.S.C. 2014(g)]*	Indexes to inflation the dollar limits on assets annually (adjusted to the nearest $100). *[Sec. 4008]*	Increases the dollar limits on assets to $3,500 ($4,500 for households with an elderly/disabled member) and indexes to inflation these dollar limits annually (adjusted down to the nearest $250). *[Sec. 4104]*	Indexes to inflation the dollar limits on assets annually (adjusted down to the nearest $250) beginning with FY2009. *[Sec. 4104(a)]*
Sec. 5(g) of the Food Stamp Act specifies differing rules as to which types of retirement plans/accounts/savings are excluded from countable liquid assets in judging eligibility. No provision is made for excluding education savings. *[7 U.S.C. 2014(g)]*	Excludes all tax-qualified retirement plans/accounts/savings and all tax-qualified education savings from countable liquid assets in judging eligibility. *[Sec. 4009 & 4010.]*	Same as the House bill. *[Sec. 4104]*	Adopts House and Senate provision to exclude all tax-qualified retirement plans/accounts/savings and education savings from countable liquid assets in judging eligibility. *[Sec. 4104 (b) & (c)]*
Sec. 6(o) of the Food Stamp Act limits the eligibility of able-bodied adults without dependents (ABAWDs) — who are (1) not working (20+ hours a week), (2) in an employment/training program, or (3) in a workfare program — to 3 months in every 36-month period. ABAWDs who gain eligibility by meeting one of the three above tests, but later fail to meet these tests, may remain eligible for an additional 3 consecutive months. *[7 U.S.C. 2015(o)]*	No comparable provision.	Lengthens the basic eligibility period for ABAWDs to 6 months in every 36-month period. Eliminates the current provision extending eligibility for ABAWDs who subsequently fail to meet work/training tests. *[Sec. 4107]*	Deletes Senate provision.
Sec. 11(s) of the Food Stamp Act allows states to provide eligibility for up to 5	No comparable provision.	Adds permission for states to provide eligibility for up to 5 months' transitional	Adopts Senate provision to expand permission for states to provide

(Continued)

Prior Law/Policy	House-Passed Bill (H.R. 2419)	Senate-Passed Substitute Amendment (H.R. 2419)	Enacted 2008 Farm Bill (P.L. 110-246)
months of transitional food assistance benefits to households that stop receiving federally funded cash aid under the Temporary Assistance for Needy Families (TANF) program. *[7 U.S.C. 2020(s)]*		food assistance benefits to households with children that cease to receive cash aid under a state-funded public assistance program. *[Sec. 4108]*	transitional benefits. *[Sec. 4106]*
No comparable provision.	No comparable provision.	Disqualifies persons found by a court or administrative agency to have intentionally (1) obtained cash by misusing program benefits to obtain money for the return of deposits on containers or (2) sold food purchased with program benefits. *[Sec. 4305]*	Adds new program disqualification rules, as in the Senate amendment. *[Sec. 4131]*
No comparable provision.	Bars those disqualified from the Food Distribution Program on Indian Reservations (the FDPIR) from the (renamed) food stamp program. *[Sec. 4004]*	Same as the House bill. *[Sec. 4501]*	Adopts House and Senate provision to bar those disqualified from the FDPIR from the (renamed) food stamp program. *[Sec. 4211]*
Sec. 6(c) of the Food Stamp Act specifies that states may require households (except for migrant/seasonal farmworkers, the homeless, and elderly/disabled households with no earned income) to file periodic reports of household circumstances. Households not required to file periodic reports must report any changes in circumstances when they occur. *[7 U.S.C. 2015(c)]*	No comparable provision.	Allows states to require periodic reporting of changes in household circumstances (versus reporting changes when they occur) by migrant/ seasonal farm workers, the homeless, and elderly/ disabled households with no earned income. Limits the frequency that these households must report changes (except changes whereby they exceed monthly income eligibility limits). *[Sec. 4105]*	Adopts Senate provision to change household reporting rules. *[Sec. 4105]*
Sec. 11(e) of the Food Stamp Act specifies that state "merit system" employees must certify households' program eligibility. *[7 U.S.C. 2020(e)(6)]*	Limits states' ability to "privatize" administrative functions. Specifies only state agency merit system employees are authorized to make eligibility certifications or: (1) represent the state in communications with client households about their application/participation, (2) participate in making any determinations regarding a household's substantive or procedural compliance with program	No comparable provision.	Deletes House provision.

(Continued)

Prior Law/Policy	House-Passed Bill (H.R. 2419)	Senate-Passed Substitute Amendment (H.R. 2419)	Enacted 2008 Farm Bill (P.L. 110-246)
	requirements, or (3) make any other required program determinations. Exemptions include nonprofit agencies assisting in outreach and contracts with automated systems, benefit issuance services, and program information activities. Prohibits federal funds for any contract that does not comply with these requirements, and bars waivers of rules that do not comply with requirements. Allows use of employees not in compliance with these requirements in cases of disasters. *[Sec. 4015]*		
No comparable provision.	No comparable provision.	Requires that USDA develop standards for identifying major changes in state agency administrative operations (e.g., substantial increases in reliance on automated systems or potential increases in administrative burdens on applicant/ recipient households). If a state implements a major change in operations, it must notify USDA and collect any information needed to identify/correct any adverse effects on program integrity or access. *[Sec. 4211]*	Adopts Senate provision adding requirements with regard to major changes in state agency administrative operations. *[Sec. 4116]*
No comparable provision.	No comparable provision.	Requires that computerized systems for state program operations must (1) be tested adequately before and after implementation, and (2) be operated under a plan for continuous updating (to reflect changed policies/circumstances) and testing (for effects on households and payment accuracy). *[Sec. 4212]*	Adopts Senate provision adding requirements for computerized state systems. *[Sec. 4121]*
Sec. 13(b) of the Food Stamp Act specifies that states must attempt to collect all over-issued benefits from recipients, unless the state agency demonstrates that collection is not cost-effective. *[7 U.S.C. 2022(b)]* There is	Permits USDA to make a determination that a state agency has over-issued benefits to a substantial number of households because of a "major systemic error" by the state. If this determination is made, USDA may (1) prohibit the state	Same as the House bill. *[Sec.4301]*	Adopts House and Senate provision permitting USDA to determine that a state has over-issued benefits because of major systemic error, to then prohibit the state from collecting resulting over-issuances from recipients, and also to

(Continued)

Prior Law/Policy	House-Passed Bill (H.R. 2419)	Senate-Passed Substitute Amendment (H.R. 2419)	Enacted 2008 Farm Bill (P.L. 110-246)
no specific provision for collection of the cost of erroneously issued benefits from state coffers.	agency from collecting any resulting over-issuances from recipients, and (2) collect the cost of the over-issuances from the state. *[Sec. 4018]*		collect the cost of over-issuances from the state. *[Sec. 4133]*
Sec. 12 of the Food Stamp Act specifies that participating retailers may be disqualified for violation of program rules (or, in cases where disqualification would cause hardship to food stamp households, receive a money penalty of up to $10,000 per violation). Disqualification is: (1) 6 months to 5 years (first violation), (2) 10 months to 10 years (second violation), and (3) permanent on a third violation, a disqualification based on trafficking in benefits, or a finding of the sale of firearms, ammunition, explosives, or controlled substances. *[7 U.S.C. 2021]*	Increases the maximum money penalty to $100,000 per violation. Removes the provision that limits the imposition of money penalties in lieu of disqualification to cases where it would cause hardship to recipient households. Permits the imposition of *both* a disqualification and a money penalty. Removes the 6-month and 10-month minimum disqualification requirements. Requires the establishment of procedures whereby participating retailers may be immediately suspended for "flagrant violations," pending appeal. *[Sec. 4017]*	Same as the House bill. Also eases conditions under which bonds are required of a participating retailer who has violated food stamp rules and wishes to be re-approved for participation. The provision limits the period of time for which bonding is required. *[Sec. 4303]* Bars charging retailers "interchange fees" in connection with implementation of electronic benefit transfer (EBT) systems. *[Sec. 4202]*	Adopts Senate provision that changes money penalty and bonding rules for retailers and bars interchange fees on retailers. *[Sec. 4132 & Sec. 4115]*
No comparable provision.	No comparable provision.	Requires USDA to issue regulations defining dietary supplements: multi-vitamin-mineral supplements providing prescribed minimum amounts of essential vitamins, minerals and certain prescribed amounts of folic acid and calcium. Requires proposed regulations within 1 year of enactment, and final regulations within 2 years of enactment. No dietary supplements may be purchased with program benefits until final regulations have been issued or a voluntary system of labeling for the identification of eligible dietary supplements is certified by USDA. *[Sec. 4402]*	Deletes Senate provision.
Sec. 7(h) of the Food Stamp Act provides that state agencies may stagger the issuance of benefits over the course of each month and specifies that any staggered issuance procedure must	No comparable provision.	Limits "split issuance" of benefits by requiring that no staggered issuance procedure may provide for more than 1 issuance during a month — except in the case of a benefit correction. *[Sec. 4203]*	Adopts Senate provision limiting split issuance of benefits. *[Sec. 4113]*

PRIOR LAW/POLICY	HOUSE-PASSED BILL (H.R. 2419)	SENATE-PASSED SUBSTITUTE AMENDMENT (H.R. 2419)	ENACTED 2008 FARM BILL (P.L. 110-246)
ensure that no household experiences an interval between issuances longer than 40 days. Staggered issuance procedures may include splitting monthly benefits into more than one "issuance." *[7 U.S.C. 2016(h)]*			
Fruits and Vegetables			
Sec. 18(g) of the Richard B. Russell National School Lunch Act requires a fresh fruit and vegetable program. Most recently, this program was funded at a total of $18.9 million for FY2008 ($9 million of which is mandatory), and $9.9 million of which was provided by the Consolidated Appropriations Act of 2008 (P.L. 110-161). This program makes fresh fruits and vegetables available in selected elementary and secondary schools in a limited number of states and on three Indian reservations. *[42 U.S.C. 1769(g)]*	Expands the existing fresh fruit and vegetable program in elementary and secondary schools. Increases annual mandatory funding to $70 million and makes the program available nationwide in (1) 35 elementary and secondary schools in each state and (2) additional schools in each state in proportion to the state's student population. No new Indian reservations would be added specifically. Up to 1% of funds may be reserved for federal administration, and states may use up to 5% of their allocation for administrative expenses. *[Sec. 4303]*	Replaces the existing fresh fruit and vegetable program. The new program provides mandatory funding of $225 million in the first year (indexed for inflation in later years) to make free fresh fruits and vegetables available in selected (by states) elementary schools. Allocates funding among states under a formula distributing about half the money equally among states and apportioning the remainder based on population. Priority is to be given to schools with high proportions of lower-income students, and at least 100 schools chosen to participate must be on Indian reservations. Annual per-student grants would be determined by states, but must be between $50 and $75. Requires an evaluation, providing $3 million in funding. *[Sec. 4904]*	Replaces existing fresh fruit/ vegetable program, beginning with the 2008-2009 school year. Similar to Senate provision, the new program is to operate in elementary schools selected by states and has mandatory Section 32 funding (see *Sec. 14222*): $40 million on October 1, 2008; $65 million on July 1, 2009; $101 million on July 1, 2010; and $150 million on July 2011. Available money each succeeding July 1 is indexed for inflation. Allocates funds by the formula proposed by the Senate provision, but includes no special set-aside for Indian reservation schools. As in the Senate provision, gives priority to schools with high proportions of low-income students, requires that annual per-student costs be between $50-$75, and mandates a $3 million evaluation. Authorizes limited set-asides for federal and state administrative costs, and discretionary appropriations (in addition to mandatory Section 32 funds). *[Secs. 4304 & 14222]*
Sec. 10603 of the 2002 farm bill states that USDA is required to use not less than $200 million a year in Section 32 funding to purchase fruits, vegetables, and certain other specialty food crops. Of this amount, at least $50 million a year must be used to acquire fresh fruits and vegetables for schools through the	Increases the set-aside for the Department of Defense Fresh program ("DoD Fresh") to $75 million a year. *[Sec. 4301]*	Provides that, in lieu of purchases required under Sec. 10603, USDA purchase fruits, vegetables, and nuts for use in domestic food assistance programs using Section 32 funds. Minimum purchase amounts are $390 million for FY2008, rising to $406 million in FY2012 and for each year thereafter.	Provides that, in addition to the minimum ($200 million-a-year) acquisitions required by the 2002 farm bill, USDA will purchase fruits, vegetables, and nuts for use in domestic nutrition assistance programs using Section 32 funds. The added purchases required are: $190 million (FY2008), $193 million

(Continued)

PRIOR LAW/POLICY	HOUSE-PASSED BILL (H.R. 2419)	SENATE-PASSED SUBSTITUTE AMENDMENT (H.R. 2419)	ENACTED 2008 FARM BILL (P.L. 110-246)
"DoD Fresh" program (using the Defense Department as the procurement agent). ***[7 U.S.C. 612c-4]***		***[Sec. 4907]***	(FY2009), $199 million (FY2010), $203 million (FY2011), and $206 million (FY2012 and each year thereafter). Deletes direct mention of the use of the DoD Fresh program as a procurement agent for fresh fruits and vegetables for schools, but retains a $50 million-a-year requirement for USDA-sponsored fresh fruit and vegetable acquisitions for schools. ***[Sec. 4404]***
The Emergency Food Assistance Program (TEFAP)			
Sec. 27 of the Food Stamp Act to use $140 million of annual appropriations under the Food Stamp Act to purchase food commodities for distribution through TEFAP emergency feeding organizations. Sec. 204(a) of the Emergency Food Assistance Act authorizes annual appropriations of $60 million for TEFAP administrative and distribution costs. ***[7 U.S.C. 2036 and 7508(a)]***	Increases required funding for TEFAP commodities to $250 million for FY2008; for later years, this amount is indexed annually for food-price inflation. ***[Sec. 4028]*** Increases the annual authorization of appropriations for TEFAP administrative and distribution costs to $100 million. ***[Sec. 4201]***	Increases required annual funding for TEFAP commodities to $250 million. ***[Sec. 4110]*** Also increases the annual authorization of appropriations for TEFAP administrative and distribution costs to $100 million. ***[Sec. 4802]*** Authorizes competitive grants totaling $10 million annually to expand the capacity and infrastructure of food banks so as to improve their ability to handle "time-sensitive" (perishable) food products, their identification of potential food donors, and procurement of locally produced foods. ***[Sec. 4915]***	Increases required funding for TEFAP commodities. For FY2008, an immediate infusion of $50 million is directed. For FY2009, $250 million in TEFAP commodities is mandated. For FY2010 through FY2012, the $250 million provided for FY2009 is to be adjusted for food-price inflation. Increases the annual authorization of appropriations for TEFAP administrative and distribution costs to $100 million. ***[Sec. 4201]*** Authorizes appropriations ($15 million a year through FY2012) for "infrastructure grants." Grants are to be made to emergency feeding organizations (emphasizing those serving mostly rural communities) for projects that improve the availability of perishable "time-sensitive" foods, improve the security/ diversity of emergency food distribution and recovery systems, make available recovered foods to emergency food providers, improve identifying potential providers of donated food, and support construction, expansion, or repair hunger relief facilities/equipment. ***[Sec. 4202]***

(Continued)

Prior Law/Policy	House-Passed Bill (H.R. 2419)	Senate-Passed Substitute Amendment (H.R. 2419)	Enacted 2008 Farm Bill (P.L. 110-246)
Commodity Supplemental Food Program (CSFP)			
Income eligibility limits for the CSFP are set at 130% of the federal income poverty guidelines for elderly applicants and 185% of the guidelines for households with women, infants, and children. ***[7 U.S.C. 612c note; regulations under Sec. 5 of the 1973 Agriculture and Consumer Protection Act]***	Requires USDA to establish income eligibility limits for the CSFP that are the same for all applicants. The limits may not exceed 185% of the poverty guidelines. ***[Sec. 4203]***	Permits the CSFP to serve low-income elderly persons with income up to 185% of the poverty guidelines, if USDA determines that appropriations have enabled every state seeking to participate in the CSFP to participate. ***[Sec. 4802]***	Bars USDA from requiring that CSFP projects prioritize assistance to either (1) the elderly or (2) women, infants, and children, but does not change eligibility rules. ***[Sec. 4221]***
Food Distribution Program on Indian Reservations (FDPIR)			
Sec. 4 of the Food Stamp Act states that, at the request of Indian tribal organizations, a FDPIR may be operated on Indian reservations in lieu of food stamps. ***[7 U.S.C. 2013]***	Authorizes annual appropriations of $5 million for a fund to purchase "traditional and locally grown" foods for the FDPIR. At least 50% of these foods are to from Native American farmers, ranchers, and producers. Requires a report reviewing the procedures for determining the contents of FDPIR food packages, the adequacy of the packages, and any plans to revise them to conform with dietary guidelines. ***[Sec. 4004]***	Similar to the House bill, but: (1) provides authority for the purchase of bison meat for the FDPIR (subject to appropriations), (2) requires a survey of participants to determine which traditional foods are most desired, and (3) does not include a House requirement that at least 50% of distributed foods be produced by Native American farmers, ranchers, and producers. ***[Sec. 4501]***	Combines elements of both the House and Senate bills. Adopts House provisions authorizing appropriations for a traditional and locally grown food fund, mandating a minimum purchases from Native American producers, and requiring a report on FDPIR food packages. Adopts Senate provisions for bison meat purchases and a survey of FDPIR participants as to their preference among traditional foods. ***[Sec. 4211]***
Senior Farmers' Market Nutrition Program (SFMNP)			
Sec. 4402 of the Farm Security and Rural Investment Act of 2002 established the SFMNP to provide low-income seniors with vouchers for fresh, locally grown fruits, vegetables and herbs usable at farmers' markets, roadside stands, and community-supported agriculture programs. Provides annual mandatory funding of $15 million. ***[7 U.S.C. 3007]***	Authorizes additional appropriations of $20 million for FY2008, rising to $75 million by FY2012. Adds honey to the list of items that may be purchased. Provides that the value of benefits not be considered income or resources for any purposes under any federal, state, or local law. Prohibits states and localities from collecting sales taxes on SFMNP food-voucher purchases. ***[Sec. 4401]***	Provides added mandatory funding of $10 million a year. The value of SFMNP benefits cannot be considered income or resources for any purposes under any federal, state, or local law. States and localities also are prohibited from collecting sales taxes on food purchased with SFMNP vouchers. ***[Sec. 4701& 4702]***	Increases total mandatory funding to $20.6 million a year and, as in the House bill, adds honey to items that may be purchased. Adopts House and Senate provisions barring consideration of SFMNP benefits as income/resources and the collection of sales taxes on SFMNP purchases. ***[Sec. 4231 & Sec. 4406(c)]***
Community Food Projects			
Sec. 25 of the Food Stamp Act authorizes USDA to make grants to non-	Expands the list of projects to be given preference to include emergency food	Provides $10 million in annual mandatory funding. ***[Sec. 4801]***	Requires USDA to provide a grant to a nonprofit organization to establish and

(Continued)

Prior Law/Policy	House-Passed Bill (H.R. 2419)	Senate-Passed Substitute Amendment (H.R. 2419)	Enacted 2008 Farm Bill (P.L. 110-246)
profit agencies for community food projects that require a one-time infusion of assistance to become self-sustaining. Certain types of projects are to be given preference. The term of a grant may not exceed 3 years. Not more than $5 million of annual appropriations under the Food Stamp Act may be used for community food project grants. Of available funds, $200,000 must be allocated to a special project grant for addressing common community food problems. Federal cost-sharing is not to exceed 50%. *[7 U.S.C. 2034]*	infrastructure initiatives, those dealing with retail access in under-served markets, those integrating urban and metro-area food production, and those providing assistance for youth, socially disadvantaged individuals, and groups with limited resources. Increases the maximum term for grants to 5 years. Authorizes annual appropriations of $30 million and increases the set-aside for the grant for common community food problems to $500,000. Raises the federal cost-sharing limit to 75%. *[Sec. 4027]*		support a "healthy urban food enterprise development center" for activities that increase access to healthy affordable foods (including locally produced food) in "underserved communities." The center is to provide technical assistance and information to small and medium-sized agricultural producers, food wholesalers and retailers, schools, and other entities regarding "best practices" and the availability of assistance for aggregating, storing, processing, and marketing locally produced food and increasing availability in underserved communities. Provides mandatory funding of $1 million a year (FY2009-2011); authorizes appropriations of $2 million (FY2012). *[Sec. 4402]*
No comparable provision.	No comparable provision.	Requires USDA give priority to projects promoting healthy food education be replicated in schools. Authorizes a new pilot project ($10 million) in not more than 5 States to provide grants to "high-poverty" schools for initiatives with hands-on gardening. No cost-sharing is required. *[Sec. 4903]*	Adopts Senate provision with an amendment to strike the authorization of appropriations to carry out the provision. *[Sec. 4303]*
Geographic Preference (Purchase of Locally Produced Foods)			
Sec. 9(j) of the Richard B. Russell National School Lunch Act requires USDA to encourage schools to purchase locally produced foods to the maximum extent practicable and appropriate. Authorizes appropriations of $400,000 a year (including for start-up grants for equipment and incurred costs). *[42 U.S.C. 1758(j)]* By regulation, any procurement under USDA entitlement programs, including school meal and other child nutrition programs, must be	Removes provisions for start-up grant funding to help carry out the locally produced food purchase policy. Requires USDA to allow schools and other institutions receiving funds under the National School Lunch and Child Nutrition Acts to use geographic preference for the procurement of locally produced foods. The Defense Department, acting as a fresh fruit and vegetable procurement agent for schools, also is covered by this rule. *[Sec. 4304]*	Same as the House bill, except that geographic preference would be allowed only for locally grown fruits and vegetables. *[Sec. 4902]*	Adopts House and Senate provision removing provisions for start-up grant funding. Requires USDA to allow schools and other institutions receiving funds under the National School Lunch and Child Nutrition Acts (and the Defense Department acting as a fresh fruit and vegetable buying agent) to use geographic preference for the procurement of "unprocessed agricultural products, both locally grown and locally raised." *[Sec. 4302]*

(Continued)

PRIOR LAW/POLICY	HOUSE-PASSED BILL (H.R. 2419)	SENATE-PASSED SUBSTITUTE AMENDMENT (H.R. 2419)	ENACTED 2008 FARM BILL (P.L. 110-246)
conducted in a manner that prohibits the use of statutorily or administratively imposed in-state or local geographic preferences (except in cases where federal statute expressly mandates or encourages geographic preference). ***[7 CFR 3016.60(c) and 3016.36(c)(2)]***			
Special Initiatives			
No comparable provision.	Authorizes annual appropriations of $10 million for a demonstration competitive grant program ("Initiative to Address Obesity Among Low-Income Americans") to develop and implement strategies to reduce obesity in the low-income population. Project proposals would be evaluated against criteria including identification of a low-income target audience, incorporation of scientifically based strategies to improve diet quality, commitment to a rigorous evaluation, and inclusion of strategies for after-school food service and improvement of children's health. Prohibits the use of demonstration funds for projects that limit the use of benefits. ***[Sec. 4023]***	Provides total mandatory funding of $50 million for pilot projects to develop and test methods of using the (renamed) Food Stamp program to improve the dietary and health status of participants and reduce overweight, obesity, and diet-related diseases. Project initiatives are to increase the availability and purchase of healthy foods and may include expanded program benefits, greater access to farmers' markets, incentives to participating food concerns, new approval requirements for participating food concerns, point-of-purchase incentives for program recipients to buy healthy foods, and education programs. Independent evaluations of the projects are mandated, and up to 50% of the funding must be used for point-of-purchase incentive projects. ***[Sec. 4403]***	Combines House and Senate provisions to authorize pilot projects that develop and test methods of using the (renamed) Food Stamp program to improve the dietary and health status of participants and reduce overweight, obesity, and diet-related diseases. Specifies that project proposals will be evaluated against the criteria laid out in the House bill, and specifies projects may not limit the use of benefits. Project initiatives may include those listed in the Senate provision, and independent project evaluations are required. Authorizes annual appropriations (set at "such sums as are necessary" annually through FY2012) to carry out these health and nutrition promotion pilot projects. Provides mandatory funding (a one-time $20 million) for point-of-purchase incentive projects. ***[Sec. 4141]***
No comparable provision. *Note:* Puerto Rico receives an annual (indexed) nutrition assistance block grant in lieu of operating a regular Food Stamp program.	Directs a comprehensive study of extending the regular (renamed) Food Stamp program to Puerto Rico. ***[Sec. 4026]***	Same as the House bill, but also provides $1 million in mandatory funding for the study. ***[Sec. 4206]***	Adopts Senate provision to direct a study of extending the (renamed) Food Stamp program to Puerto Rico and provide mandatory funding of $1 million. ***[Sec. 4142]***
No comparable provision.	No comparable provision.	Provides mandatory funding ($5 million) for grants to projects for expanding the number of farmers' markets that accept electronic benefit transfer (EBT) cards used in the (renamed) Food Stamp	Deletes Senate provision. *Note:* **Sec. 10106** effectively adopts Senate provision to provide funding for similar programs.

(Continued)

PRIOR LAW/POLICY	HOUSE-PASSED BILL (H.R. 2419)	SENATE-PASSED SUBSTITUTE AMENDMENT (H.R. 2419)	ENACTED 2008 FARM BILL (P.L. 110-246)
		program. *[Sec. 4210]*	
No comparable provision.	No comparable provision.	Requires a periodically updated study of hunger in the U.S., assessing data and hunger-reduction measures and making policy recommendations. Authorizes matching grants (1) to food program service providers and nonprofits for collaborative efforts to assess community hunger problems and to achieve "hunger-free communities" and (2) to emergency feeding organizations for infrastructure development. Authorizes $50 million a year for these studies and grants. *[Sec. 4405]*	Similar to the Senate provision, authorizes appropriations ("such sums as are necessary" through FY2012) for matching grants (1) to food program service providers and nonprofits for collaborative efforts to assess community hunger problems and to achieve "hunger-free communities" and (2) to emergency feeding organizations for infrastructure development. Any available funding is to be divided equally between these to grant initiatives, and the federal matching percentage is limited to 80%. *[Sec. 4405]*
No comparable provision.	No comparable provision.	Requires periodic USDA surveys of foods purchased by schools in the School Lunch program and provides mandatory funding of $3 million for each survey. *[Sec. 4901]*	Similar to Senate provision, requires a USDA survey of foods purchased by schools in the School Lunch program and provides mandatory one-time funding of $3 million. *[Sec. 4307]*
Sec. 19 of the Child Nutrition Act authorizes appropriations for "Team Nutrition Network" grants to states and local education agencies to carry out nutrition education activities for schoolchildren. *[42 U.S.C. 1788]*	No comparable provision.	Provides annual mandatory funding of $3 million for Team Nutrition Network grants. *[Sec. 4905]*	Deletes Senate provision.
No comparable provision.	No comparable provision.	Establishes a pilot project to provide whole grain products as meal supplements (snacks) in after-school programs operated by a limited number of elementary and secondary schools in 6 states and on one Indian reservation. Provides mandatory funding of $4 million to carry out and evaluate project. Funds are to be derived equally from money otherwise available for the Senior Farmers' Market Nutrition program and community food projects. *[Sec. 4912]*	Similar to Senate provision, requires USDA to purchase whole grains and whole grain products for use in school meal programs and an evaluation of this initiative. From mandatory Section 32 funding made available under *Sec. 14222*, provides $4 million (in FY2009) to carry out the whole grain project. *[Sec. 4305 & Sec. 14222(d)]*

(Continued)

Prior Law/Policy	House-Passed Bill (H.R. 2419)	Senate-Passed Substitute Amendment (H.R. 2419)	Enacted 2008 Farm Bill (P.L. 110-246)
No comparable provision.	No comparable provision.	Authorizes a "food employment empowerment and development" (FEED) program with USDA providing grants to public and private nonprofit agencies to encourage the use of community resources to combat hunger through projects that recover unused food and provide job training related to food recovery and preparation. *[Sec. 4914]*	Deletes Senate provision.
Section 4404 of the Farm Security and Rural Investment Act of 2002 authorizes a Congressional Hunger Fellows program to encourage future U.S. leaders to pursue careers in humanitarian and public service and provide aid to people in need. Two types of fellowships are authorized: (1) Bill Emerson fellowships relating to domestic hunger and humanitarian needs and (2) Mickey Leland fellowships that address international hunger and other humanitarian needs. Establishment of a trust fund is authorized to support these fellowships through an agreement with the Congressional Hunger Center (which administers the fellowships). *Note:* The trust fund has not been funded. Instead, the Congressional Hunger Center typically receives $2.5 million as part of the annual USDA appropriations.	Restates Sec. 4404 with provisions similar to those contained in prior law. Provisions differ primarily by authorizing annual appropriations of $3 million and by specifically naming the Congressional Hunger Center as the administering entity for Emerson and Leland fellowships. *[Sec. 4402]*	Same as the House bill, except that the Senate amendment (1) authorizes issuance of "grants" to the Hunger Center (as opposed to "contracts") and authorizes appropriations at "such sums as are necessary." *[Sec. 4404]*	Adopts Senate provision restating and revising Sec. 4404. *[Sec. 4401]*
Title V: Agricultural Credit			
USDA Farm Ownership Loans			
Requires borrowers for USDA direct farm ownership loans to have training, or farming and ranching experience, that "is sufficient to assure reasonable prospects of success in the proposed farming operations." *[7 U.S.C. 1922(a)(2)]* Same	No comparable provision.	Expands and clarifies the qualification criteria to allow USDA to take "into consideration all farming experience...without regard to any lapse between farming experiences." *[Sec. 5001]* Same provision for farm operating	Adopts Senate provision. *[Sec. 5001]*

Prior Law/Policy	House-Passed Bill (H.R. 2419)	Senate-Passed Substitute Amendment (H.R. 2419)	Enacted 2008 Farm Bill (P.L. 110-246)
provision for farm operating loans. *[7 U.S.C. 1941(a)(2)]*		loans *Sec. 5101(1)-(2)]*	
Allows direct farm ownership loans to be used for buying a farm or ranch, capital improvements, closing costs, soil/water conservation projects, and refinancing bridge loans. *[7 U.S.C. 1923(a)(1)]*	No comparable provision.	Adds another loan purpose to allow beginning farmers and ranchers to use a direct farm ownership loan to refinance a guaranteed farm ownership loan. *[Sec. 5002]*	Deletes Senate provision.
Allows direct and guaranteed loans for soil and water conservation projects. Gives priority to conservation structures or practices that comply with 16 U.S.C. 3812. Maximum loan size is $50,000. *[7 U.S.C. 1924]*	Replaces prior law provision with a new conservation loan guarantee program. Provides loan guarantees or interest rate subsidies for qualified conservation projects to farmers, ranchers or other entities primarily engaged in farming. Gives priority to beginning farmers and ranchers, socially disadvantaged farmers or ranchers, conversion to sustainable or organic production, or structures or practices that comply with 16 U.S.C. 3812. Maximum loan size is $1 billion, for up to 10 years. Allows guarantees up to $1 million on between 80% to 90% of the loan amount. Interest rate subsidies range from 3-5%, depending on loan. Authorizes appropriations of such sums as necessary (FY2008-12). *[Sec. 5001]*	Adds additional purposes for the conservation loan to include (1) conversion to organic production and (2) certain environmental quality practices. Gives priority to beginning farmers or ranchers, socially disadvantaged farmers or ranchers, and management practices that comply with 16 U.S.C. 3812. Eliminates the maximum loan size provision. *[Sec. 5003]*	Adopts House provision, with modification. Allows USDA to make loans or issue loan guarantees. Does not specify a maximum amount or duration, but says that guarantees should be 75% of the loan principal. Does not include interest subsidies. Managers statement says loan limits for other Farm Service Agency operating loans should apply. *[Sec. 5002]*
Sets a maximum per borrower of $200,000 for direct farm ownership loans. *[7 U.S.C. 1925]*	Raises to $300,000 the maximum loan per borrower. Requires USDA to establish plans to encourage borrowers to graduate to private or commercial credit. *[Sec. 5002]*	Similar to House provision, except does not have clause about graduation to other sources of credit. *[Sec. 5004]*	Adopts Senate provision. *[Sec. 5003]*
Authorizes down payment loans on farm real estate to beginning farmers and ranchers up to 40% of the purchase price or appraised value. Maximum property value is $250,000. Loan duration up to 15 years. Interest rate equals 4%. Requires borrower down payment of	Expands eligibility to include socially disadvantaged farmers and ranchers. Raises loan limit to 45% of the lowest of the purchase price, appraised value, or $500,000. Removes limit on property value. Extends loan duration up to 20 years. Interest rate equals the greater of	Similar to House provision, except interest rate equals the greater of 2% or the regular direct farm ownership interest rate minus 4%. *[Sec. 5005]*	Adopts House provision, with modification. The interest rate equals the greater of 1.5% or the regular direct farm ownership interest rate minus 4%. *[Sec. 5004]*

(Continued)

PRIOR LAW/POLICY	HOUSE-PASSED BILL (H.R. 2419)	SENATE-PASSED SUBSTITUTE AMENDMENT (H.R. 2419)	ENACTED 2008 FARM BILL (P.L. 110-246)
10%. *[7 U.S.C. 1935]*	1% or the regular direct farm ownership interest rate minus 4%. Reduces borrower down payment requirement to 5%. Requires USDA to establish annual performance goals. *[Sec. 5003]*		
Creates a pilot program to guarantee loans made by a private seller of a farm or ranch to a beginning farmer or rancher on a contract land sales basis. Available in 5 States, up to 5 loans per state in each of fiscal years 2003-07. *[7 U.S.C. 1936]*	Makes permanent and nationwide the guarantee program for privately financed contract land sales. Expands eligibility to socially disadvantaged farmers/ranchers. Requires a 5% down payment. Sets maximum purchase price at $500,000 and a 10-year maximum guarantee. Seller chooses a guarantee plan of either 3 amortized annual installments, or 90% of the outstanding principal. *[Sec. 5004]*	Similar to House provision, except does not include socially disadvantaged farmers and ranchers, and does not include seller's choice for guaranteed payment of 90% of outstanding principal. *[Sec. 5006]*	Adopts House provision, with modification. Provides a transition period until 2011. *[Sec. 5005]*
Authorizes USDA loans to any Indian Tribe or tribal corporation to acquire lands within the tribe's reservation or a community in Alaska. *[25 U.S.C. 488]*	Authorizes loans to purchasers of highly fractioned land. *[Sec. 5005]*	Similar to House provision. *[Sec. 5401]*	Adopts Senate provision. *[Sec. 5501]*
USDA Farm Operating Loans			
Limits direct farm operating loans to (1) beginning farmers or ranchers, or (2) farmers or ranchers who have received direct operating loans for 6 or fewer years. *[7 U.S.C. 1941(c)(1)]*	No comparable provision.	Clarifies that USDA may consider all farming experience for eligibility. Extends by one year (to 7 years) the eligibility for a direct operating loan. *[Sec. 5101(3)]*	Adopts Senate provision, but does not extend the duration of eligibility. *[Sec. 5101]*
Sets a maximum per borrower of $200,000 for direct farm operating loans. *[7 U.S.C. 1943]*	Raises to $300,000 the maximum per borrower. *[Sec. 5011]*	Same as House provision. *[Sec. 5102]*	Adopts House provision. *[Sec. 5102]*
Limits eligibility for guaranteed operating loans to 15 years (a.k.a., term limits). The 2002 farm bill and successive legislation suspended application of this limit. *[7 U.S.C. 1949(b)(1)]*	Extends the suspension of term limits for guaranteed operating loans through Jan. 1, 2008. *[Sec. 5012]*	Repeals the term limits provision on guaranteed operating loans. *[Sec. 5103]*	Adopts House provision, except extends the suspension of term limits through Dec. 31, 2010. *[Sec. 5103]*

PRIOR LAW/POLICY	HOUSE-PASSED BILL (H.R. 2419)	SENATE-PASSED SUBSTITUTE AMENDMENT (H.R. 2419)	ENACTED 2008 FARM BILL (P.L. 110-246)
USDA Emergency Loans			
Authorizes Farm Service Agency emergency disaster loans for farmers, ranchers, and aquaculture operations under certain conditions. *[7 U.S.C. 1961(a)]*	No comparable provision.	Adds eligibility for emergency loans to equine farmers and ranchers. *[Sec. 5404]*	Adopts Senate provision. Managers' statement notes horses for racing, showing, recreation, or pleasure are not eligible. *[Sec. 5201]*
Administrative Provisions			
No comparable provision.	No comparable provision.	Creates a pilot program of individual development accounts for beginning farmers and ranchers in at least 15 States. Producers make deposits into savings accounts and receive a matching deposit ranging between 3 to 1 per dollar deposited (maximum $9,000 matching per year). Withdrawals allowed to buy farmland; make mortgage payments; buy equipment, breeding stock, or trees; harvest timber; or pay for training. Withdrawals must be made within 2 years after the last match. Program delivered through private entities that apply for grants up to $300,000; requires non-federal matching of 25%. Duration is 5 years, plus 2 years for withdrawals. Authorizes appropriations of $10 million annually in FY2008-12. *[Sec. 5201]*	Adopts Senate provision, with modification. Reduces matching deposit to between 2 to 1 per dollar deposited (maximum $6,000 matching per year), reduces grants to $250,000 maximum, increases non-federal matching to 50%, limits administrative cost to 10%, reduces authorization for appropriation to $5 million annually. *[Sec. 5301]*
Requires USDA to include beginning farmers and ranchers in the process when selling or leasing USDA-acquired property. *[7 U.S.C. 1985(c)]*	Gives first priority to socially disadvantaged farmers and ranchers when USDA sells or leases property. Adds socially disadvantaged farmers to beginning farmers when discussing steps in the process. *[Sec. 5021]*	Similar to House provision, except does not give priority to socially disadvantaged farmers over beginning farmers. *[Sec. 5202(a)]*	Adopts Senate provision. *[Sec. 5302(a)]*
Reserves loan amounts for beginning farmers and ranchers: 70% of direct farm ownership loans, and of this amount, 60% for down payment loans. Reserves	Increase the reservation for beginning farmers and ranchers: 75% of direct farm ownership loans, and of this amount ⅔ for the down payment and participation	Same reservations as the House provision, except refers to "joint financing arrangements" instead of "participation loans." *[Sec. 5202(b)]*	Adopts Senate provision. *[Sec. 5302(b)]*

(Continued)

Prior Law/Policy	House-Passed Bill (H.R. 2419)	Senate-Passed Substitute Amendment (H.R. 2419)	Enacted 2008 Farm Bill (P.L. 110-246)
35% of direct operating loans in FY2003-07. Reserves 25% of guaranteed farm ownership loans, and 40% of guaranteed operating loans. *[7 U.S.C. 1994(b)(2)]*	loans. Reserves 50% of direct operating loans in FY2008-12. Reserves 40% of guaranteed farm ownership loans. *[Sec. 5022]*		
Sets loan authorization levels totaling $3.796 billion for each of FY2003-07, including $770 million for direct loans ($205 million for farm ownership loans, and $565 million for operating loans), and $3.026 billion for guaranteed loans ($1 billion farm ownership loans, and $2.026 billion for operating loans). *[7 U.S.C. 1994(b)(1)]*	No comparable provision.	Raises loan authorization levels to $4.226 billion for each of FY2008-12, including $1.2 billion for direct loans ($350 million for farm ownership loans and $850 million for operating loans). No increases for guaranteed loans. *[Sec. 5204]*	Adopts Senate provision. *[Sec. 5303]*
No comparable provision.	Requires USDA to promote the goal of moving borrowers to commercial or other sources of credit when writing regulations. *[Sec. 5023]*	Same as House provision. *[Sec. 5203]*	Adopts House provision. *[Sec. 5304]*
Allows a borrower-owner who is occupying a homestead that is in the possession of USDA because of foreclosure or bankruptcy to have the right of first refusal to reacquire the homestead property at any time during occupancy. *[7 U.S.C. 2000(c)(4)(B)]*	Extends the right of first refusal to reacquire homestead property during the occupancy period to members of the immediate family of a borrower-owner who is a socially disadvantaged farmer or rancher. *[Sec. 5024]*	No comparable provision.	Adopts House provision. *[Sec. 5305]*
Requires an interest rate reduction program for guaranteed loans. *[7 U.S.C. 1999(a)]*	No comparable provision.	Clarifies that the interest rate reduction program be available for new guaranteed operating loans or restructured guaranteed operating loans. *[Sec. 5205]*	Deletes Senate provision.
Allows re-amortization of the repayment of a shared appreciation agreement that is delinquent. *[7 U.S.C. 2001(e)(7)(D)]*	No comparable provision.	Clarifies that deferral is an available loan servicing tool and that deferral may not exceed 1 year. *[Sec. 5206]*	Deletes Senate provision.
No comparable provision.	Prohibits USDA from studying or entering into a contract for competitive sourcing in the rural development or farm loan programs. *[Sec. 5025]*	Similar to House provision. *[Sec. 5207]*	Adopts House provision. *[Sec. 5306]*

(Continued)

PRIOR LAW/POLICY	HOUSE-PASSED BILL (H.R. 2419)	SENATE-PASSED SUBSTITUTE AMENDMENT (H.R. 2419)	ENACTED 2008 FARM BILL (P.L. 110-246)
Defines "farmer" and "farming" to include fish farming. *[7 U.S.C. 1991(a)(1)-(2)]*	No comparable provision.	Adds commercial fishing to the definition of farmer and farming. *[Sec. 6020(c)]*	Deletes Senate provision.
Farm Credit System (FCS)			
Allows voting stock in Banks for Cooperatives to be held only by (1) cooperative associations that are eligible to borrow from the banks; and (2) other banks for cooperatives. *[12 U.S.C. 2124(c); 12 U.S.C. 2154a(c)(1)(D)]*	Allows other categories of persons and entities who are eligible to borrow from Banks for Cooperatives to hold voting stock, as determined by the board of directors. *[Sec. 5031]*	No comparable provision.	Adopts House provision. *[Sec. 5403]*
Sets the par value of capital stock in the Bank for Cooperatives. *[12 U.S.C. 2124(b)]*	No comparable provision.	Makes a technical correction to refer to par value. *[Sec. 5302]*	Adopts Senate provision. *[Sec. 5402]*
Requires that members of the Farm Credit Administration Board be appointed by the President and confirmed by the Senate. The President designates one as chairman. *[12 U.S.C. 2242(a)]*	No comparable provision.	Requires Senate confirmation of the President's choice of chairman. *[Sec. 5303]*	Deletes Senate provision.
Defines what types of loans are eligible for the agricultural mortgage secondary market (Farmer Mac). *[12 U.S.C. 2279aa(9)]* Establishes a risk-based capital test for the Federal Agricultural Mortgage Corporation (Farmer Mac) to determine a sufficient level of capitalization. *[12 U.S.C. 2279bb-1(a)]*	Allows rural utility (electric or telephone facility) loans to qualify for the agricultural mortgage secondary market (Farmer Mac). Does not apply certain requirements for guarantees and standards to rural utility loans. *[Sec. 5032]* Provides for separate consideration of rural utility (electric and telephone) loans when determining credit risk. *[Sec. 5034]*	Similar to House provision, except (1) it recasts loan standards in terms of agricultural mortgages, and (2) removes congressional review of standards (in 12 U.S.C. 2279aa-8). *[Sec.5306(a)-(c)]* Similar to House provision. *[Sec. 5306(d)]*	Adopts Senate provision. *[Sec. 5406(a)-(c)]* Adopts Senate provision. *[Sec. 5406(d)]*
Allows any Farm Credit System bank to charge associations in its district for the cost of premiums due to the Farm Credit System Insurance Corporation (FCSIC)	Replaces the formula for computing assessments with a requirement that the assessment be computed in an equitable manner. *[Sec. 5033(a)]*	Similar to House provision, except it specifies that the equitable manner be determined by the FCS Insurance Corporation. *[Sec. 5301(a)]*	Adopts Senate provision. *[Sec. 5401(a)]*

Prior Law/Policy	House-Passed Bill (H.R. 2419)	Senate-Passed Substitute Amendment (H.R. 2419)	Enacted 2008 Farm Bill (P.L. 110-246)
by using a specified formula based on accrual status and various government guarantees. *[12 U.S.C. 2020(b)]*			
Gives the FCSIC Board of Directors power to issue rules and regulations. *[12 U.S.C. 2277a-7(10)]*	Gives rule-making authority for "authority to pass along cost of insurance premiums" to FCSIC Board of Directors. *[Sec. 5033(h)]*	Same as House provision. *[Sec. 5301(b)]*	Adopts House provision. *[Sec. 5401(b)]*
Allows the FCSIC to collect premiums annually. When the Insurance Fund does not exceed the secure base amount, premiums equal different rates multiplied by outstanding principal of accrual (0.0015), non-accrual (0.0025), and various government-guaranteed (0.00015-0.0003) loans. Allows reduced premiums when the Fund exceeds the secure base amount. The base amount is 2% of outstanding insured obligations of all insured System banks reduced by 90% of federally-insured loans and 80% of state-insured loans. Excess premiums collected in the Fund over the secure base amount are allocated to insurance reserve accounts for each bank and the Financial Assistance Corporation using average daily balances. *[12 U.S.C. 2277a-4]*	Removes specification of "annual" when referring to FCSIC premiums. Changes formula for premiums by using different rates and classes of assets: insured obligations after deducting 90% of federally-insured accrual loans and investments and 80% of state-insured accrual loans and investments (0.002), non-accrual loans (0.001), and other than temporarily-impaired investments (0.001). Revises the secure base amount definition to include federal- and state-guaranteed investments. Calculate outstanding principal to include investments. Calculate excess funds using year-end balances and simplify formula to allocate among banks. Terminates reserve fund for Financial Assistance Corp. *[Sec. 5033(b)-(f)]*	Similar to House provision. *[Sec. 5304]*	Adopts Senate provision. *[Sec. 5404]*
Requires annual certification, reporting of loan balances, and payment of premiums by FCS banks to the FCSIC. *[12 U.S.C. 2277a-5]*	Gives discretion to FCSIC for timing of certification. Premiums are not to be collected more than quarterly. *[Sec. 5033(g)]*	Similar to House provision. *[Sec. 5305]*	Adopts Senate provision. *[Sec. 5405]*
No comparable provision.	No comparable provision.	Changes lending authorities of FCS associations (in AL, MS, and LA) by allowing Federal Land Bank Associations to make short-and intermediate-term loans, and Production Credit Associations to make long-term loans. Requires board of directors and	Adopts Senate provision. *[Sec. 5407]*

(Continued)

PRIOR LAW/POLICY	HOUSE-PASSED BILL (H.R. 2419)	SENATE-PASSED SUBSTITUTE AMENDMENT (H.R. 2419)	ENACTED 2008 FARM BILL (P.L. 110-246)
		stockholder approval. Effective Jan. 1, 2010. ***[Sec. 5307]***	
TITLE VI: RURAL DEVELOPMENT			
Defining Rural Eligibility			
Sec. 343(a) of the Consolidated Farm and Rural Development Act (Con Act), as amended, defines rural as any area other than a city or town with a population greater than 50,000 and the urbanized area contiguous and adjacent to such a city or town. ***[7 U.S.C. 1991(a)]***	Directs USDA to submit a report to the Agriculture Committees: (1) assessing the varying definitions of rural used by USDA; (2) describing the effect of varying definitions on USDA's programs; and (3) recommending changes to better target funds through rural development programs. ***[Sec. 6001]***	Creates a standard definition of rural area that *excludes* (1) cities of 50,000 or more, (2) any urbanized area contiguous and adjacent to a city of 50,000 or more, and (3) any collection of contiguous census blocks with a specific housing density, or adjacent to a city of 50,000 or urban area. ***[Sec. 6020]***	Adopts the Senate provision but deletes the housing density criterion from the definition of "rural." Permits USDA to include "areas rural in character" if they meet certain non-urban criteria (excluding Honolulu, HI, and San Juan, PR). Does not change eligibility for water and waste water funding. ***[Sec. 6018]***
No comparable provision.	Authorizes USDA to review socio-economic variables as factors in awarding rural development loans and grants, and issue regulations. ***[Sec. 6014]***	No comparable provision.	Deletes the House provision.
Rural Infrastructure: Water and Waste Disposal			
The 2002 farm bill amended the Con Act to authorize USDA to make water and wastewater grants for development projects for the storage, treatment, purification, or distribution of water or the collection, treatment, or disposal of waste in rural areas. Authorizes $30 million in annual appropriations for FY2002-07. ***[7 U.S.C. 1926(a)(2)(B)(vii)]***	Reauthorizes through 2012. ***[Sec. 6002]***	Reauthorizes through 2012. ***[Sec. 6001]***	Reauthorizes through 2012. ***[Sec. 6001]***
The Con Act, as amended, sets interest rate levels not to exceed current market yields for outstanding municipal obligations. Low-income rural residents receive interest rates below this maximum level. ***[7 U.S.C. 1927 (a)(3)(A)]***	No comparable provision	Senate provision establishes interest rates for water and waste water loans based on a market index of loans to ensure that interest rates for intermediate and poverty rate loans are tied to the current market rate. The poverty rate is set at 60% of the market rate and the intermediate rate is set at 80% of market rate. ***[Sec. 12602]***	Adopts the Senate provision. ***[Sec. 6011]***

PRIOR LAW/POLICY	HOUSE-PASSED BILL (H.R. 2419)	SENATE-PASSED SUBSTITUTE AMENDMENT (H.R. 2419)	ENACTED 2008 FARM BILL (P.L. 110-246)
The 2002 farm bill amended the Con Act to authorize appropriations for a water/wastewater circuit rider program, providing technical assistance based on a National Rural Water Association program. *[7 U.S.C. 1926(a)(22)(C)]* Authorizes appropriations of $15 million for FY2003 and each year thereafter.	Reauthorizes through 2012, authorizing $25 million for FY2008. *[Sec. 6004]*	Reauthorizes through 2012, authorizing $20 million for FY2008. *[Sec. 6004]*	Adopts the House provision. *[Sec. 6006]*
The 2002 farm bill amended the Con Act to authorize USDA to provide Emergency and Imminent Community Water Assistance Grants to rural areas and small communities comply with the Water Pollution Control Act or Safe Drinking Water Act. *[7 U.S.C. 1926a(i)(2)]* Authorizes an appropriation of $35 million annually for FY2003-07.	Reauthorizes through FY2012. *[Sec. 6006]*	Reauthorizes through FY2012. *[Sec. 6011]*	Reauthorizes through FY2012. *[Sec. 6008]*
No comparable provision.	No comparable provision.	Provides $135 million in mandatory spending for pending water/wastewater loans, grants and emergency community assistance grants, to be available until expended. *[Sec. 6033]*	Adopts the Senate provision. Provides $120 million in mandatory spending. *[Sec.6029]*
The Con Act, as amended, authorizes USDA to make grants for water systems for rural and native villages in Alaska. Authorizes appropriations of $30 million annually for FY2001-07. *[7 U.S.C. 1926d(d)(1)]*	Reauthorizes through FY2012. *[Sec. 6007]*	Reauthorizes through FY2013. Amends program to provide grants to the Denali Commission for solid waste management and for rural drinking water sites in Alaska (not more than 5% of total program funding). *[Sec. 6012]*	Adopts House provision with an amendment to provide $1.5 million annually FY2008-2012 to the Denali Commission for solid waste management. *[Sec. 6009]*
The 2002 farm bill amended the Con Act to authorize USDA to make grants to private nonprofits for loans to eligible low-income individuals to construct, refurbish, and service household water well systems in rural areas (giving priority to certain applicants). Authorizes appropriations of $10 million annually for FY2003-07. *[7 U.S.C. 1926e(d)]*	Reauthorizes through FY2012. Amends program to authorize USDA to make grants to private non-profits for loans to eligible low-income individuals. *[Sec. 6008]*	Reauthorizes through FY2012. *[Sec. 6013]*	Adopts House provision with changes, striking consideration of matching funds and increasing the assistance limit for each well from $8,000 to $11,000. *[Sec. 6010]*

(Continued)

Prior Law/Policy	House-Passed Bill (H.R. 2419)	Senate-Passed Substitute Amendment (H.R. 2419)	Enacted 2008 Farm Bill (P.L. 110-246)
No comparable provision.	No comparable provision	Amends the ConAct to authorize grants to develop wells in isolated rural areas. Provides $10 million annually FY2008-2012. *[Sec. 6013]*	Deletes Senate provision.
The 2002 farm bill amended the Con Act to establish a SEARCH grant program, providing technical assistance for water and waste disposal facilities. ***[7 U.S.C. 2009ee]***	No comparable provision.	Reauthorizes SEARCH grants and amends program. Provides up to 4% of funds available for water, waste disposal, and essential community facilities to financially distressed communities. Directs USDA to develop a simplified application for applicants. ***[Sec. 6010]***	Adopts Senate provision. ***[Sec. 6002]***
Rural Broadband and Telecommunications Development			
The 2002 farm bill amended the Con Act to authorize grants to acquire radio transmitters to increase rural coverage by all-hazards weather radio broadcasts of the National Oceanic and Atmospheric Administration. Authorizes appropriation of such sums as needed for FY2002-07. ***[7 U.S.C. 2008p]***	Reauthorizes grants through FY2012. ***[Sec. 6018]***	Reauthorizes grants through FY2012. ***[Sec. 6026]***	Reauthorizes grants through FY2012. ***[Sec. 6021]***
The 2002 farm bill amended the Rural Electrification Act (REA) of 1936 ***[7 U.S.C. 901 et seq.]*** by authorizing USDA to provide loans and loan guarantees for the costs of providing broadband service to rural areas, as part of the Enhancement of Access to Broadband Service in Rural Areas provisions. ***[7 U.S.C. 950bb]***	Reauthorizes through FY2012. Redefines eligibility and prioritizes loan applications to areas based on number of service providers. Amends definition of rural area. Limits loan terms to 35 years. Extends authority to provide loans to FY2012. Authorizes a National Center for Rural Telecommunications Assessment and authorizes $1 million in appropriations annually. ***[Sec. 6023]***	Reauthorizes through FY2012. Redefines eligibility and prioritizes loan applications. Prohibits loans to areas served by 3 or more providers. Authorizes $25 million annually (FY2008-12). Authorizes a National Center for Rural Telecommunications Assessment and authorizes $1 million in appropriations annually. ***[Sec. 6110]*** Establishes which areas are eligible for REA assistance. ***[Sec. 6105]***	Adopts Senate provision with changes. Permits assistance to areas with more than 3 providers under certain conditions. Gives highest priority to applicants serving the most rural residents. Prohibits eligibility to providers res serving more than 20% of the market. Permits USDA to require cost-share funding. ***[Sec. 6110]*** Adopts House and Senate measure to authorizes National Center for Rural Telecommunications. ***[Sec. 6111]***. Adopts Senate provision defining REA eligibility. Eligible rural areas exclude town of 20,000 or more. ***[Sec. 6104]***
Sec. 601(a) *et seq*. of the REA, as amended, authorizes USDA to provide	Authorizes the Community Connect Grant Program to provide broadband	Authorizes the Connect the Nation Act and creates a competitive matching grant	No change to current law.

(Continued)

PRIOR LAW/POLICY	HOUSE-PASSED BILL (H.R. 2419)	SENATE-PASSED SUBSTITUTE AMENDMENT (H.R. 2419)	ENACTED 2008 FARM BILL (P.L. 110-246)
loans and loan guarantees to electric utilities to serve customers in rural areas. *[7 U.S.C. 950bb(b)]*	service for education, public safety, and health care in rural areas. Authorizes appropriations of $25 million annually (FY2008-12). *[Sec. 6024]*	program to encourage state initiatives for public-private partnerships *[Sec. 6201]* and authorizes grants to encourage state initiatives *[Sec. 6202]* to provide broad-band service to rural areas. Authorizes appropriations of $40 million annually (FY2008-12).	
Sec. 2333 of the 1990 farm bill (Food, Agriculture, Conservation, and Trade Act, P.L. 101-624) provides grants to non-commercial television that serve rural areas. *[7 U.S.C. 950aaa-2]* Also, Sec. 2335A authorizes Telemedicine and Distance Learning Services in Rural Areas by providing loans/grants to schools and medical facilities for telecom technologies. *[7 U.S.C. 950aaa5]*	Reauthorizes the grant program to assist rural public television stations in making the transition from analog to digital broadcast equipment. *[Sec. 6028]* Reauthorizes Telemedicine and Distance Learning Services in Rural Areas through FY2012. *[Sec. 6029]*	Reauthorizes appropriations through FY2012. Amends provision: (1) adds library connectivity and public television station digital conversion into the notification; (2) species requirements on how financial assistance is to be used and criteria for prioritizing; (3) requires USDA to notice the amount of financial assistance available to applicants, among other provisions. Renames program the Telemedicine, Library Connectivity, Public Television, and Distance Learning in Rural Areas, and reauthorizes through FY2012. *[Sec. 6302]*	Adopts Senate provision with changes that only libraries are added as eligible entities. Makes public television stations eligible for funding for high-speed telecommunications for educational programming in rural areas. *[Sec. 6201]*
The Con Act, as amended, authorizes appropriations for grants for broadcasting systems, funded at $5 million annually (FY2002-07). *[7 U.S.C. 1932(f)]*	No comparable provision.	Reauthorizes through FY2012. *[Sec. 6016]*	Reauthorizes through FY2012 *[Sec. 6014]*
No comparable provision.	Directs USDA to prepare a report that develops a comprehensive national broadband strategy. *[Sec. 6031]*	Directs the Federal Communications Commission, in coordination with USDA, to submit a report to Congress describing a comprehensive rural broadband strategy. *[Sec. 6111]* Instructs the U.S. Comptroller General of to conduct a study of the Rural Utilities Service administration and of Federal assistance for broadband programs, with recommendations. *[Sec. 6113]*	Adopts Senate provision. Requires an update of the report in the third year following enactment. *[Sec. 6112]* Adopts Senate provision striking an obsolete reference to dial-up Internet and place the provision in a separate section. *[Sec. 6005]*. Deletes Senate provision for a GAO study.
The 2002 farm bill amended the REA to authorize USDA to expand 911 access	Reauthorizes through FY2012. *[Sec. 6022]*	Reauthorizes through FY2012. Expands eligibility to emergency communication	Adopts the Senate provision changes to make emergency communication

(Continued)

Prior Law/Policy	House-Passed Bill (H.R. 2419)	Senate-Passed Substitute Amendment (H.R. 2419)	Enacted 2008 Farm Bill (P.L. 110-246)
and make telephone loans for rural emergency services. *[7 U.S.C. 940e]*		providers. Authorizes USDA to use funds made available for telephone or broadband loans; requires USDA promulgate regulations *[Sec. 6107]*	equipment providers eligible for loans. *[Sec. 6107]*
Agricultural-Based Rural Economic Development			
The 2002 farm bill amended the Con Act to authorize appropriations for direct and guaranteed loans for rural business development. *[7 U.S.C. 1926(a)(11)(D)]*	Establishes new criteria for loans and loan guarantees, directing USDA to favor projects that support local/regionally produced agricultural products. *[Sec. 6010]*	Similar to the House bill, but also: defines "under-served community;" establishes priorities for projects that support community development and marketing, distributing, storing, aggregating, or processing a locally-produced product; sets a per-facility limit of up to $250,000 in loan/loan guarantees to modify/update facilities; and requires USDA to submit an annual report to Congress. *[Sec. 6017]*	Adopts Senate provision with changes that extend the distance a product can travel and be eligible. Defines "underserved community" and gives priority to entities providing products to these communities. *[Sec. 6015]*
No comparable provision.	Authorizes appropriations for Appropriate Technology Transfer for Rural Areas at $5 million annually (FY2008-12). *[Sec. 6011]*	Similar to the House bill. *[Sec. 6018]*	Adopts the Senate provision with minor changes to elaborate on the purpose of the program. *[Sec. 6016]*
No comparable provision. Sec. 231 of the Agricultural Risk Protection Act of 2000 *[7 U.S.C. 1621]*, as amended by the 2002 farm bill, authorizes USDA to make Value-Added Agricultural Product Development Grants to assist agricultural producers to establish businesses to produce value-added agricultural products, and provide for technical assistance and planning.	No comparable provision. Authorizes $6 million subject to appropriations annually (FY2008-12). *[Sec. 6027]*	Establishes Artisanal Cheese Centers to provide educational technical assistance to eligible cheese manufacturing and marketing businesses. *[Sec. 6023]* Expands the definition of value-added products. Reduces the maximum grant amount to $300,000. *[Sec. 6401]*	Deletes the provision. Adopts the Senate provision with modification. Reserves 10% of funds for projects benefitting beginning farmers and ranchers and socially disadvantaged farmers and ranchers, and 10% of funds for projects to develop mid-tier value chains. Provides $15 million in mandatory funding. *[Sec. 6202]*
No comparable provision.	No comparable provision.	Amends the Con Act to provide reimbursement payments to	Adopts Senate provision with technical changes. *[Sec. 1621]*

(Continued)

Prior Law/Policy	House-Passed Bill (H.R. 2419)	Senate-Passed Substitute Amendment (H.R. 2419)	Enacted 2008 Farm Bill (P.L. 110-246)
		"geographically disadvantaged farmers" (e.g., AK, HI) for costs associated with transporting or producing an agricultural commodity. *[Sec. 6021]*	
The 2002 farm bill amended the Con Act authorizing a program to restore historic barns. *[7 U.S.C. 2008o(c)]*	Reauthorizes through FY2012 and gives priority to projects that identify, document, and conduct research on historic barns and develop and evaluate appropriate techniques or best practices for protecting historic barns. *[Sec. 6017]*	Similar to the House bill; also establishes a grant may be made for projects that rehabilitate or repair historic barns; preserve historic barns; and identify, document, survey, and conduct research on historic barns/structures. *[Sec. 6025]*	Adopts House provision with technical changes. *[Sec. 6020]*
The 2002 farm bill authorized an Agricultural Innovation Center Demonstration program. *[7 U.S.C. 1621]*	Reauthorizes the program and provides $6 million annually (FY2008-12). *[Sec. 6025]*	No comparable provision	Adopts the House provision. *[Sec. 6203]*
Regional Economic Development and Planning			
The Con Act, as amended, authorizes a Delta Regional Authority, providing funds for 240 counties in 8 states in Mississippi Delta. *[7 U.S.C. 2009aa-1]*	Reauthorizes appropriations through FY2012. *[Sec. 6019]*	Reauthorizes appropriations through FY2012; amends program to allow for grants for health care facility development. *[Sec. 6029]*	Adopts House provision with changes. Adds counties to be eligible. Establishes separate Health Care Services section and defines eligibility to mean Mississippi River Delta region. *[Secs. 6024-6025]*
The 2002 farm bill amended the Con Act authorizing the Northern Great Plains Regional Authority to make grants and loans and implement a regional development plan. *[7 U.S.C. 2009bb-1]*	Amends program to eliminate prioritization of activities to be funded. Modifies federal share of administrative expenses. Eliminates Isolated Areas of Distress designation. *[Sec. 6020]*	Eliminates requirement for a federal commission member, unless appointed. Broadens list of eligible organizations. Provides assistance to states in providing regional plans for renewable energy and transportation. *[Sec. 6030]*	Adopts Senate provision with changes. Requires the Commission to coordinate with tribal leaders if no federal co-chair is names. Defines organizations that may serve in the capacity of federal co-chair. *[Sec. 6026]*
No comparable provision.	No comparable provision.	Amends the Con Act to authorize a new Northern Border Economic Development Commission (VT, NY, NH, and ME); authorizes appropriations of $40 million annually (FY2008-12). *[Sec. 6034]*	Adopts Senate provision. Authorizes 3 additional regional commissions and provides them $30 million each FY2002-2012. *[See Sec. 14217]*
The Con Act, as amended, authorizes grants to multi-jurisdictional regional planning and development organizations; $30 million annually through FY2007.	No comparable provision.	Reauthorizes through FY2012. *[Sec. 6005]*	Deletes Senate provision.

(Continued)

PRIOR LAW/POLICY	HOUSE-PASSED BILL (H.R. 2419)	SENATE-PASSED SUBSTITUTE AMENDMENT (H.R. 2419)	ENACTED 2008 FARM BILL (P.L. 110-246)
[7 U.S.C. 1926(a)(23)]			
The Con Act, as amended, authorizes loans and grants for business and community development. *[7 U.S.C. 1932]*	No comparable provision.	Directs USDA to continue Rural Economic Area Partnership Zones (NY, ND, and VT) with areas of high unemployment/poverty. *[Sec. 6019]*	Adopts Senate provision. *[Sec. 6017]*
No comparable provision.	Directs USDA, in coordination with the Department of Transportation, to prepare a report on railroad issues regarding the movement of agricultural products, renewable fuels, and economic development. *[Sec. 6032]*	No comparable provision.	Adopts House provision. Includes other transportation systems in addition to rail. *[Sec. 6206]*
Rural Entrepreneurship and Business Investment Programs			
The 2002 farm bill amended the Con Act to authorize USDA to make Rural Business Opportunity Grants for business development or labor training in rural areas. Authorizes appropriations of $15 million annually through FY2007. *[7 U.S.C. 1926(a)(11)]*	Reauthorizes through FY2012. *[Sec. 6003]*	Similar to the House bill. *[Sec. 6002]*	Adopts Senate provision. *[Sec. 6003]*
The Con Act, as amended, authorizes appropriations for grants to cooperative development centers. *[7 U.S.C. 1932(e)(5)]*	Authorizes USDA to give preference to grant applications that establish centers for rural cooperative development that demonstrate specified requirements. Authorizes $50 million annually (FY2008-12). *[Sec. 6009]*	Similar to House bill. Allows USDA to award multi-year grants to programs as deemed by the Secretary; establishes a cooperative research program; and creates a reserve for socially disadvantaged communities. *[Sec. 6015]*	Adopts Senate provision with minor modifications. *[Sec. 6013]*
No comparable provision.	Authorizes a new Rural Entrepreneur and Microenterprise Assistance Program. Authorizes $20 million annually (FY2008-12). *[Sec. 6013]*	Authorizes a new Rural Microenterprise Assistance Program, with mandatory spending of $40 million for FY2008, available until expended. *[Sec. 6022]*	Adopts House provision with changes. Provides $15 million in mandatory funding. *[Sec. 6022]*
The 2002 farm bill amended the Con Act, authorizing the Rural Strategic Investment Program, providing an equity generating program for rural business development modeled on the Small	Limits discretionary funding of not more than $25 million annually FY2008-2012. Adds planning grant eligibility for "rural heritage sites." *[Sec. 6021]*	Authorizes a new Rural Collaborative Investment Program, with mandatory spending of $135 million for grants and administrative activities. *[Sec. 6032]*	Adopts Senate provision with changes to include adding rural heritage as a goal of the program. Authorizes $135 million in discretionary funds for the period FY2009-2012. *[Sec. 6028]*

(Continued)

Prior Law/Policy	House-Passed Bill (H.R. 2419)	Senate-Passed Substitute Amendment (H.R. 2419)	Enacted 2008 Farm Bill (P.L. 110-246)
Business Investment Companies of the Small Business Administration. *[7 U.S.C. 2099dd et seq.]*			
The 2002 farm bill amended the Con Act, authorizing the Rural Business Investment Program to make loans/grants through regional investment boards. *[7 U.S.C. 2009cc-5]*	No comparable provision.	Reauthorizes through FY2012, with changes: debentures may be pre-paid at any time; distributions may be made to cover tax liability; USDA fees are limited to an application fee of $500; and USDA will not be required to operate program with other federal agencies. *[Sec. 6031]*	Adopts the Senate provision with modifications. Removes provision allowing distributions to cover tax liability. Limits on funding from certain financial institutions is raised to 25%. *[Sec. 6027]*
Community Development Programs			
The 2002 farm bill authorizes grants for units of general local government, Indian tribes, to pay the cost of training firefighters and emergency medical personnel. *[7 U.S.C. 1621]*	Reauthorizes the Rural Firefighters and Emergency Personnel Grant Program, appropriations up to $30 million annually (FY2008-12). Expands the types of eligible entities. *[Sec. 6026]*	No comparable provision.	Adopts House provision with minor modifications. *[Sec. 6204]*
The 2002 farm bill amended the Con Act to authorize the National Rural Development Partnership, a state-federal partnership of community rural development entities. *[7 U.S.C. 2008m]*	Reauthorizes through FY2012. *[Sec. 6016]*	Reauthorizes through FY2012. *[Sec. 6024]*	Reauthorizes through FY2012. *[Sec. 6019]*
The 2002 farm bill amended the Con Act to authorize loans and grants for "essential community facilities" (incl. child day care). *[7 U.S.C. 1926(a)(19)]*	No comparable provision.	Authorizes $40 million in mandatory spending for loans, grants, and loan guarantees to construct child day care facility grants. *[Sec. 6003*]	Adopts Senate provision, but specifies that the program not receive mandatory funding. *[Sec. 6004]*
The Con Act, as amended, authorizes the Community Facility Grants Program, limited to $10 million per fiscal year for grants to local governments, nonprofits, and Indian tribes to provide the federal share of the cost of developing specific essential community facilities authorizes funds for essential community facilities. *[7 U.S.C. 1926(a) 19]*	No comparable provision	Reserves 0.5% of the funds for community facilities to eligible entities located in freely associated states or outlying areas as defined in the Elementary and Secondary Education Act of 1965. *[Sec. 6008]*	Deletes Senate provision.

PRIOR LAW/POLICY	HOUSE-PASSED BILL (H.R. 2419)	SENATE-PASSED SUBSTITUTE AMENDMENT (H.R. 2419)	ENACTED 2008 FARM BILL (P.L. 110-246)
The Con Act, as amended, authorizes funds for essential community facilities; the maximum amount of a community facility grant cannot exceed 75% of the project costs. *[7 U.S.C. 1926(a) 19B]*	No comparable provision	Amends to give priority for community facility projects that are carried out with a non-Federal share of funds that is substantially greater than the minimum requirement, as determined by USDA regulation. *[Sec. 6009]*	Deletes Senate provision
The 2002 farm bill amended the Con Act to authorize USDA to provide cost-share grants to tribal colleges and universities for developing essential community facilities in rural areas and universities, as defined in the Higher Education Act of 1965 for developing essential community facilities in rural areas. Authorizes $10 million in annual appropriation for FY2003-07. *[7 U.S.C.1926(a)(25)]*	Reauthorizes through FY2012. Amends program to direct USDA to establish a maximum percentage of the cost of a facility covered by a grant. Caps non-federal support to no more than 5% of the facility's total cost. *[Sec. 6005]*	Reauthorizes through FY2012. Increases the maximum federal grant tribal colleges and universities receive for the cost of developing essential community facilities in rural areas to 95%. *[Sec. 6007]*	Adopts House provision. *[Sec. 6007]*
No comparable provision.	No comparable provision	Makes technical changes to address funding for cooperative organizations by allowing for business guarantees of loans. *[Sec. 6014]*	Adopts Senate provision with technical changes. *[Sec. 6012]*
No comparable provision.	No comparable provision.	Provides grants to expand rural employment opportunities for individuals with disabilities. Authorizes $2 million annually (FY2008-12). *[Sec. 6028]*	Adopts Senate provision with minor changes. *[Sec. 6023]*
The Con Act authorizes grants to train farm workers in new technologies and in specialized skills for higher value crops; authorizes appropriations of $10 million annually through FY2007 *[7 U.S.C. 1981q(c)]*	No comparable provision.	Reauthorizes through FY2012. *[Sec. 6027]*	Deletes Senate provision.
No comparable provision.	Amends the Con Act to authorize grants to improve the technical infrastructure of rural health care facilities at $30 million annually (FY2008-12). Identifies types of eligible rural health facilities. *[Sec. 6012]*	Authorizes loans and loan guarantees to improve the technical infrastructure of rural health care facilities. Provides $50 million in mandatory spending (with at least $25 million for hospitals with less than 50 acute care beds). *[Sec. 6006]*	Deletes both House and Senate provisions.

(Continued)

Prior Law/Policy	House-Passed Bill (H.R. 2419)	Senate-Passed Substitute Amendment (H.R. 2419)	Enacted 2008 Farm Bill (P.L. 110-246)
No comparable provision.	No comparable provision.	Authorizes the Secretary of Housing and Urban Development to provide financial assistance for community-based housing development entities and affordable housing projects, and other requirements. Authorizes appropriations of $10 million (FY2008); $15 million annually (FY2009-10). *[Secs. 6501-6505]*	Adopts Senate provision with changes to permit GAO to use private audits for review of the Housing Assistance Council. *[Secs. 6301-6305]*
Other Rural Development Provisions			
The REA, as amended, authorizes USDA to make loans for rural electrification and telephone services and to assist borrower in implementing improvements to electrical and telephone service. *[7 U.S.C. 901 et seq.]*	No comparable provision.	Inserts "efficiency and" before "conservation;" makes technical changes to loan and grants for electric generation; establishes fees for baseload generation loan guarantees; defers loan payments for improved energy efficiency; defines "rural" and "farm" for borrowing eligibility; and specifies procedures for borrowers. *[Secs. 6101-6104; 6109]*	Adopts Senate provision authorizing energy efficiency program. *[Sec. 6101]*. Deletes provision for loans and grants for electric generation. *[Sec. 6102]*. Deletes Senate provision on fees for loan guarantees, but requires a study of electric generating needs in rural areas. *[Sec. 6113*]. Adopts Senate provision to allow energy audits. *[Sec. 6104]*. Makes technical changes for certain financing.
The REA, as amended, authorizes USDA to issue bonds for rural electrical generation or telephone purposes. *[7 U.S.C. 940c-1(f)]* The REA, as amended, authorizes USDA to make loans for electrical generation in rural areas. *[7 U.S.C. 940f]*	Reauthorizes through FY2012. *[Sec 6030]* No comparable provision.	Similar to the House bill, but limits guarantees to no more than $1 billion and establishes technical provisions for bond guarantees. *[Sec. 6106]* Defines "qualified energy source" and permits loans for electrical generation from renewable sources sold to non-rural residents at sets loan rates. *[Sec. 6108]*	Adopts Senate provision. *[Sec. 6106]* Adopts Senate provision with modifications. Defines "renewable energy source." *[Sec. 6108]*
The REA, as amended, authorizes USDA to make exceptions for electrification borrowers to relieve them of regulatory requirements. *[7 U.S.C. 936e]*	No comparable provision.	Defines "substantially underserved trust areas" and to authorize USDA to make loan rates as low as 2% to qualified utilities serving these areas. *[Sec. 6112]*	Adopts Senate provision with changes. *[Sec. 6105]*
Title III of the REA establishes agency funding procedures for direct loans and loan guarantee. *[7 U.S.C. 940-c 1]*	No comparable provision	Requires the Rural Utility Service (RUS) follow new procedures in dealing with borrowers. Allows USDA to adjust population limitations related to digital	Adopts Senate provision regarding bonding requirements and strikes the other provisions. *[Sec. 6109]*

(Continued)

PRIOR LAW/POLICY	HOUSE-PASSED BILL (H.R. 2419)	SENATE-PASSED SUBSTITUTE AMENDMENT (H.R. 2419)	ENACTED 2008 FARM BILL (P.L. 110-246)
		mobile wireless service; requires USDA to review bonding requirements for all programs administered by RUS. *[Sec 6109]*	
The 1990 farm bill, as amended, authorizes a rural electronic commerce extension program to provide assistance to rural businesses. *[7 U.S.C. 59239e]*	No comparable provision.	Reauthorizes through FY2012. *[Sec. 6301]*	Deletes Senate provision.
The Housing Act of 1949, as amended, authorizes a loan and grant program to provide housing construction and assistance to farm labor. *[42 U.S.C. 1484(f)(3)]*	No comparable provision.	Amends program to include aquacultural workers. *[Sec. 6420]*	Adopts Senate provision. *[Sec. 6205]*
TITLE VII: AGRICULTURAL RESEARCH			
Structure and Funding of Research, Education, and Extension			
Existing policy functionally categorized Cooperative State Research, Education, and Extension Service (CSREES) programs for state-level research, education, or extension activities as "formula funded" or "competitive."	Formally categorizes each existing CSREES program as a "capacity program" or "competitive program," and designates the current level of funding for each category as "critical base funding." *[Sec. 7101]*	Formally categorizes each existing CSREES program as an "infrastructure program" (i.e., capacity program) or "competitive program," as in the House provision. *[Sec. 7401]*	Defines the terms capacity, infrastructure, and competitive programs *[Sec. 7511]*; critical base funding; Hispanic-serving institutions; non-land grant colleges of agriculture *[Sec. 7101]*; and 1862, 1890, and 1994 institutions for the purposes of restructuring the Research, Extension, and Economics mission area as outlined in this act. *[Sec. 7501]*
The 1994 USDA reorganization act merged the Extension Service with the Cooperative State Research Service to establish CSREES. The Agricultural Research Service (ARS) was established in its current form in 1953. Both agencies are under the jurisdiction of the Undersecretary for Research, Extension, and Economics. *[7 U.S.C. 6971]*	Establishes a National Agricultural Research Program Office (NARPO) under the Under Secretary, with six sub-offices organized by research focus. The provision integrates the administrative functions of ARS's and CSREES's respective National Program Leaders, but retains the separate agencies. *[Sec. 7104]*	Directs the Undersecretary to coordinate the programs under the authority of the ARS and CSREES national program leaders, as well as the Director of the National Institute of Food and Agriculture. *[Sec. 7402]*	Reflects the House provision but changes the name to the Research, Extension, and Education Office (REEO). Establishes a National Institute of Food and Agriculture as of October 1, 2009. The new Institute comprises all the programs of the former CSREES. *[Sec. 7511]*

(Continued)

Prior Law/Policy	House-Passed Bill (H.R. 2419)	Senate-Passed Substitute Amendment (H.R. 2419)	Enacted 2008 Farm Bill (P.L. 110-246)
The National Research Initiative (NRI) grant program is an expansion of a 1990 farm bill program initially authorized in 1965, funded by annual appropriations 7 U.S.C. 450i). The Initiative for Future Agriculture and Food Systems (IFAFS) is a mandatory-funded grant program, which was first authorized in the Agricultural Research, Extension, and Education Reform Act of 1998 (AREERA, P.L. 105-185), reauthorized in the 2002 farm bill. *[7 U.S.C. 7621]*	Merges the existing NRI and IFAFS grant programs and groups them with all other competitive grant programs currently administered by CSREES to become the National Institute of Food and Agriculture (NIFA). *[Sec. 7105]*	CSREES is terminated as an agency; all of its competitive and infrastructure programs are to be administered under the National Institute of Food and Agriculture. *[Sec. 7401]*	Amends the statute authorizing the NRI to create a new Agriculture and Food Research Initiative to make competitive grants for fundamental and applied research, and for purchasing research equipment. Repeals 7 U.S.C. 7621 (IFAFS) Authorizes $700 million annually from FY2008 through FY2012. *[Sec. 7406]*
Matching funds are not required for NRI and IFAFS grants; they may be required for certain applied research grants. *[IFAFS 7 U.S.C. 7621; NRI 7 U.S.C. 450(b)]*	Creates two categories of research grants; fundamental research (to be called NRI grants); and applied research (to be called IFAFS grants). Matching funds are required for IFAFS grants, not for NRI grants. Authorizes $500 million in annual appropriations, and transfers mandatory IFAFS funds to support the merged NRI/IFAFS program. *[Sec. 7106]*	No comparable provision.	Requires 50% matching funds for equipment grants and 100% matching funds for applied research grants on topics that are commodity-specific and not of national scope. *[Sec. 7406]*
Most research and extension programs appear as individual line items in the President's annual budget request.	Requires the President to submit an annual budget making a single line item request for capacity programs and a single line item request for competitive programs. *[Sec. 7102]*	Directs the Under Secretary for Research, Education, and Economics to submit a "roadmap" that, among other things, describes recommended funding for competitive programs and infrastructure programs as unified categories, with some flexibility in implementation. *[Sec. 7402]*	Requires the President to submit an annual budget making a single line item request for all programs under the Research, Extension, and Economics mission area. Retains Senate language regarding a "roadmap." Recommends that budget emphasis within the request for capacity/infrastructure base funding be on certain institutions, and within competitive base funding, on emerging problems. *[Sec. 7504 and 7506]*
Sec. 1403 of the National Agricultural Research, Extension, and Teaching Policy Act of 1977 (NARETPA, Title XIV of P.L. 97-98) sets out the purposes of agricultural research and extension. *[7 U.S.C. 3101]*	Adds integrating and organizing all USDA research, extension, and education programs; minimizing duplication; and maximizing cooperation to the purposes of the Department's research mission. *[Sec. 7103]*	No comparable provision.	Deletes House provision.

(Continued)

Prior Law/Policy	House-Passed Bill (H.R. 2419)	Senate-Passed Substitute Amendment (H.R. 2419)	Enacted 2008 Farm Bill (P.L. 110-246)
Prior law limited eligibility for formula funded and other non-competitive research, extension, and education programs to 1862 and 1890 institutions, 1994 institutions (tribally controlled colleges), and Hispanic-serving institutions. *[7 U.S.C. 361a, 343]*	Expands eligibility for these programs to public non-land grant institutions offering 4-year degrees in agriculture (American Association of State Colleges of Agriculture and Renewable Resources, or ASCARR institutions). *[Sec. 7102]*	Requires the "roadmap" to include consideration of the needs of ASCARR institutions in addition to those of the 1862, 1890, 1994, and Hispanic-serving institutions. *[Sec. 7402]*	Adopts Senate provision. *[Sec. 7504]*
Prior law authorized capacity-building grant programs for 1890 and 1994 institutions, and Hispanic-serving institutions. *[7 U.S.C. 3152]*	Authorizes the appropriation of such sums as necessary for a new capacity-building grant program for ASCARR institutions. *[Sec. 7107]*	No comparable provision.	Authorizes capacity-building grants to ASCARR institutions *[Sec. 7138]*; and changes the term from ASCARR institutions to non-land grant college of agriculture (NLGCA) institutions. *[Sec. 7101]*
Sec. 1417 of NARETPA makes land grant and non-land grant, high minority enrollment, and secondary and post-secondary institutions eligible for higher education grants. *[7 U.S.C. 3152]*	Expands eligibility for grants/fellowships for food and agricultural sciences education to include Agriculture in the K-12 Classroom programs, nonprofit organizations, and other institutions of higher education. *[Sec. 7206]*	Similar to the House bill. *[Sec. 7007]*	Adopts House provision. *[Sec. 7109]*
The Hatch Act and Smith-Lever Acts set the minimum amount of federal formula funds spent on multi-state research and extension projects at either 25% or twice the percentage of state matching funds, whichever is less. *[7 U.S.C. 361a, 343]*	Sets 25% as the minimum amount of federal formula funds that must be expended on multistate research and extension projects. *[Sec. 7603]*	No comparable provision.	Deletes House provision.
Annual work plans are required from states to receive federal funds under NARETPA, the Hatch Act of 1887 (for research), and the Smith-Lever Act (for extension). *[7 U.S.C. 3101, 361a, 343]*	Requires USDA to prepare a report for Congress on how to streamline the submission, reporting requirements, and implementation of plan-of-work requirements. *[Sec. 7602]*	Similar to the House bill. *[Sec. 7503]*	Requires a review of annual plans of work, but does not require a report on how to streamline the plan-of-work process. *[Sec. 7505]*
The 1998 reseach act (AREERA) requires CSREES to conduct merit reviews of non-competitive research and extension grants, and peer reviews of grants. *[7 U.S.C. 7613]*	Provides for the continuation, under NIFA, of the mandatory merit and peer review processes for non-competitive and competitive grants. *[Sec. 7601]*	No comparable provision.	Deletes House provision.

(Continued)

Prior Law/Policy	House-Passed Bill (H.R. 2419)	Senate-Passed Substitute Amendment (H.R. 2419)	Enacted 2008 Farm Bill (P.L. 110-246)
Sec. 1462(a) of NARETPA limits the amount of indirect costs that USDA reimburses to institutions that are awarded research, extension, and teaching grants to 19%. ***[7 U.S.C. 3310]***	Amends Sec. 1462(a) to apply the 19% limitation on indirect cost reimbursement to any grant the institution receives, not just competitively awarded grants. ***[Sec. 7225]***	Raises the limitation on reimbursement of indirect costs from 19% to 30% for competitively awarded research, extension, and teaching grants. ***[Sec. 7027]***	Raises the limitation on reimbursement of indirect costs from 19% to 22%. ***[Sec. 7132]***
Sec. 1408(g)(1) of NARETPA authorizes the National Agricultural Research, Extension, Education, and Economics Advisory Board. ***[7 U.S.C. 3123]***	Increases authorized appropriations for the Advisory Board from $350,000 to $500,000 annually. ***[Sec. 7201]***	Similar to the House bill. ***[Sec. 7002]***	Increases the maximum annual appropriations for the Advisory Board to $500,000. Reduces the number of Board members from 31 to 24. Adds representation from crop, livestock, aquaculture producer organizations. ***[Sec. 7102]***
Subtitle K of NARETPA authorizes miscellaneous programs. ***[7 U.S.C. 3311]***	No comparable provision.	Authorizes $19 million annually through FY2012 for grants to the Consortium for Agricultural and Rural Transportation Research and Education. ***[Sec. 7051]***	Adopts Senate provision. ***[Sec 7529]***
Sec. 103(a) of AREERA authorizes peer and merit review of research and extension grant applications. ***[7 U.S.C.7613(a)]***	No provision.	No provision.	Amends AREERA to prohibit consideration of the availability of matching funds in reviewing grant proposals. ***[Sec. 7301]***
Provisions Affecting Certain Research Institutions			
Sec. 1417 of NARETPA authorizes grants/fellowships to land grant colleges and universities for food and agricultural sciences education. ***[7 U.S.C. 3152]***	Specifies that the University of the District of Columbia (UDC) be eligible to receive education grants and fellowships. ***[Sec. 7204]***	Similar to the House bill. ***[Sec. 7004]***	Makes UDC eligible to compete for food and agricultural sciences education grants. ***[Sec. 7106]***
Sec. 1447 of NARETPA authorizes grants to upgrade agriculture and food science facilities at the 1890 institutions. ***[7 U.S.C. 3222b]***	Extends eligibility to UDC for grants to upgrade facilities, and authorizes appropriations of $750,000 annually through FY2012. ***[Sec. 7212]***	Similar to the House bill. ***[Sec. 7020]***	Authorizes $750,000 in annual appropriations to upgrade agriculture and food science facilities at UDC. ***[Sec. 7124]***
The District of Columbia Public Post-secondary Education Reorganization Act (P.L. 93-471) designates UDC as a land grant institution, with certain exceptions. ***[88 Stat. 1423]***	Makes UDC eligible to receive funds for the Expanded Food and Nutrition Education Program (EFNEP). ***[Sec. 7512]***	Similar to the House bill. ***[Sec. 7313]***	Makes UDC eligible for EFNEP grants. ***[Sec. 7417]***

(Continued)

Prior Law/Policy	House-Passed Bill (H.R. 2419)	Senate-Passed Substitute Amendment (H.R. 2419)	Enacted 2008 Farm Bill (P.L. 110-246)
The Hatch Act of 1887 authorizes federal funding for agricultural research in the states/U.S. territories. *[7 U.S.C. 361a]*	Exempts UDC from the matching funds requirement for Hatch Act funds. *[Sec. 7513]*	Requires 50% matching funds from the District of Columbia in order for UDC to be eligible for Hatch Act funds. *[Sec. 7304]*	Adopts Senate provision. *[Sec. 7404]*
The 2002 farm bill amended NARETPA Sec. 1444(a)(2) to require that (1) federal support for extension at the 1890 institutions be at least 15% of the amount appropriated for extension at 1862 universities *[7 U.S.C. 3221]*, and (2) federal support for research at the 1890 institutions be at least 25% of the amount appropriated for research at the 1862 universities. *[7 U.S.C. 3222]*	Increases the level of required federal support for (1) extension at the 1890 institutions to 20% of the amount appropriated for 1862 extension programs *[Sec. 7215]*, and (2) research at the 1890 schools to 30% of the amount appropriated for research at the 1862 schools *[Sec. 7216]*	Identical to the House bill. *[Sec. 7017-7018]*	Adopts House and Senate provision. *[Sec. 7121-7122]*
Sec. 1434(b) of NARETPA defines which institutions are eligible for animal health/disease funding. *[7 U.S.C. 3196]*	No comparable provision.	Specifies that 1890 institutions are eligible to receive funding for animal health and disease research. *[Sec. 7016]*	Adopts Senate provision. *[Sec. 7120]*
The 2002 farm bill amended NARETPA Sec. 1449 to phase in increasing state matching funds for federal formula funds for research and extension at the 1890 institutions, reaching the 100% matching level in FY2007. *[7 U.S.C. 3222d]*	Makes permanent the requirement that states provide a 100% match to federal funds for research and extension at the 1890 colleges. *[Sec. 7220]*	Similar to the House bill. *[Sec. 7022]*	Extends 100% matching funds requirement for research and extension at the 1890 colleges through FY2012 and amends NARETPA to update permanent law and clarify current requirement of providing equal matching funds from non-federal sources. *[Sec. 7127]*
Sec. 3 of the Smith-Lever Act (extension) authorizes federal funds for cooperative state extension programs. *[7 U.S.C. 343]*	No comparable provision.	Makes 1890 institutions eligible for the Children, Youth, and Families Education and Research Network funds (CYFERNet). *[Sec. 7303]*	Adopts Senate provision. Also stipulates that funds for Smith-Lever 3(d) programs (of which CYFERNet is one) are competitively awarded. *[Sec. 7403]*
The McIntire-Stennis Cooperative Forestry Act (P.L. 87-788) authorizes certain forestry research programs. *[16 U.S.C. 582]*	No comparable provision.	Makes the 1890 institutions eligible for McIntire-Stennis forestry research funds. *[Sec. 7310]*	Adopts Senate provision. *[Sec. 7412]*
Sec. 1447 of NARETPA authorizes grants to upgrade facilities at 1890 institutions. *[7 U.S.C. 3222]*	Authorizes $8 million annually for grants to land grant institutions in insular areas to upgrade agriculture and food science	No comparable provision.	Amends NARETPA to authorize assistance to insular area land grant institutions, and authorizes $8 million in

(Continued)

Prior Law/Policy	House-Passed Bill (H.R. 2419)	Senate-Passed Substitute Amendment (H.R. 2419)	Enacted 2008 Farm Bill (P.L. 110-246)
	facilities. *[Sec. 7237]*		annual appropriations through FY2012. *[Sec. 7125]*
Funding for tropical and subtropical research is provided under the Special Research grant program to the land grant institutions in U.S. insular areas. *[7 U.S.C. 450i]*	No comparable provision.	Within miscellaneous programs, adds authority for a Tropical and Subtropical Agricultural Research competitive grant program limited to the insular area land grant institutions and divided equally between the Caribbean and Pacific basins. Authorizes appropriations of such sums as necessary. *[Sec. 7038]*	Adds Tropical and Subtropical Agricultural Research to the list of high-priority research and extension initiatives under Sec. 1672 of the 1990 farm bill. Extends authorization of Sec. 1672 through FY2012. *[Sec. 7204]*
Subtitle K of NARETPA authorizes miscellaneous programs. *[7 U.S.C. 3311]*	No comparable provision	Establishes a grant program for research, extension, and education programs at land grant institutions in the American Pacific region (AK, HI). *[Sec. 7041]*	Adds Agricultural Development in the American-Pacific Region to the list of high-priority research and extension initiatives. *[Sec. 7204]*
Sec. 1425(c)(2)(B) of NARETPA sets a minimum distribution level and a formula for distribution of any annual appropriation for the Expanded Food and Nutrition Education Program (EFNEP) in excess of the previous year's level. *[7 U.S.C. 3175]*	Authorizes annual appropriations of $90 million; sets a minimum $100,000 annual distribution in EFNEP funds to each land grant institution; sets minimum funding percentages for the 1890 institutions through FY2013; establishes a formula to distribute funds to states after FY2013; and establishes UDC's eligibility to receive EFNEP funds. *[Sec. 7604]*	Establishes UDC's eligibility to receive EFNEP funds. *[Sec. 7313]* The bill does not contain other comparable provisions in the House bill.	UDC is eligible to receive EFNEP funds *[Sec. 7417]*. Amends NARETPA to increase the authorization of appropriations to $90 million annually. Authorizes each institution to receive $100,000as a base amount. *[Sec. 7116]*
Sec. 1417(b)(4) of NARETPA authorizes capacity-building grants for research and teaching at high minority-enrollment institutions. *[7 U.S.C. 3152]*	Adds extension capacity-building as one of the purposes for which grants and fellowships may be made to high minority-enrollment institutions. *[Sec. 7605]*	Similar to the House bill. *[Sec. 7005]*	Adopts House provision. *[Sec. 7107]*
Sec. 1455 of NARETPA authorizes grants to Hispanic-serving institutions to strengthen educational capacity, and authorized $20 million annually in appropriations through FY2007. *[7 U.S.C. 3241]*	Authorizes the grant program through FY2012 at $20 million annually. *[Sec. 7221]*	Amends NARETPA to require that Hispanic-serving institutions compete for strengthening grants; and increases the authority for appropriations to $40 million annually through FY2012. *[Sec. 7023]*	Adopts Senate provision. *[Sec. 7128]*

(Continued)

Prior Law/Policy	House-Passed Bill (H.R. 2419)	Senate-Passed Substitute Amendment (H.R. 2419)	Enacted 2008 Farm Bill (P.L. 110-246)
No comparable provision.	Establishes an endowment fund for Hispanic-serving institutions for basic institutional support. Authorizes appropriations through FY2012 for grant programs and for collaborative extension projects at 1862 institutions. *[Sec. 7222]*	Similar to the House bill. *[Sec. 7024]*	Amends NARETPA to establish an endowment fund, institutional capacity-building program, and a competitive grant program to benefit Hispanic-serving agricultural colleges and universities; authorizes necessary funds to be appropriated through FY2012. *[Sec. 7129]*
The definition of "Hispanic-serving institution" in NARETPA is based on a formula found in Sec. 316(b)(1) of the Higher Education Act of 1965. *[20 U.S.C. 1059]*	Changes the definition of a Hispanic-serving institution in NARETPA from one based on a formula to one based on the total enrollment of students being at least 25% Hispanic, as it is in the Higher Education Act of 1965. *[Sec. 7234]*	Similar to the House bill. *[Sec. 7001]*	Adopts House and Senate provision. *[Sec. 7101]*
Sec. 1458 of NARETPA authorizes USDA agencies and land grant institutions to participate in international research, extension, and teaching programs. *[7 U.S.C. 3291]*	Extends eligibility for participation in international research, extension, and teaching programs to Hispanic-serving institutions. Gives priority to institutions with existing cooperative agreements with federal or state agencies. *[Sec. 7223]*	Similar to the House bill. *[Sec. 7025]*	Adopts House and Senate provision and adds anti-hunger, nutrition, and food availability to the purposes of international programs. *[Sec. 7130]*
Organic Agricultural Research			
Sec. 1672B of the 1990 farm act, as amended, provides $3 million annually in mandatory funds to support an organic agriculture research and extension initiative. *[7 U.S.C. 5925]*	Adds emphasis on the environmental impact of organic farming and on new plant varieties suited to organic farming. Authorizes annual appropriations through FY2012 of $25 million; and provides $25 million annually in mandatory funds through FY2012. *[Sec. 7310]*	Provides $16 million annually in mandatory funds through FY2012 to support the organic research and extension initiative. *[Sec. 7104]*	Reflects the House bill language concerning new areas of program emphasis. Provides a total of $78 million in mandatory funds in Fiscal Years 2009-2012. *[Sec. 7206]*
No comparable provision.	Sense of Congress that in -house funding for ARS research on organic agriculture be at least commensurate with its share of the U.S. food market. *[Sec. 7608]*	Similar to the House bill. *[Sec. 7505]*	Deletes House and Senate provisions.
Sec. 7405 of the 2002 farm bill established the Beginning Farmer and	No comparable provision.	Authorizes annual appropriations of $30 million through FY2012; permits grants	Provides $18 million in mandatory funds for the program in FY2009, and $19

(Continued)

PRIOR LAW/POLICY	HOUSE-PASSED BILL (H.R. 2419)	SENATE-PASSED SUBSTITUTE AMENDMENT (H.R. 2419)	ENACTED 2008 FARM BILL (P.L. 110-246)
Rancher Development program. *[7 U.S.C. 3319f]*		for farmers/ranchers who convert to certified organic production. *[Sec. 7309]*	million in mandatory funds in FY2010-2012. Authorizes $30 million in annual appropriations (FY2008-12). *[Sec. 7410]*
Specialty Crops Research			
The Specialty Crop Competitiveness Act of 2004 (P.L. 108-465) established a specialty crop committee to inform the Advisory Board on research needs. *[7 U.S.C. 3123a(c)]*	Expand information that the specialty crop committee provides the Advisory Board to include a comprehensive analysis of the specialty crop sector. *[Sec. 7204]*	No comparable provision.	Adopts House provision. *[Sec. 7103]*
Sec. 1672 of the 1990 farm act authorizes research and extension grants on specialty crops as a high-priority research area. *[7 U.S.C. 5925]*	Adds new specialty crop research initiative to the 1998 research act (AREERA). Authorizes annual appropriations of $100 million through FY2012; provides $215 million annually in mandatory funds. *[Sec. 7411]*	Similar to the House bill, but authorizes $16 million annually in mandatory funds through FY2012. *[Sec. 7211]*	Provides $230 million in mandatory funds over 5 years for a specialty crop research and extension initiative; provides $25 million for grants to research fresh produce food safety; and authorizes an additional $100 million in appropriated funds annually. *[Sec. 7311]*
Food safety research is part of USDA's research, extension, and education mission area, and is included in both intra- and extramural programs.	Authorizes USDA to make competitive grants to universities/others to design and implement programs to improve the safety of fresh-cut produce; authorizes appropriation as necessary, with $25 million annually in mandatory funds through FY2012. *[Sec. 7511]*	No comparable provision.	Includes a food safety research emphasis in Sec. 7311, above.
Section 1419A of NARETPA authorizes the Secretary to enter into a wide variety of grants and other collaborative agreements with private and public educational institutions, corporations, and individuals to conduct independent research and public policy analysis on food and agriculture. *[7 U.S.C. 3155]*	Requires the Food and Agricultural Policy Research Institute (a university-based economic research institute, in part supported by federal funds) to establish a corollary institute specializing in specialty crop policy research (drawing on university expertise in states with specialty crop production). *[Sec. 7235]*	Amends NARETPA to specify four university-based policy research centers as eligible to receive grants under existing authority. Adds specialty crops policy issues as a priority focus for these centers' activities. *[Sec. 7009]*	Adopts Senate provision. *[Sec. 7111]*
Bioenergy Research			
No comparable provision.	Establishes a renewable energy committee to report to the Advisory	No comparable provision.	Amends NARETPA to authorize the establishment of a permanent renewable

(Continued)

Prior Law/Policy	House-Passed Bill (H.R. 2419)	Senate-Passed Substitute Amendment (H.R. 2419)	Enacted 2008 Farm Bill (P.L. 110-246)
	Board concerning research needs and budget recommendations. *[Sec. 7203]*		energy subcommittee to the Advisory Board. *[Sec. 7104]*
Sec. 404 of the 1998 research act (AREERA) authorizes public-private cooperative agreements to conduct pilot projects to develop biobased products with commercial potential, and authorizes the appropriation of such sums as necessary. *[7 U.S.C. 7624]*	Extends this authority through 2012. *[Sec. 7403]*	Contains an identical provision. *[Sec. 7204]*	Adopts House and Senate provision. *[Sec. 7304*
Section 1419 of NARETPA authorizes grants for research on production and marketing of alcohols and industrial hydrocarbons from agricultural commodities and forest products. *[7 U.S.C. 3154]*	Establishes a bioenergy and biobased products research initiative, coordinated by the National Agricultural Research Program Office, focused on the conversion of biomass to renewable fuels. Authorizes $50 million annually in appropriations through FY2012. *[Sec. 7410]*	No comparable provision.	Repeals the authority under NARETPA *[Sec. 7110]* and amends the 1998 research act (AREERA) to authorize a bioenergy and biobased products research initiative as in the House provision. Incorporates several additional provisions from House bill and Senate amendment energy titles authorizing research on: 1) on-farm renewable energy ***[H. Sec. 9010 and S. Sec. 9011]***; 2) using sweet sorghum and switchgrass to supplement corn as an ethanol feedstock ***[H. Sec. 9020]***; 3) regional biomass crop experiments ***[S. Sec. 9010]***; 4) renewable energy at a laboratory in Colorado ***[S. Sec. 9022]***; and 5) farm energy demonstration projects ***[S. Sec. 9025]***. *[Sec. 7207]*
No comparable provision.	Establishes a New Era Rural Technology Program that makes grants available to rural community colleges and technical centers to support training a workforce in bioenergy, renewable energy, and pulp and paper manufacturing; authorizes appropriation as necessary. *[Sec. 7312]*	Similar to the House bill. *[Sec. 7043]*	Adopts House provision. *[Sec. 7137]*

(Continued)

Prior Law/Policy	House-Passed Bill (H.R. 2419)	Senate-Passed Substitute Amendment (H.R. 2419)	Enacted 2008 Farm Bill (P.L. 110-246)
Other Research Provisions Related to Bioenergy Programs — see section on Energy Programs (below)			
Other Research Provisions			
No comparable provision.	Authorizes USDA to establish animal disease laboratories, conduct research on diseases that constitute a threat to the livestock industry, and gives USDA discretion over the importation and movement of live viruses. *[Sec. 7108]*	Requires USDA to issue a permit to the Department of Homeland Security (DHS) for work on live Foot and Mouth Disease virus at the National Bio- and Agro-defense Lab; clarifies only the Secretary of Agriculture has the authority to grant and revoke such permits. *[Sec. 11016]*	Adopts Senate provision, but replaces the term National Bio- and Agro-defense Lab with more general language. *[Sec. 7524]*
Sec. 1433(a) of NARETPA authorizes animal health and disease research. *[7 U.S.C. 3195(a)]*	Requires the Secretary to encourage cooperation among institutions eligible to receive these funds. *[Sec. 7213]*	No comparable provision.	Adopts House provision. *[Sec. 7118]*
The ARS National Animal Disease Center in Ames, Iowa, is currently the highest bio-security lab for animal disease research.	No comparable provision.	Authorizes $16 million annually through FY2012 to construct a higher-level bio-secure ARS animal health and disease facility in Bozeman, MT. *[Sec. 7508]*	Deletes Senate provision.
Sec. 1415A of NARETPA authorizes a program to defray the school loans of veterinary medical school graduates who agree to serve for limited time periods in under-served areas. *[7 U.S.C. 3151]*	No comparable provision.	Amends program to require USDA to favor large and mixed animal practitioner shortages in rural areas in initial phases of program implementation. *[Sec. 7003]*	Adopts Senate provision and clarifies that large and mixed animal practitioner shortages have priority. Requires USDA to return to the Food Safety and Inspection Service fund that had been transferred to CSREES. *[Sec. 7105]*
Sec. 1672(e) of the 1990 farm bill, as amended, authorizes USDA to make grants for research and extension projects in a variety of high priority topic areas. *[7 U.S.C. 5925]*	Adds nine new subjects to the list of high priority research and extension areas. *[Sec. 7305]*	Adds 14 subjects to the list of high priority research and extension areas. *[Sec. 7102]*	Adopts House and Senate lists of priority areas to add 23 subjects. Eliminates certain specified areas from the list enacted in the 1990 farm bill; includes some of these in the larger priority area categories in the new law. *[Sec. 7204]*
Sec. 410 of the 1998 research act (AREERA) provides a one-time transfer of $8 million in mandatory funds to make grants to four national youth groups to support pilot projects in rural areas; also provides authority for appropriations	Amends the 1998 act to encourage flexibility in making grants to youth organizations; allows the organizations to redistribute grant funds among themselves; authorizes the appropriation of such sums as necessary through	Reauthorizes appropriations through FY2012. *[Sec. 7209]*	Adopts House provision. *[Sec. 7309]*

(Continued)

Prior Law/Policy	House-Passed Bill (H.R. 2419)	Senate-Passed Substitute Amendment (H.R. 2419)	Enacted 2008 Farm Bill (P.L. 110-246)
through FY2007. *[7 U.S.C. 7630]*	FY2012. *[Sec. 7408]*		
A number of international agricultural research exchange opportunities currently are available under several different statutes.	Authorizes appropriations to establish a Borlaug International Agricultural Science and Technology Fellowship Program to promote collaboration between U.S. and foreign agricultural professionals and international research systems. *[Sec. 7606]*	Similar to the House bill. *[Sec. 7042]*	Adopts Senate provision, which amends NARETPA to authorize the program and annual appropriations. *[Sec. 7139]*
The Act of March 4, 1927, authorizes the establishment of the National Arboretum. *[20 U.S.C. 191]*	Authorizes the construction of a Chinese garden at the National Arboretum, using federal appropriations. *[Sec. 7509]*	Also authorizes construction of a Chinese garden at the arboretum, but forbids use of appropriated funds. *[Sec. 7312]*	Adopts House provision. *[Sec. 7415]*
Subtitle K of NARETPA authorizes miscellaneous programs. *[7 U.S.C. 3311]*	No comparable provision.	Authorizes appropriations for grants to nonprofits to distribute donated vegetable seeds to community food projects in low-income areas. *[Sec. 7046]*	Adopts Senate provision. *[Sec. 7523]*
	No comparable provision.	Authorizes appropriations for grants to support farm safety education/outreach. *[Sec. 7047]*	Included in Sec. 7204 (high priority research and extension areas).
	No comparable provision.	Authorizes appropriations for grants to increase participation of women/under-represented minorities from rural areas in science, technology, engineering, and math education/careers. *[Sec. 7408]*	Included in Sec. 7204 (high priority research and extension areas).
	No comparable provision.	Establishes a National Farm Management Center to create a public benchmarking database and to improve farm financial management training. *[Sec. 7037]*	Authorizes competitive research and extension grants to improve farm management skills and create a financial management database. *[Sec. 7208]*
	No comparable provision.	Authorizes appropriations for a research program to develop pharmaceuticals and agrichemicals from plant, marine, and microbial sources. *[Sec. 7049]*	Adopts Senate provision. *[Sec. 7525]*
	No comparable provision.	Authorizes a $1 million annual appropriation to support nonprofit	Adds the purposes of the Senate provision to the authorization for

(Continued)

Prior Law/Policy	House-Passed Bill (H.R. 2419)	Senate-Passed Substitute Amendment (H.R. 2419)	Enacted 2008 Farm Bill (P.L. 110-246)
		research on international anti-hunger and nutrition activities. *[Sec. 7050]*	international agricultural research, extension, and education. *[Sec. 7130]*
	No comparable provision.	Authorizes appropriations for competitive grants to establish regional centers of excellence in food systems veterinary medicine. *[Sec. 7052]*	Adds food systems veterinary medicine to the list of high priority research and extension areas. *[Sec. 7204]*
	No comparable provision.	Authorizes appropriations for land grant institutions to establish regional centers of excellence for agricultural commodities (incl. poultry sustainability); requires matching funds. *[Sec. 7039]*	Adds regional centers of excellence to the list of high priority research and extension areas. *[Sec. 7204]*
	No comparable provision.	Establishes a farm and ranch stress assistance network to provide behavioral programs to U.S. producers. *[Sec. 7044]*	Adopts Senate provision. *[Sec. 7522]*
	No comparable provision.	Establishes a grant program focusing on critical rural and agricultural transportation and logistics issues facing producers and rural businesses. *[Sec. 7051]*	Adopts Senate provision with additional language to give priority to collaborative efforts. *[Sec. 7529]*
No comparable provision.	Prohibits USDA from disposing of land or facilities at the Grazinglands Research Laboratory in El Reno, OK. *[Sec. 7109]* Authorizes USDA to lease land at the El Reno facility to the University of Oklahoma. *[Sec. 7111]*	No comparable provision.	Adopts House provisions. *[Sec. 7502 and 7503]*
No comparable provision.	Requires scientists conducting research on biotech crops to receive training in USDA's biotech regulatory regime. Authorizes the certification of third-party providers of such training. *[Sec. 7110]*	No comparable provision	Deletes provision.
No comparable provision.	Authorizes appropriations for a grant to update USDA's Nutrient Composition Handbook for Beef. *[Sec. 7112]*	No comparable provision.	Deletes provision.

(Continued)

PRIOR LAW/POLICY	HOUSE-PASSED BILL (H.R. 2419)	SENATE-PASSED SUBSTITUTE AMENDMENT (H.R. 2419)	ENACTED 2008 FARM BILL (P.L. 110-246)
No comparable provision.	Sense of Congress that there should be greater support for USDA human nutrition research. *[Sec. 7113]*	No comparable provision.	Deletes provision.
Sec. 1433 of NARETPA authorizes animal health and disease research. *[7 U.S.C. 3195]*	Encourages setting priorities for animal health/disease research through regular regional/national meetings. *[Sec. 7213]*	No comparable provision.	Contains the House provision. *[Sec. 7119]*
Sec. 1424(b) of NARETPA authorizes a human nutrition research initiative. *[7 U.S.C. 3174]*	Adds as a focus of human nutrition research examination of the efficacy of current agriculture policies in promoting the health and welfare of economically disadvantaged populations. *[Sec. 7236]*	No comparable provision.	Adopts House provision. *[Sec. 7113]*
Sec. 1672(d) of the 1990 farm bill, as amended, encourages USDA to give funding priority to high-priority collaborative research proposals. *[7 U.S.C. 5925]*	Requires USDA to give funding priority to collaborative research grants. *[Sec. 7303]*	No comparable provision.	Adopts House provision. *[Sec. 7203]*
The 1990 farm bill authorizes research and extension on aflatoxin. *[7 U.S.C. 5925]*	Changes the focus of aflatoxin research and extension from controlling aflatoxin to improving and commercializing control technologies. *[Sec. 7304]*	No comparable provision.	Adds the House bill language to the authority for high priority research and extension projects on aflatoxin. *[Sec. 7204]*
The 1990 farm bill, as amended, authorizes a nutrient management research and extension initiative. *[7 U.S.C. 5925]*	Adds dairy cattle waste and regional concerns to the purposes of the nutrient management research and extension initiative. *[Sec. 7307]*	Establishes a consortium of northeast colleges for research on dairy nutrient management and energy production. *[Sec. 9023]* Establishes a southwest regional dairy, environment, and private land research program. *[Sec. 11092]*	Adds House bill language to section reauthorizing the nutrient management research and extension initiative. *[Sec. 7205]* Deletes Senate provisions.
Sec. 1417(i) of NARETPA authorizes a National Food and Agricultural Sciences Teaching Awards. *[7 U.S.C. 3152]*	No comparable provision.	Adds extension and research to the award program and requires at least one cash award be made per year. *[Sec. 7006]*	Adopts Senate provision. *[Sec. 7108]*
Sec. 604 of AREERA, the 1998 research act, authorizes the Food Animal Residue Avoidance Databank. *[7 U.S.C. 7642]*	No comparable provision.	Authorizes annual appropriations of $2.5 million through FY2012. *[Sec. 7213]*	Adopts Senate provision. *[Sec. 7312]*

PRIOR LAW/POLICY	HOUSE-PASSED BILL (H.R. 2419)	SENATE-PASSED SUBSTITUTE AMENDMENT (H.R. 2419)	ENACTED 2008 FARM BILL (P.L. 110-246)
USDA's FY2000 appropriations act (P.L. 106-7) authorizes grants for education at Alaska Native- and Native Hawaiian-serving institutions. *[7 U.S.C. 3242]*	No comparable provision.	Permits the Alaskan consortia of institutions to designate fiscal agents for each member institution, and to allocate funds among members. *[Sec. 7308]*	Adds Senate bill language to reauthorization of the education grants program at Alaska Native- and Native Hawaiian-serving institutions. *[Sec. 7112]*
In 1994, the Federal Crop Insurance Reform and Department of Agriculture Reauthorization Act (P.L. 103-354) reorganized USDA, including the merger and realignment of certain research agencies. *[7 U.S.C. 6971]*	No comparable provision.	Gives USDA authority to exchange, sell, or otherwise dispose of ARS' animals, animal products, plants, and plant products (except for seeds, germplasm). *[Sec. 7314]* Authorizes a pilot program to allow property at the ARS research center and the National Agricultural Library to be leased out. *[Sec.7316]*	Adopts Senate provision. *[Sec. 7408]*
Research on antibiotic-resistant bacteria in livestock is authorized under general statutory authority for federal and state agricultural research.	No comparable provision.	Authorizes a competitive grant program for targeted research on antibiotic-resistant bacteria in livestock. *[Sec. 7317]*	Adopts Senate provision. *[Sec. 7521]*
No comparable provision.	No comparable provision.	Directs USDA to prepare a report, in coordination with other federal agencies, on the prevalence of areas in the U.S. with limited access to affordable and nutritious food, and to make recommendations. *[Sec. 7504]*	Adopts Senate provision. *[Sec. 7527]*
No comparable provision.	No comparable provision.	Requires USDA to prepare a report on: (1) domestic and international markets for products from cloned animals; and (2) the safety of foods from cloned animals (particularly milk). *[Sec. 7507]*	Deletes provision.
No comparable provision.	No comparable provision.	Provision concerning the recruitment and hiring processes for ARS and the Forest Service, but does not directly affect research policy. *[Sec. 7502]*	Adopts Senate provision. *[Sec. 7528]*
No comparable provision.	No comparable provision.	Provision concerning the National Finance Center and National Information Technology Center, but not related to USDA's research mission.*[Sec. 7502]*	No comparable provision.

(Continued)

Prior Law/Policy	House-Passed Bill (H.R. 2419)	Senate-Passed Substitute Amendment (H.R. 2419)	Enacted 2008 Farm Bill (P.L. 110-246)
Extended Program Authorizations			
Program under prior law in some cases.	Comparable provision in some cases.	Comparable provision in some cases.	— Human nutrition intervention and health promotion research program *[Sec. 7114]* — Pilot research program to combine medical/agricultural research *[Sec. 7115]* — Continuing animal health and disease research *[Sec.7117]* — Grants to upgrade agricultural and food sciences facilities at 1890 land grant colleges, incl. Tuskegee University *[Sec. 7123]* — National research and training virtual centers *[Sec. 7126]* — Competitive grants for international agricultural science and education *[Sec. 7131]* — Equipment grants *[Sec. 7133]* — University research *[Sec. 7134*] — Extension Service *[Sec. 7135]* — Supplemental and alternative crops *[Sec. 7136]* — Aquaculture assistance programs *[Sec. 7140]* — Rangeland grants *[Sec. 7141]* — Authorization for biosecurity planning/response *[Sec. 7142*] — Resident instruction & distance education grants program for insular area institutions of higher education *[Sec. 7143]* — National genetic resources program *[Sec. 7201]* — National Agricultural Weather Information System *[Sec. 7202]* — Assistive technology program for farmers with disabilities *[Sec. 7210]*

Prior Law/Policy	House-Passed Bill (H.R. 2419)	Senate-Passed Substitute Amendment (H.R. 2419)	Enacted 2008 Farm Bill (P.L. 110-246)
			— National Rural Information Center Clearinghouse *[Sec. 7212]* — Integrated research, education, and extension competitive grants program *[Sec. 7306]* — Fusarium graminearum grants *[Sec. 7307]* — Bovine Johne's disease control program *[Sec. 7308]* — Agricultural biotechnology research and development for developing countries *[Sec. 7310]* — Office of pest management policy *[Sec. 7313]* — Critical Agricultural Materials Act *[Sec. 7401]* — Equity in Educational Land-Grant Status Act of 1994 *[Sec. 7402]* — Agricultural Experiment Station Research Facilities Act *[Sec. 7405]* — Agricultural Risk Protection Act of 2000 *[Sec. 7407]* — National Aquaculture Act of 1980 *[Sec. 7414]* — National Agricultural Research, Extension, and Teaching Policy Act Amendments of 1985 *[Sec. 7416]*
Repeal of Program Authorizations			
Program under prior law in some cases.	Comparable provision in some cases.	Comparable provision in some cases.	— Agricultural telecommuni-cations program *[Sec. 7209]* — Research on honey bee diseases *[Sec. 7211]* — Partnerships for high value agricultural product quality research *[Sec. 7302]* — Precision agriculture *[Sec. 7303]* — Biobased products *[Sec. 7304]* — Public education regarding the use

(Continued)

Prior Law/Policy	House-Passed Bill (H.R. 2419)	Senate-Passed Substitute Amendment (H.R. 2419)	Enacted 2008 Farm Bill (P.L. 110-246)
			of biotechnology in producing food for human consumption *[Sec. 7411]*
Title VIII: Forestry			
Cooperative Forestry Programs			
The Forest Land Enhancement Program provided financial aid for private forest practices (mandatory spending of $100 million for FY2002-07). Only about half of $100 million was spent; the remainder was borrowed for firefighting or cancelled by Congress. ***[16 U.S.C. 2103]*** General authority is provided for under the Cooperative Forestry Assistance Act of 1978 (CFAA, P.L. 95-313), as amended, authorizes USDA to establish a variety of cooperative programs to protect and manage nonfederal forest lands. ***[16 U.S.C. 2101-2114]***	No reauthorization provision, allowing program to terminate.	No reauthorization provision, allowing program to terminate.	No reauthorization provision, allowing program to terminate.
No comparable CFAA provision.	Adds new priorities: (1) conserving working forests, (2) protecting forests from natural threats and restoring forests, and (3) enhancing public benefits from private forests. ***[Sec. 8001]***	Similar to the House bill, but with subtle differences in priorities for protecting and restoring forests and for enhancing benefits. ***[Sec. 8001]***	Adopts House provision with modifications. ***[Sec. 8001]***
No comparable CFAA provision.	No comparable provision.	Authorizes new cost-share grants for local governments, tribes, and non-profits to acquire lands threatened by conversion to non-forest uses and provide public benefits. ***[Sec. 8002]***	Adopts Senate provision, creating the Community Forest and Open Space Conservation Program. ***[Sec. 8003]***
No comparable CFAA provision.	Adds requirements for financial assistance: state-wide forest assessment of conditions, trends, threats, and priorities and strategies to address threats and describe resources. ***[Sec. 8002]***	Adds new requirements for financial assistance: statewide forest plan to identify critical areas; address regional needs; and plan for managing and monitoring forests, achieving national priorities. ***[Sec. 8004]***	Adopts House provision with modifications. ***[Sec. 8002]***

(Continued)

Prior Law/Policy	House-Passed Bill (H.R. 2419)	Senate-Passed Substitute Amendment (H.R. 2419)	Enacted 2008 Farm Bill (P.L. 110-246)
Sec. 2109(d)(1) of the CFAA defines "State" to include "Trust Territory of the Pacific Islands." *[16 U.S.C. 2109]*	Replaces "Trust Territory of the Pacific Islands" with "the Federated States of Micronesia, the Republic of the Marshall Islands, the Republic of Palau." *[Sec. 8003]*	Same as the House bill. *[Sec. 8005]*	Adopts House provision. *[Sec. 8004]*
Subsections of the CFAA provide for a USDA Coordinating Committee, to coordinate among agencies, and for State Coordinating Committees, to coordinate with state foresters and other interested parties. *[16 U.S.C. 2113]*	Replaces USDA Committee with new Forest Resource Coordinating Committee, to coordinate among agencies, state agency representatives, and others. *[Sec. 8004]* Modifies state committee duties to include recommendations concerning the new state-wide forest assessment and strategies. *[Sec. 8005]*	Exempts projects proposed by Indian tribes from State Coordinating Committee recommendations. *[Sec. 8003]*	Adopts House provision with modifications. *[Sec. 8005 & 8006]*
No comparable CFAA provision.	Requires a Secretary-determined portion of funds to be allocated competitively among states. *[Sec. 8006]*	No comparable provision.	Adopts House provision. *[Sec. 8007]*
No comparable CFAA provision.	Authorizes up to 5% of funding for cost-shared competitively-allocated innovative education, outreach, or technology transfer projects. *[Sec. 8006]*	No comparable provision.	Adopts House provision. *[Sec. 8008]*
CFAA permanently authorized an Emergency Reforestation program. *[16 U.S.C. 2106a]* It has not been funded since FY1993.	Authorizes a new Emergency Reforestation program as part of the Emergency Conservation program (16 U.S.C. 2201-2204). *[Sec. 8102]*	Establishes new emergency landscape restoration program to rehabilitate croplands, grasslands, and private non-industrial forests following natural disasters. *[Sec. 2398]*	Adopts House provision with changes, creating the Emergency Forest Restoration Program. Provides definitions of disaster; and authorizes such sums as needed. *[Sec. 8203]*
Other Forestry Provisions			
No comparable provision.	No comparable provision.	Includes definitions and makes tribes eligible for Forest Legacy funding *[16 U.S.C. 2103c]* and forest management assistance. *[Sec. 8101-8112]*	Deletes provision.
No comparable provision.	No comparable provision.	Authorizes Cultural and Heritage Cooperation, with purposes, definitions, and prohibition on disclosing information, and provides for reburial of human remains and cultural items; for temporary area closures for traditional and cultural	Adopts Senate provision. *[Sec. 8101-8107]*

(Continued)

PRIOR LAW/POLICY	HOUSE-PASSED BILL (H.R. 2419)	SENATE-PASSED SUBSTITUTE AMENDMENT (H.R. 2419)	ENACTED 2008 FARM BILL (P.L. 110-246)
		purposes; and for free use of forest products for traditional and cultural purposes. *[Sec. 8121-8127]*	
The Healthy Forests Restoration Act of 2003 (P.L. 108-148) authorized easements through FY2008 to protect private forests for endangered species and biodiversity. *[16 U.S.C. 6578]*	Extends program with $10 million from the CCC annually through FY2012. *[Sec. 8101]*	Moves section to Title II (Conservation) and authorizes such sums as necessary. Expands purposes to support endangered species, carbon storage, and biodiversity. Replaces 99-year easement option with permanent easement, and encourages tribes to participate. *[Sec. 2331]*	Adopts House provision with changes. Provides $9.75 million annually (FY2009-2012). Provides permanent easements instead of 99-year easements. Encourages tribal participation. *[Sec. 8205]*
The Renewable Resources Extension Act of 1978 (P.L. 95-306) authorized educational and technical aid via state extension agencies and eligible universities and colleges. *[16 U.S.C. 1671-1676]*	Extends the program through FY2012. *[Sec. 7507]*	Extends the program through FY2012. *[Sec. 8201]*	Extends the program through FY2012. *[Sec. 7413]*
The Global Climate Change Prevention Act of 1990 within the 1990 farm bill authorized the Forest Service Office of International Forestry through FY2007. *[7 U.S.C. 6704(d)]*	Extends the program through FY2012. *[Sec. 8103]*	Extends the program through FY2012. *[Sec. 8202]*	Extends the program through FY2012. *[Sec. 8202]*
The 1990 farm bill, as amended, authorized Rural Revitalization Through Forestry via technology transfer, business assistance, and local training, through FY2008. *[7 U.S.C. 6601(d)(2)]*	Extends the program through FY2012. *[Sec. 8104]*	No comparable provision.	Adopts House provision. *[Sec. 8201]*
No comparable provision.	Authorizes competitive forestry grants to Hispanic-serving institutions to recruit, retain, and train "Hispanics and other under-represented groups." *[Sec. 8201]*	No comparable provision.	Adopts House provision. *[Sec. 8402]*
No comparable provision.	No comparable provision.	Amends the Lacey Act Amendments of 1981 (P.L. 97-79; 16 U.S.C. 3371-78) to expand restrictions on and penalties for importing wild plants/parts (e.g., lumber, logs) removed in violation of U.S. or foreign laws. *[Sec. 8204]*	Adopts Senate provision with changes. Provides definition of *plant*; excludes recycled materials; clarifies impact on exports; and specifies need for regulations to further define *plant*. *[Sec. 8204]*
No comparable provision.	No comparable provision.	Expands boundary of Green Mountain National Forest (VT) to allow additional	Adopts Senate provisions. *[Sec. 8301 & 8303]*

Prior Law/Policy	House-Passed Bill (H.R. 2419)	Senate-Passed Substitute Amendment (H.R. 2419)	Enacted 2008 Farm Bill (P.L. 110-246)
		land acquisition. Authorizes exchange/ sale of specific forest lands to Bromley Mountain Ski Resort, with directions on proceeds use. *[Sec. 8203 & 8205]*	
No comparable provision.	No comparable provision.	Authorizes certain land conveyance in New Mexico. *[Sec. 11075]*	Adopts Senate provision, amended to authorize a land conveyance in Virginia. *[Sec. 8302]*
No comparable provision.	No comparable provision.	For non-salvage timber sale contracts awarded between 7/1/04 and 12/31/06, the purchasers may request that the contract be cancelled, the contract payment rate be redetermined, or a substitute Producer Price Index be used. USDA may agree to the contract modification, if the specified terms and limitations are met. *[Sec. 8301]*	Adopts Senate provision with modifications. *[Sec. 8401]*
No comparable provision.	No comparable provision.	Sense of Senate that the President should act to ensure that imports of softwood lumber from Canada are consistent with the U.S.-Canada Softwood Lumber Agreement. *[Sec. 11903]*	Amends the Tariff Act of 1930 to require a softwood lumber importer declaration program to verify and reconcile data on softwood lumber imports, to assure implementation of U.S.-Canada Softwood Lumber Agreement. *[Sec. 3301]*
Title IX: Energy			
Farm and Community Energy Systems			
Sec. 9005 of the 2002 farm bill authorizes appropriations for the Energy Audit and Renewable Energy Development Program to provide grants to state agencies and organizations to assist farmers and rural businesses to be energy efficient. *[7 U.S.C. 8105]*	Extends program through FY2012. Funding continues of such sums as necessary for FY2008-12. *[Sec. 9004]*	Extends program through FY2012, but folds it into the new Rural Energy for America Program where mandatory funding is available (see next section). *[Sec. 9007]*	Folds the Energy Audit and Renewable Energy Development Program into the Rural Energy for America Program. Amended section 9007. *[Sec. 9001]*
Sec. 9006 of the 2002 farm bill authorizes the Renewable Energy Systems and Energy Efficiency Program to assist farmers, ranchers, and rural	Renames program "Rural Energy for America Program." Adds production and sale of electricity by renewable energy systems to loan purposes, and assistance	Renames program similar to House bill. Implements energy audits for state agencies, coops, educational institutions and utilities. Provides grants, loan	Establishes the Rural Energy for America Program to promote energy efficiency and renewable energy development for agricultural producers and rural small

(Continued)

Prior Law/Policy	House-Passed Bill (H.R. 2419)	Senate-Passed Substitute Amendment (H.R. 2419)	Enacted 2008 Farm Bill (P.L. 110-246)
businesses with renewable energy systems and improving energy efficiency. $23 million annual mandatory CCC funding for FY2003-07, but with annual spending, shortfalls lapse. *[7 U.S.C. 8106]*	for feasibility studies. Reserves 15% for small projects. Mandatory CCC funds: $50 million (FY2008), $75 million (FY2009), $100 million (FY2010), $125 million and $125 million (FY2011-FY2012), respectively. *[Sec. 9005]*	guarantees and incentive payments for energy efficiency and renewable energy, and manure-to-energy projects. Reserves 20% for small projects. Establishes the Rural Energy Star program. Mandatory funds of $230 million (FY2008), available until expended. *[Sec. 9007]*	businesses. Provides grants up to 25% of cost for energy audits, renewable energy development, and financial assistance (20% of funds for grants less than $20,000). Total assistance limited to 75% of cost. Provides mandatory funds: $55 million (FY2009), $60 million (FY2010), $70 million (FY2011), and $70 million (FY2012). Authorizes $25 million in discretionary funds annually (FY2009-12). Amended section 9007. *[Sec. 9001]*
No comparable provision.	Establishes the Farm Energy Production Pilot Program, with matching grants (up to 75% of cost) to demonstrate energy-neutral farms with existing technologies. Authorizes appropriations of $5 million (FY2008-12). *[Sec. 9010]*	No comparable provision.	Deletes provision. The conferees state the goals of this provision are addressed in Section 7207 of Title VII (Research).
No comparable provision.	Establishes Future Farmsteads Program to equip and demonstrate five farms and farm households in five regions with energy efficient/producing technologies. Authorized appropriations of such sums as necessary. *[Sec. 9015]*	Similar to the House bill. *[Sec. 9025]*	Deletes provision. The conferees state the goals of this provision are addressed in Section 7207 of Title VII (Research).
No comparable provision.	Establishes Community Wood Energy Program, with matching grants (up to 50%) for use of local, sustainable wood sources for a community wood energy system. Authorizes appropriations of such sums as necessary. *[Sec. 9019]*	Provides matching grants for use of local, sustainable wood sources for a community wood energy system based on a Community Wood Energy Plan. Authorizes appropriations of $5 million annually (FY2008-12). *[Sec. 9014]*	Establishes the Community Wood Energy Program, providing grants of up to $50,000 for up to 50% of the cost for communities to plan and install wood energy systems in public buildings. Authorizes $5 million in discretionary funds annually (FY2009-12). Amended section 9013. *[Sec. 9001]*
No comparable provision.	No comparable provision.	Establishes New Century Farm Project to develop and operate sustainable and integrated biomass — biofuel production systems. Authorizes one-time funding of $15 million (FY2008-FY2012) until spent. *[Sec. 9011]*	Deletes provision. The conferees state the goals of this provision are addressed in Section 7207 of Title VII (Research).

(Continued)

PRIOR LAW/POLICY	HOUSE-PASSED BILL (H.R. 2419)	SENATE-PASSED SUBSTITUTE AMENDMENT (H.R. 2419)	ENACTED 2008 FARM BILL (P.L. 110-246)
No comparable provision.	Establishes the Rural Energy Self-Sufficiency Initiative, providing grants to increase community energy self-sufficiency. Provides discretionary appropriations of $5 million authorized for FY2008 and such sums as necessary for FY2009-12). ***[Sec. 9013]***	Establishes the Rural Energy Systems Renewal program, with matching grants (up to 50% of project cost) to fund community energy renewal projects. Authorizes annual appropriations of $5 million (FY2008-12). ***[Sec. 9015]***	Establishes the Rural Energy Self-Sufficiency Initiative, providing cost- share (up to 50%) grants for rural communities to assess energy systems and to make improvements. Authorizes $5 million in discretionary funds annually (FY2009-12). Amended section 9009. ***[Sec. 9001]***
No comparable provision.	No comparable provision.	Establishes program to provide grants to Northeast land-grant universities for research, extension, and demonstration projects for dairy nutrient management and energy development in the Northeast. Authorizes discretionary funds of such sums as are necessary. ***[Sec. 9023]***	Deletes provision. The conferees state that the nutrient management research and extension initiative are addressed in Section 7204 of Title VII (Research).
No comparable provision.	No comparable provision.	Establishes Voluntary Renewable Biomass Certification Program to certify biomass feedstocks meet standards designed to reduce greenhouse emissions and improve soil carbon content while protecting wildlife habitat/environment. No funds authorized. ***[Sec. 9016]***	Deletes provision.
No comparable provision.	No comparable provision.	Creates Rural Energy Self Sufficiency program (five annual matching grants up to 75% of cost) for rural communities with other institutions to increase energy self-sufficiency through integrated renewable energy systems. Authorizes $5 million (FY2008) and such sums as necessary (FY2009-12). ***[Sec. 9011]***	Deletes provision. The conferees state the goals of this provision are addressed in section 7204 of Title VII (Research).
Biofuel Feedstocks			
Sec. 9008 of the 2002 farm bill extended the Biomass Research and Development Act of 2000 (P.L. 106-224), providing grants and financial assistance from the Department of Energy (DOE) and USDA to establish a technical advisory committee and a Biomass Research and	Extends the Biomass Research & Development Act of 2000 from 2007 to 2012. Mandatory CCC funding of $35 million (FY2008), $60 million (FY2009), $75 million (FY2010), $100 million (FY2011), $150 million (FY2012). Authorizes appropriations of $200	Extends the Biomass Research & Development Act of 2000 from 2007 to 2012, and incorporates it into the farm bill. Mandatory CCC funding of $15 million (FY2008), $25 million (FY2009), $35 million (FY2010). Authorizes appropriations of $85 million annually	USDA and DOE shall continue to coordinate biofuels and biobased products research and development programs under the Biomass Research & Development Act of 2000 through the Biomass Research and Development Board. Provides competitive grants,

(Continued)

PRIOR LAW/POLICY	HOUSE-PASSED BILL (H.R. 2419)	SENATE-PASSED SUBSTITUTE AMENDMENT (H.R. 2419)	ENACTED 2008 FARM BILL (P.L. 110-246)
Development Board. Mandatory funds of $5 million (FY2002); $14 million annually (FY2003-07). Authorized additional annual appropriations of $49 million (FY2002-07). *[7 U.S.C. 8101]*	million annually (FY2006-15). *[Sec. 9008]*	(FY2008-12). *[Sec. 9008]*	contracts, and financial assistance for research, development, and demon-stration of biofuels and biobased products Provides mandatory funding: $20 million (FY2009), $28 million (FY2010), $30 million (FY1022), and $40 million (FY2012). Authorizes $35 million in discretionary funds (FY2009-12). Amended section 9008. *[Sec. 9001]*
No comparable provision.	Establishes the Feedstock Flexibility Program, authorizing the use of CCC funds to purchase sugar to be resold as a biomass feedstock to produce bioenergy. *[Sec. 9013]*	Similar to the House bill. *[Sec. 1501]*	Adopts the House provision with modifications. Amended section 9010. *[Sec. 9001]*
No comparable provision.	Establishes the Renewable Woody Biomass for Energy Program, providing Forest Service grants to encourage the use of woody renewable biomass for energy. Mandatory funds of $15 million annually (FY2008-12). *[Sec. 9019]*	Similar to the House bill. Authorizes $5 million annually (FY2008-12). *[Sec. 9013]*	Requires the Forest Service to conduct a competitive research and development program to encourage use of forest biomass for energy. Adopts the House provision with amendments. Authorizes $15 million per year for FY2009-12. Amended section 9012. *[Sec. 9001]*
No comparable provision.	Establishes the Biomass Energy Reserve (BER). Provides financial and technical assistance to landowners and operators to grow dedicated energy crops as feedstock for cellulosic ethanol and other energy production. Incentives cover harvesting, storing, and transporting of biomass to bioenergy facilities. BER projects would be within a 50-mile radius of a bioenergy facility. Authorizes mandatory funding of such sums as necessary. *[Sec. 9018]*	Establishes the Biomass Crop Transition Program, providing technical/financial assistance to agricultural producers for production, collection, harvest, storage and transportation of biomass crops for use in a biorefinery. Directs USDA to collect and disseminate information on production potential, environmental impacts, and best practices. Authorizes mandatory funds: $130 million for FY2008 and $10 million annually for FY2009- 11, available until expended. *[Sec. 9004]*	Establishes the Biomass Crop Assistance Program (BCAP). Provides USDA contracts with producers to promote the production of bioenergy feedstock crops not primarily grown for food or animal feed. Also provides payments to eligible entities through contracts for collection, harvest, storage, and transportation of renewable biomass material for use in a biomass conversion facility. Provides discretionary CCC funding of such sums as necessary annually for FY2008-12. Amended section 9011. *[Sec. 9001]*
No comparable provision.	No comparable provision.	Establishes Regional Biomass Crop Experiment Grants for regional biomass production experiments by land grant institutions. Mandatory CCC funds: $10	Establishes a program to award competitive grants for projects with a focus on supporting on-farm biomass crop research and the dissemination of

(Continued)

Prior Law/Policy	House-Passed Bill (H.R. 2419)	Senate-Passed Substitute Amendment (H.R. 2419)	Enacted 2008 Farm Bill (P.L. 110-246)
		million (FY2008), $20 million (FY2009), $10 million (FY2010); also authorizes appropriations of such sums as necessary (FY2008-12). *[Sec. 9010]*	results to enhance the production of biomass energy crops and the integration of such with the production of bioenergy. Moves this provision to Title VII (Research). *[Sec. 7207]*
Sec. 9002 of the 2002 farm bill requires federal agencies to purchase biobased products and authorizes a voluntary biobased labeling program. USDA sets procurement regulations. Provides $1 million in mandatory CCC annual funding (FY2002-07) for testing biobased products. *[7 U.S.C. 8102]*	Extends the federal procurement program; sets biobased component of product at 5% minimum; and revises the deadline for USDA to set labeling requirements. Authorizes appropriations of $1 million for procurement provisions and $1 million for labeling provisions (FY2008-13). *[Sec. 9002]*	Extends the program and refines federal procurement rules for biobased products. Requires federal agencies to maximize procurement of biobased products and submit reports to Congress. Continues voluntary labeling. Establishes testing centers and education grants. Authorizes $3 million in annual mandatory funds (FY2008-12). *[Sec. 9002]*	Adopts Senate provision with changes. Extends the federal biobased procurement program. Eliminates the 5% minimum biobased content. Continues voluntary labeling. Authorizes mandatory funding of $1 million for FY2008 and $2 million annually for FY2009-12. Authorizes discretionary appropriations of $2 million annually for FY2009-2012 for testing and labeling. Amended section 9002. *[Sec. 9001]*
Sec. 9003 of the 2002 farm bill provides grants to help finance biorefineries and biofuel plants for demonstration. No mandatory funds were authorized; no discretionary funds were appropriated. Implementing regulations have been developed. *[7 U.S.C. 8103]*	Renames as "Biorefinery and Repowering Assistance." Provides cost sharing grants; adds loan guarantees for new and developing technologies for advanced cellulosic bioenergy production and biorefinery repowering. Mandatory CCC funds: $75 million (FY2008); $100 million (FY2009); $125 million (FY2010); $200 million (FY2011); and $300 million (FY2012). *[Sec. 9003]*	Renames as "Biorefinery and Repowering Assistance." Provides cost sharing grants (up to 50%) and loan guarantees (up to 80%) to assist new and developing technologies focusing on advanced cellulosic bioenergy production and biorefinery repowering. Authorizes mandatory CCC funds of $300 million for FY2008 to be made available until expended. *[Sec. 9005]*	Provides funds for repowering assistance to reduce or eliminate their use of fossil fuels for biorefineries in existence at enactment. Provides mandatory CCC funding of $35 million, available until expended. Provides discretionary funding of $15 million annually for FY2009-12. Amended section 9004. *[Sec. 9001]*
No comparable provision.	No comparable provision.	Establishes E-85 fuel program, providing grants for E-85 fuel infrastructure. Authorizes discretionary funding of $20 million (FY2008-12) to be available until expended. *[Sec 9021]*	Deletes provision.
Biofuel Research and Education			
The 2002 farm bill amended the Sun Grant Research Initiative Act of 2003 to establish bioenergy research programs through grants to land grant institutions and five regional centers. Authorizes	Extends program through FY2012, and adds new center in Hawaii. *[Sec. 9008]*	Similar to the House bill. Provides grants to other land grant institutions within each region, in addition to designated centers. Authorizes mandatory funds of $5 million (FY2008) and $10 million	Continues sun grant program. Provides matching grants to sun grant centers to develop, distribute, and implement biobased energy technologies and to promote diversification and sustainability

(Continued)

PRIOR LAW/POLICY	HOUSE-PASSED BILL (H.R. 2419)	SENATE-PASSED SUBSTITUTE AMENDMENT (H.R. 2419)	ENACTED 2008 FARM BILL (P.L. 110-246)
appropriations: $25 million (2005), $50 million (2006), and $75 million annually (FY2007-10). *[7 U.S.C. 8109]*		(FY2009-10) until expended, and discretionary annual appropriations of $70 million (FY2008-15). *[Sec. 9009]*	of agricultural production, and economic diversification in rural areas through biobased energy and product technologies. Establishes a Sun Grant Information Analysis Center. Requires annual reports. Discretionary funds of $75 million for FY2008-12 are authorized. *[Sec. 7526]*
Sec. 7134 of the 2002 farm bill provides research grants on production/marketing of alcohols and industrial hydrocarbons from agricultural and forest products. Authorized annual appropriations of $20 million (FY1991-2007). *[7 U.S.C. 3154]*	No comparable provision.	The House bill amends Section 1417(j) of NARETPA by adding agriculture programs for grades K-12 to the purposes of these grants. *[Sec. 7008]*	Adopts Senate provision with amendment to repeal this section from prior law. *[Sec. 7110]*
Sec. 9004 of the 2002 farm bill provides for the Biodiesel Fuel Education Program with grants to educate fleet operators and the public on biodiesel benefits. Annual mandatory CCC funding of $1 million (FY2003-07). *[7 U.S.C. 8104]*	Extends the Biodiesel Fuel Education Program. Provides annual mandatory funding of $2 million (FY2008-12). *[Sec 9017]*	Similar to the House bill. *[Sec 9003]*	Adopts House and Senate provision except provides mandatory CCC funding of $1 million annually (FY2008-12). Amended section 9006. *[Sec. 9001]*
Sec. 9009 of the 2002 farm bill provides for Cooperative Research and Extension projects, encouraging research on carbon sequestration in soils and plants, and the role of agriculture in greenhouse gas emissions. *[7 U.S.C. 6711]*	The House bill extends the section 221of the Agricultural Risk Protection Act of 2000 (ARPA) through 2012. (Section 7506)	No comparable provision, but authorizes USDA to continue to provide for grants to the Consortium for Agricultural Soils Mitigation of Greenhouse Gases to develop, analyze, and implement carbon cycle and greenhouse gas management research through cooperative research at the land grant universities. Authorizes appropriations of $15 million annually (FY2008-12). *[Sec. 7315]*	Adopts House provision. *[Sec. 7407]*
Sec. 9010 of the 2002 farm bill provides for the Bioenergy Program, continuing incentive payments to biofuel producers based on year-to-year production increases. Annual mandatory CCC funding of $150 million (FY2002-06), but no funding authorized for FY2007. *[7 U.S.C. 8108]*	Renews and extends the "Bioenergy Program for Advanced Biofuels" through FY2012. Excludes ethanol produced from corn starch; expands eligibility for production incentives for combined heat and power production using biomass at biofuels plants. Provides mandatory CCC funding of $225 million (FY2008),	Similar to the House bill. Bases payments on: biofuel production; feedstock prices; and net non-renewable energy fuel content. Benefits purchasers of feedstocks for cellulosic biofuels and biodiesel; excludes those who claim a biofuel production tax credit or who produce more than 150 million gallons	Establishes the Bioenergy Program for Advanced Biofuels. Provides payments to producers to support and expand production of advanced biofuels. Provides mandatory funding of $55 million (FY2009), $55 million (FY2010), $85 million (FY2011), and $105 million (FY2012). Authorizes additional

(Continued)

Prior Law/Policy	House-Passed Bill (H.R. 2419)	Senate-Passed Substitute Amendment (H.R. 2419)	Enacted 2008 Farm Bill (P.L. 110-246)
	$250 million (FY2009), $275 million (FY2010), $300 million (FY2011), and $350 million (FY2012). *[Sec. 9007]*	per year. One-time mandatory funding of $245 million (FY2008-12), available until expended. *[Sec. 9006]*	appropriations of $25 million annually (FY2009-12)Amended section 9005. *[Sec. 9001]*
No comparable provision.	Establishes grants for Biochar Research, Development, and Demonstration. Annual authorized appropriations of $3 million (FY2008-12). *[Sec. 9012]*	No comparable provision.	Deletes provision. Includes biochar research as a high-priority area in Title VII (Research). *[Sec. 7203]*.
No comparable provision.	Sense of Congress on renewable energy production, use, and benefits. *[Sec. 9016]*	No comparable provision.	Deletes provision.
No comparable provision.	No comparable provision.	Sense of Congress on ethanol blends, and need for inter-agency cooperation to encourage increased production of intermediate ethanol blends between E-10 and E-85. *[Sec 9002]*	Deletes provision.
No comparable provision.	No comparable provision.	Sense of Congress regarding research, extension and education programs on biofuels and bioproducts. *[Sec. 9004]*	Deletes provision. The managers encourage USDA to continue to allow and support these activities.
No comparable provision.	Authorized appropriations for a pilot program for academic internships with government, private, or non-profit institutions. *[Sec 9015]*	No comparable provision.	Deletes this provision.
No comparable provision.	No comparable provision.	Directs USDA and other agencies to submit reports: growth potential for cellulosic material *[Sec. 9024]*; biofuels infra-structure *[Sec, 9018]*; rural nitrogen fertilizers *[Sec. 9019]*; and life-cycle analysis of biofuels *[Sec. 9020]*.	Requires joint USDA, DOE, EPA studies: (1) infrastructure needs associated with significant expansion in biofuels production and use *[Sec. 9002]*; and (2) rural nitrogen fertilizer study, with appropriations authorized at $1 million (FY2009) *[Sec. 9003]*
No comparable provision.	No comparable provision.	Directs USDA and other agencies on Research and Development of Renewable Energy, including farm to fuel research on biofuel production; and research on small scale wind and solar energy production and fuel cells. Annual authorized appropriations of $5 million (FY2008-12), and annual appropriations of $110 million for cellulosic biofuel	Deletes provision. The conferees state the goals of this provision are addressed in section 7207 of Title VII (Research).

(Continued)

Prior Law/Policy	House-Passed Bill (H.R. 2419)	Senate-Passed Substitute Amendment (H.R. 2419)	Enacted 2008 Farm Bill (P.L. 110-246)
		research (USDA) and $110 million for smaller-scale biorefineries and plant research (DOE). *[Sec. 9022]*	
Other Renewable Energy Provisions			
No comparable provision.	No comparable provision.	Directs USDA report on the potential economic issues (including costs) associated with animal manure used in normal agricultural operations and as a bioenergy feedstock. *[Sec. 10307]*	Adopts Senate provision. *[Sec. 11014]*
Sec. 9001of the 2002 farm bill provides for definitions, including biomass, biobased product, biomass, renewable energy and small rural business *[7 U.S.C. 8101]*	Same as prior law.Keeps prior law definitions, but adds definitions for advanced biofuels, biofuel, biomass conversion facility, biorefinery, feedstock, among others. Adds ocean and hydroelectric to the renewable energy definition. *[Sec. 9001]*	Keeps prior law definitions, but adds definitions for advanced biofuels, biofuel, biomass conversion facility, biorefinery, feedstock, among others. Adds ocean and hydroelectric power to the renewable energy definition. *[Sec. 9001]*	Adopts Senate provision with changes. Adds definitions for advanced biofuels to include home heating fuels and aviation and jet fuels from cellulosic biomass. Amended section 9001. *[Sec. 9001]*
No comparable provision.	Establishes the USDA Energy Council to coordinate energy related activities within USDA and between USDA and other agencies. *[Sec. 9017]* Establishes an entity at USDA to provide oversight and coordination, liaison activities, biobased product evaluation, and database maintenance. *[Sec. 9009]*	Similar to the House bill, in that it would establish an entity at USDA to provide oversight and coordination, liaise with other federal departments, evaluate new biobased products, and maintain a database of biobased research and best practices. *[Sec. 9017]*	Deletes the House and Senate provision. The managers express their support of a single entity at USDA to coordinate activities in the Senate bill.
No comparable provision.	No comparable provision.	No comparable provision.	Directs the Secretary of the Treasury, in consultation with other agencies, to request that the National Academy of Sciences produce an analysis of current scientific findings relating to the future production of biofuels and its domestic impacts. *[Sec. 15322]*
Title X: Horticulture and Organic Production			
Marketing and Trade Promotion, Consumer Access			
No comparable provision.	No comparable provision.	Sets definitions to apply throughout one of the bill's subtitles the terms "specialty	Adopts Senate provision with an amendment to remove the definition of

(Continued)

Prior Law/Policy	House-Passed Bill (H.R. 2419)	Senate-Passed Substitute Amendment (H.R. 2419)	Enacted 2008 Farm Bill (P.L. 110-246)
		crop", "state", and "state department of agriculture." *[Sec. 1801]*	the term "State." *[Sec. 10001]*
No comparable provision.	Requires an independent evaluation of the commodity purchasing processes and the importance of increasing purchases of specialty crops. *[Sec. 10104]*	No comparable provision.	Adopts House provision with changes to require the Secretary to arrange to have performed an independent evaluation of the purchasing processes used by USDA to implement the requirement that funds available under Section 32 be principally devoted to perishable agricultural commodities. *[Sec. 10101]*
The Specialty Crops Competitiveness Act of 2004 *[P.L. 108-465, 7 U.S.C. 1621 note]* established a program of block grants to states to support projects in marketing, research, pest management, and food safety, among other purposes. Authorizes $44.5 million annually through FY2009.	Reauthorizes the block grant program through FY2012 and provides mandatory funding starting at $60 million in FY2008, rising to $95 million in FY2012. Increases the number of U.S. insular areas eligible to receive grants. *[Sec. 10102]*	Contains an identical provision, except that funding ends after FY2011. Specifies that turfgrass sod and herbal crops also are specialty crops. *[Sec. 1841]*	Adopts House provision with changes to specify that any funds made available for a fiscal year under the program that are not expended by certain date, to be determined by USDA, will be reallocated to other States; change the minimum grant amount to $100,000 or one-third of 1% of the overall funding allocated to the program in a given fiscal year (whichever is higher). Provides mandatory funding: $10 million (FY2008); $49 million (FY2009); and $55 million annually (FY2010-2012). *[Sec. 10109]*
The Farmer-to-Consumer Direct Marketing Act established a Farmers' Market Promotion Program promote farmers' markets, authorizing annual appropriations for grants to local governments, and nonprofit organizations. *[7 U.S.C. 3005]*	Expands the types of activities that are eligible for funding. Renames program the Farmer Marketing Assistance Program. Provides annual mandatory funds of $5 million (FY2008-10) and $10 million (FY2011-12). *[Sec. 10404]*	Reauthorizes the Farmers' Market Promotion Program and provides mandatory funds of $5 million annually in FY2008-11, and $10 million in FY2012. *[Sec. 1812]*	Adopts Senate provision with an amendment to specify that 10% of the funds available to carry out the program be used to implement electronic benefit transfer systems at farmers' markets; and to specify mandatory funding: $3 million (FY2008); $5 million annually (FY2009-2010); $10 million annually (FY2011-2012). *[Sec. 10106]*
No comparable provision.	Authorizes grants to a variety of public and private entities to improve transporting specialty crops to markets.	Similar to the House bill, but with minor technical differences. *[Sec. 1842]*	Adopts House provision with changes to allow national/state/regional organizations of producers, shippers or carriers to be eligible for grants. *[Sec. 10403]*

(Continued)

PRIOR LAW/POLICY	HOUSE-PASSED BILL (H.R. 2419)	SENATE-PASSED SUBSTITUTE AMENDMENT (H.R. 2419)	ENACTED 2008 FARM BILL (P.L. 110-246)
No comparable provision.	No comparable provision.	Requires the Government Accountability Office (GAO) to investigate the impact on specialty crops of lowering foreign trade barriers and to prepare a strategy for addressing the issue. *[Sec. 1831]*	Deletes Senate provision.
No comparable provision.	No comparable provision.	Encourages USDA and the U.S. Trade Representative to increase attention to sanitary and phytosanitary trade issues, and to develop a strategic risk management framework. *[Sec. 1833]*	Deletes Senate provision.
No comparable provision.	Establishes a grant program entitled the Healthy Food Urban Enterprise Program to support feasibility studies on improving the access of underserved communities to affordable, locally produced, nutritious food. Authorizes annual appropriations for this purpose. *[Sec. 10405]*	Authorizes a grant program to establish a Healthy Food Enterprise Development Center, providing information and technical assistance to entities to make affordable, locally produced, nutritious food available in underserved communities. Provides $1 million in mandatory funds (FY2009); $2 million annually (FY2010-12). *[Sec. 1843]*	Adopts Senate provision with changes to place language for the Healthy Urban Food Enterprise Development Center within the Community Food Projects statute. Clarifies that subgrants may be used to establish and facilitate enterprises that process, distribute, aggregate, store, and market healthy affordable foods. Limits allocations for administrative expenses. Provides $1 million in funding annually (FY2009-2011) and authorizes $2 million (FY2012). *[Sec. 4402]*
Specialty Crop Provisions Related to Nutrition Programs — see section on Nutrition Programs (below)			
Organic Agriculture Production			
The 2002 farm bill established a cost-share program to help producers and handlers of organic products obtain certification under the National Organic Program (NOP), and provided a one-time transfer of $5 million in mandatory crop insurance funds. *[7 U.S.C. 1524]*	Provides a one-time transfer (FY2008) of $22 million in mandatory funds to continue the cost-share program, caps the federal share of certification cost at no more that 75%, and raises the maximum amount a producer can receive from $500 to $750. *[Sec. 10301]*	Similar to the House bill, and requires an annual report to the House and Senate Agriculture Committees on cost-share expenditures in each state. *[Sec. 1823]*	Adopts Senate provision but deletes the cap on the federal cost share. *[Sec. 10301]*
The 2002 farm bill required USDA to keep segregated data on organic production and marketing. *[7 U.S.C. 5925c]*	Provides $3 million in mandatory CCC funds to support data collection and analysis on organic production, marketing, pricing, and crop loss risk.	Similar to the House bill and provides a one-time transfer of $5 million for segregated data collection and analysis. *[Sec. 1821]*	Adopts Senate provision with changes to clarify USDA's data collection, analysis, and survey development requirements, and to specify the contents of USDA's

(Continued)

PRIOR LAW/POLICY	HOUSE-PASSED BILL (H.R. 2419)	SENATE-PASSED SUBSTITUTE AMENDMENT (H.R. 2419)	ENACTED 2008 FARM BILL (P.L. 110-246)
	[Sec. 10302]		report to the House/Senate Agriculture Committees. Provides $5 million in mandatory funding, with an additional authorization of appropriations of $25 million (FY2008-2012), further specifying that $3.5 million in funding be allocated to AMS to collect and distribute comprehensive reporting of prices relating to organically produced agricultural products. *[Sec. 10302]*
No comparable provision.	Authorizes $50 million subject to appropriations over the life of the farm bill to provide technical assistance and cost-sharing grants to producers seeking to convert from conventional to organic production. *[Sec. 10303]*	Provides for technical assistance and cost-sharing under the Environmental Quality Incentives Program (EQIP) to producers seeking to convert to organic production. *[Sec. 2361]* Authorizes grants for this purpose under the Beginning Farmer and Rancher Development Program. *[Sec. 7309]*	Deletes House provision, but includes language addressing the goal of providing technical assistance to farmers transitioning to organic farming under the EQIP program *[Sec. 2501]* of the conservation title.
Assessments are exempted under marketing orders for conventionally-grown fruits and vegetables, for producers whose operations are 100% organic. *[7 U.S.C. 7401]*	No comparable provision.	Allows producers who have part of their farm certified organic under the NOP to receive the exemption. *[Sec. 1822]*	Deletes Senate provision.
The Organic Foods Production Act of 1990 authorizes appropriations of such sums as necessary for the National Organic Program. *[7 U.S.C. 6522]*	No comparable provision.	Specifies increased authorized annual funding levels for the NOP, starting at $5 million in FY2008 and rising to $11 million in FY2012. *[Sec. 1824]*	Adopts Senate provision with changes to provide such additional sums as are necessary. *[Sec. 10303]*
No comparable provision.	Provides grants using Section 32 funds (7 U.S.C. 612c) to help urban gardening and greenhouse projects to purchase and operate organic fruit and vegetable gardens and greenhouses. *[Sec. 10103A]*	No comparable provision.	Deletes House provision.
Pest and Disease Control			
No comparable provision.	Establishes a cooperative program with state agriculture departments to conduct	Similar to the House bill, with technical differences. Provides mandatory funds	Adopts Senate provision with changes to describe the application process; prohibit

PRIOR LAW/POLICY	HOUSE-PASSED BILL (H.R. 2419)	SENATE-PASSED SUBSTITUTE AMENDMENT (H.R. 2419)	ENACTED 2008 FARM BILL (P.L. 110-246)
	early pest detection and surveillance activities and create action plans, among other things. Provides mandatory funds starting at $10 million (FY2008), rising to $70 million (FY2012). ***[Sec. 10201]***	starting at $10 million (FY2008), rising to $64 million (FY2012). ***[Sec. 12101(f)]***	USDA from considering nonfederal funds; direct USDA to consider risk factors when considering an application; and express disapproval of a cost-sharing rule for animal and health emergency programs. Specifies mandatory funds: $12 million (FY2009); $45 million (FY2010); $50 million (FY2011); and $50 million (FY2012). ***[Sec. 10201]***
No comparable provision.	Authorizes the appropriation of $15 million for the construction of a sterile fruit fly rearing facility in Waimanalo, Hawaii, and the appropriation of $1 million annually thereafter. ***[Sec. 10202]***	No comparable provision.	Deletes House provision.
No comparable provision.	Authorizes the appropriation of necessary funds through FY2012, in addition to $20 million annually in mandatory funds, to create a National Clean Plant Network where the specialty crop industry can obtain pest- and disease-free planting stock. ***[Sec. 10404]***	Similar to the House bill, with technical differences. ***[Sec. 1851]***	Adopts Senate provision with an amendment to add NLGCA institutions to the list of entities that USDA shall consult with to carry out the program. Specifies mandatory funding: $5 million annually (FY2009-2012). ***[Sec. 10202]***
The Plant Protection Act (PPA) ***[7 U.S.C. 7701 et seq.]*** authorizes USDA's Animal and Plant Health Inspection Service (APHIS) to cooperate with states, localities and others to prevent the spread of and eradicate invasive pests and diseases.	No comparable provision.	Modifies penalties in the PPA as follows: $500,000 for each violation adjudicated in a single proceeding; $1,000,000 for each violation adjudicated in a single proceeding involving a genetically modified organism. Requires an action, suit or proceeding regarding a violation of the PPA to be considered no later than 5 years after the date the violation is initially discovered by USDA. ***[Sec. 11017]***	Adopts Senate provision with changes to strike the change to the statute of limitations, to expand the penalties to cover any willful violation of the PPA, and to clarify subpoena authorities of USDA under the PPA. Modifies the ability of the executive branch to delay the provision of compensation for economic losses. ***[Sec. 10203]*** Identical amendments were made to the Animal Health Protection Act in the livestock title. ***[Sec. 11012]***
The PPA ***[7 U.S.C. 7701 et seq.]*** authorizes USDA's activities under the Federal Coordinated Framework for the Regulation of Biotechnology. APHIS	No comparable provision.	Requires USDA, no later than 18 months after enactment, to take action on each issue identified in the document "Lessons Learned and Revisions under Consider-	Adopts Senate provision, with modification. ***[Sec. 10204]***

(Continued)

Prior Law/Policy	House-Passed Bill (H.R. 2419)	Senate-Passed Substitute Amendment (H.R. 2419)	Enacted 2008 Farm Bill (P.L. 110-246)
regulated the importation, interstate movement, and field testing of genetically engineered organisms that may pose a plant risk.		ation for APHIS' Biotechnology Framework," dated October 4, 2007; and as USDA considers appropriate, to promulgate regulations to improve the management and oversight of articles regulated under the PPA, among other specified action items. *[Sec. 11077]*	
Food Safety Provisions			
The Agricultural Adjustment Act governs the terms and conditions of marketing orders applicable to specified commodities. *[7 U.S.C. 608c(6)]*	Authorizes the implementation of quality-related food safety programs under marketing orders for specialty crops. *[Sec. 10106]*	No comparable provision.	Deletes House provision.
No comparable provision.	Authorizes appropriation of necessary sums to implement a program to educate fresh produce industry personnel and consumers about ways to reduce pathogens in fresh produce. *[Sec. 10110]*	Similar to the House bill, and authorizes $1 million in annual appropriations for that purpose. *[Sec. 1813]*	Adopts Senate provision, authorizing appropriations of $1 million annually (FY2008-2012) to remain available until expended. *[Sec. 10105]*
Disaster Assistance			
The 2002 farm bill established the Tree Assistance Program to compensate commercial orchardists for losses due to natural disasters and authorized annual appropriations for the program. *[7 U.S.C. 8201]*	Makes nursery tree growers eligible for disaster assistance under the program, increases the limitation on annual assistance from $75,000 to $150,000, and continues appropriations authority. *[Sec. 10101]*	Makes nursery tree growers eligible for disaster assistance, increases the limit on annual assistance to $100,000, adds reimbursement for orchard management to repair losses, and provides necessary mandatory funding over the life of the farm bill. *[Sec. 1210(e)]*	Adopts Senate provision with changes to modify the reimbursement cost of replanting trees lost from a natural disaster; amend the Federal Crop Insurance Act with a provision identical to that in the Trade Act of 1974; and to make other technical changes. The Manager's report clarifies the insurance requirement for eligibility applies only to insurance on crops and not underlying vines/trees. *[Secs. 12033 and 15101]*
Specialty Crop Sector Data Collection			
No comparable provision.	Authorizes necessary funds through FY2012 to support the collection and dissemination of market news for specialty crops. *[Sec. 10402]*	Authorizes $9 million annually in funds to support market news and price information on specialty crops. *[Sec. 1811]*	Adopts Senate provision with changes to authorize appropriations $9 million annually (FY2008-2012) to remain available until expended, in addition to

(Continued)

PRIOR LAW/POLICY	HOUSE-PASSED BILL (H.R. 2419)	SENATE-PASSED SUBSTITUTE AMENDMENT (H.R. 2419)	ENACTED 2008 FARM BILL (P.L. 110-246)
			available annual appropriations for market news services. *[Sec. 10107]*
The 1997 Census of Agriculture Act (P.L. 105-113) authorizes a Census of Agriculture to be taken every 5 years. *[7 U.S.C. 2204g(a)]*	Amends the 1997 law to include a census of specialty crops as part of the Census of Agriculture. *[Sec. 10107]*	Allows USDA to include a census of specialty crops in the Census or to conduct a separate census of specialty crops not later than the end of FY2008 and every 5 years thereafter. *[Sec. 1814]*	Adopts House provision. *[Sec. 10103]*
Other Commodity-Specific Provisions			
No comparable provision.	Requires USDA to submit a report on the investigation of honey bee colony collapse disorder and strategies to combat the problem. *[Sec. 10001]*	No comparable provision.	Deletes House provision.
The Honey Research, Promotion, and Consumer Information Act (P. L. 98-590), as amended, provides for coordinated research, promotion, and consumer information to expand their markets for honey. *[7 U.S.C. 4601 note]*	No comparable provision.	Amends the Honey Research, Promotion, and Consumer Information Act with provisions regarding the Honey Board and referenda on the honey research and promotion order. *[Sec. 1854]*	Adopts Senate provision with changes to discontinue the current Honey Board after USDA conducts a referendum for honey producers or honey packers, importers and handlers. Requires USDA to act as a fiduciary in conducting the referenda. *[Sec. 10401]*
No comparable provision.	No comparable provision.	Amends 7 U.S.C. 1622(h) to require the USDA grading or inspection mark be located close to the country of origin label on packaged honey. *[Sec. 1855]*	Adopts Senate provision with changes to specify that violations of the labeling requirements of this section, with respect to honey, may be deemed by USDA as sufficient cause for debarment from the benefits of the Agricultural Marketing Act of 1946. *[Sec 10402]*
The Agricultural Adjustment Act requires that imported commodities that are under marketing orders in the U.S. meet the order's standards. *[7 U.S.C. 608e-1(a)]*	Adds clementines to the list of commodities to which this requirement applies. *[Sec. 10105]*	Similar to the House bill. *[Sec. 3207]*	Adopts the House and Senate provision. *[Sec. 10102]*
The Agricultural Marketing Act of 1946 governs research and marketing programs for agricultural products. *[7 U.S.C. 1641]*	Adds a section to the 1946 act to regulate the minimum maturity of all Hass avocados sold in the United States. *[Sec. 10108]*	Authorizes USDA to initiate the process of establishing a marketing order regulating the grades and standards of Hass avocados, if a U.S. organization	Adopts Senate provision. *[Sec. 10108]*

(Continued)

PRIOR LAW/POLICY	HOUSE-PASSED BILL (H.R. 2419)	SENATE-PASSED SUBSTITUTE AMENDMENT (H.R. 2419)	ENACTED 2008 FARM BILL (P.L. 110-246)
		submits such a proposal. *[Sec. 1856]*	
A 1990 law contains the terms and conditions of the mushroom marketing order. *[7 U.S.C. 6104]*	Makes changes to the geographic composition of the Mushroom Board, and other provisions. *[Sec. 10109]*	Similar to the House bill, with technical differences. *[Sec. 1853]*	Adopts House provision with changes to clarify that the mushroom council may develop and propose to USDA programs for good agricultural and good handling practices and related activities for mushrooms. *[Sec. 10104]*
No comparable provision.	No comparable provision.	Establishes a program to compensate asparagus growers for losses in 2004 — 2007 due to imports. Provides $15 million in mandatory funds for this purpose. *[Sec. 1852]*	Adopts Senate provision. *[Sec. 10404]*
The 2002 farm bill did not include a separate title for horticultural products or organic production.	Creates new farm bill title, Horticulture and Organic Agriculture (Title X).	No new title; includes most horticulture and organic agriculture provisions as part of the Commodity Title I.	Creates new farm bill title, Horticulture and Organic Agriculture (Title X).
TITLE XI: LIVESTOCK			
Livestock Mandatory Reporting			
The Livestock Mandatory Reporting Act of 1999 *[7 U.S.C. 1635-1636h]* established a program of mandatory reporting of information regarding the marketing of live cattle, boxed beef, swine, and lambs. Requires packers, processors, and importers to provide periodic reporting of price, volume, contract, and demand information to USDA. The information is used to create price reports for livestock producers.	No comparable provision.	Changes the time of the afternoon swine report. After an economic study of wholesale pork product prices, USDA is authorized to establish mandatory packer reporting of wholesale pork product sales, specifying that USDA will make this information publicly available. *[Sec. 10001]*	Directs USDA to conduct a study of the economic impacts of requiring plants to report pork product sales, focusing on wholesale pork cuts. Also directs USDA to improve electronic reporting and publishing under the program. *[Sec. 11001]*
Meat and Poultry Inspection			
The Federal Meat Inspection Act (FMIA) *[21 U.S.C. 601 et seq.]* and the Poultry Products Inspection Act (PPIA) *[21 U.S.C. 451 et seq.]* permit states to operate their own meat and poultry inspection programs, if they are at least	Requires USDA to report to Congress on the effectiveness of each state inspection program and on the changes necessary to ensure enforcement of federal requirements. Replaces current federal-state cooperative inspection program	Provides for a new opt-in program for state-inspected plants with 25 or fewer employees, which subjects them to federally-directed inspection using state employees. During the first 3 years, state plants with 26-35 employees may also	State inspection provisions generally the same as the Senate bill, without the provision to provide 100% reimbursement for programs with pathogen testing that exceed federal testing. *[Sec. 11015]* Amends the FMIA

(Continued)

Prior Law/Policy	House-Passed Bill (H.R. 2419)	Senate-Passed Substitute Amendment (H.R. 2419)	Enacted 2008 Farm Bill (P.L. 110-246)
"equal to" (but not necessarily identical to) the federal program. State-inspected meat and poultry cannot be shipped in interstate commerce.	with a new program whereby USDA would approve the shipment of state-inspected meat and poultry from a state where key program requirements are identical to federal requirements; permits many plants currently under federal inspection to shift to state inspection; raises the federal reimbursement maximum from 50% to 60% for poultry programs only; among other things. *[Sec. 11103]* No comparable provisions regarding reportable food registries, recall plans, *E. coli* reassessment, or sanitary food transportation.	apply. Sets federal reimbursement at not less than 60% for both meat and poultry programs and permits 100% reimbursements if pathogen testing exceeds typical federal testing, among other provisions. *[Sec. 11067]* Requires USDA to establish "reportable food registries" for meat and poultry and their products. Requires all establishments to include recall plans in their safety prevention (i.e., HACCP) plans, and requires all beef establishments to have an *E. coli* reassessment. Directs HHS and USDA to issue sanitary food transportation regulations. *[Sec. 11087]*	and PPIA to require all establishments to promptly notify USDA if they have reason to believe an adulterated or misbranded product has entered commerce. Requires establishments to prepare, and maintain in writing, a product recall plan. *[Sec. 11017]*
Seafood Grading and Inspection			
The 2002 farm bill identifies the market and common name for catfish for labeling purposes. ***[21 U.S.C. 321d]*** Sec. 203(c) of the Agricultural Marketing Act (AMA) of 1946 ***[7 U.S.C. 1622]*** authorizes USDA to develop standards to encourage uniformity and consistency in commercial marketing. Sec. 1(w) of FMIA ***[21 U.S.C. 601 et seq.]*** defines "amenable species" subject to mandatory inspection.	No comparable provision.	Authorizes the establishment of a voluntary USDA grading program for catfish. Requires USDA to provide inspection activities for catfish, by adding catfish to the list of "amenable species" under the FMIA. Specifies that new catfish grading and certification programs shall not duplicate, impede, or undermine similar activities conducted by the Department of Commerce or by the Food and Drug Administration. *[Sec. 10002]*	Makes "catfish," as defined by the Secretary, an amenable species under the FMIA, thus subjecting catfish products to mandatory inspection; authorizes USDA to take into account the conditions under which catfish are raised and transported. Amends the 1946 AMA to authorize USDA to establish a voluntary fee-based grading program for catfish and to permit other farm-raised seafood industries to apply for such grading. *[Sec. 11016]*
Country of Origin Labeling (COOL)			
Sec. 10816 of the 2002 farm bill amended the AMA of 1946 by requiring retailers (excl. restaurants) to provide country of origin labeling (COOL) for beef, lamb, pork, seafood, peanuts, and perishable agricultural commodities. Specifies requirements on labeling USA products, on recordkeeping, certification, and on enforcement and fines for non-	Continues to require COOL implementation by 2008 for red meats and other covered commodities. Adds meat produced from goats. Makes changes to the labeling requirements for fresh red meats, by creating a new labeling system for red meats with new designation categories, e.g., defines U.S. origin as a product from an animal exclusively born,	Similar to the House bill, but further makes macadamia nuts and chicken covered commodities. *[Sec. 10003]* Creates a separate program for ginseng for country of harvest labeling. *[Sec. 10004]*	Continues to require COOL implementation by Sept. 30, 2008, for covered commodities, to which are added goat meat, chicken, macadamia nuts, pecans, and ginseng. Changes the labeling requirements for fresh red meats by creating a new labeling system for red meats with new designation categories, e.g., defines U.S. origin as a product

(Continued)

Prior Law/Policy	House-Passed Bill (H.R. 2419)	Senate-Passed Substitute Amendment (H.R. 2419)	Enacted 2008 Farm Bill (P.L. 110-246)
compliance. Annual appropriations bills delayed implementation of mandatory COOL for all covered commodities until Sept. 30, 2008 (except wild and farm-raised fish and shellfish, which went into effect in 2005.) *[7 U.S.C. 1621 et seq.]*	raised and slaughtered in the U.S. (or present in the U.S. before Jan. 1, 2008). For all covered commodities, eases record-keeping, certification requirements, and reduces fines for noncompliance. *[Sec. 11104]*		from an animal exclusively born, raised and slaughtered in the U.S. (or present in the U.S. before July 15, 2008). For all covered commodities, eases record-keeping, certification requirements, and reduces fines for noncompliance. *[Sec. 11002]*
Agricultural Fair Practices Act			
The Agricultural Fair Practices Act (AFPA) of 1967 (P.L. 90-288) allows farmers to file complaints with USDA if a processor refuses to deal with them because they are members of a bargaining or marketing association of producers. Makes it unlawful for handlers to coerce, intimidate, or discriminate against producers because they belong to such groups. *[7 U.S.C. 2301 et seq.]*	No provision.	Amends AFPA as follows: — Expands the definition of "association of producers" to also include general livestock, poultry and farm groups. *[Sec. 10101]* — Broadens the types of prohibited practices. *[Sec. 10102]* — Amends the enforcement provisions; clarifies civil actions against handlers, providing for preventive relief, damage, and attorneys fees. *[Sec. 10103]* — Directs USDA to promulgate rules/regulations. *[Sec. 10104]*	Amends AFPA to modify the previous definition of "association of producers" to include organizations with membership exclusively limited to agricultural producers and dedicated to promoting their products. Also modifies the definition of "handler." *[Sec. 11003]*
Packers and Stockyards Act			
The Packers and Stockyards Act (P&S Act) of 1921 (P.L. 67-51), as amended, provides USDA with the basic authority to regulate marketing practices in the livestock, poultry, and meat industries. The law is to prevent unfair, deceptive, and monopolistic trade practices, focusing on livestock terminal and auction markets, livestock marketing agencies, dealers, meat packers, and live poultry dealers. *[7 U.S.C. 181 et seq.]*	Amends the P&S Act to direct USDA to establish regulatory standards for arbitration provisions in livestock and poultry contracts. Among other things, such regulations are intended to permit a producer to seek relief in a small claims court, if within the court's jurisdiction, regardless of a contract's arbitration clause. *[Sec. 11102]*	Amends the P&S Act as follows: — Creates a new special counsel at USDA to investigate/prosecute violations of competition laws. *[Sec. 10201]* — Strengthens USDA enforcement authorities over live poultry dealers. *[Sec. 10202]* — Specifies conditions regarding cancelling and securing contracts. Provides for producer choice of jurisdiction and venue, including arbitration. *[Sec. 10203]* — Allows growers to discuss contract terms. *[Sec. 10204]*	Amends the P&S Act as follows: — Requires an annual report from USDA on detailed investigations into possible violations of the P&S Act; *[Sec. 11004]* — Permits poultry and swine producers to cancel their contracts up to 3 business days after signing, unless a later date is specified in the contract; and requires clear disclosure in contracts of cancellation terms; — Requires poultry/swine contracts to contain a conspicuous statement that additional large capital

PRIOR LAW/POLICY	HOUSE-PASSED BILL (H.R. 2419)	SENATE-PASSED SUBSTITUTE AMENDMENT (H.R. 2419)	ENACTED 2008 FARM BILL (P.L. 110-246)
		— Allows producers to seek remedy for violations. *[Sec. 10205]* — Allows USDA to seek outside counsel to aid in investigations and civil cases. *[Sec. 10206]* — Prohibits major packers from owning, feeding, or controlling livestock more than 14 days prior to slaughter. *[Sec. 10207]* — Directs USDA to promulgate regulations. *[Sec. 10208]*	investments may be required during the term of the contract; — Contains provisions intended to assist producers deal with contract disputes, including arbitration terms, venue for any litigation. *[Sec. 11005]* — Requires USDA to issue rules on such criteria as, for example, the reasonable period of time a producer should be given to remedy a breach of contract before it is cancelled. *[Sec. 11006]*
Animal Pest and Disease Programs			
Sec. 2506(d) of the 1990 farm bill authorizes appropriations and directs USDA to carry out pseudorabies eradication in U.S. swine populations. Current concerns are that this disease persists in feral populations and may be reintroduced. *[21 U.S.C. 114i]*	Sense of Congress regarding pseudorabies eradication program that USDA recognize the threat feral swine pose to the domestic swine population, and the need for a surveillance program for monitoring and eradication. *[Sec. 11101]*	Similar to the House bill, and also recognizing the threat to the entire livestock industry. *[Sec. 10301]*	Similar to the House bill, also recognizing the threat to the entire livestock industry. *[Sec. 11007]*
Sec. 10409 of the Animal Health Protection Act (AHPA), enacted as part of the 2002 farm bill, directs USDA to carry out operations and measures to detect, control, or eradicate any livestock pest or disease, incl. animals at slaughterhouse, stockyard, or other point of concentration. *[7 U.S.C. 8308]*	No comparable provision.	Directs USDA to establish and implement a trichinae certification program. Authorizes appropriations of $1.25 million annually for FY2008-12. *[Sec. 10304]*	Directs USDA to establish and implement a voluntary trichinae certification program. Requires USDA to use not less than $6.2 million for the program, subject to availability of appropriations. Authorizes annual appropriations of $1.5 million for FY2008-2012. *[Sec. 11010]*
USDA has authority to cooperate with states on laws that exclude, eradicate, and/or control agricultural pests within the AHPA *[7 U.S.C. 8301 et seq.]* and the Talmadge-Aiken Act *[7 U.S.C. 450]*. Sections of 21 U.S.C., Title 21 (Food and Drugs) also cover the prevention and spread of contagion. Current concerns are about pesticide-resistant populations	Sense of Congress regarding the cattle fever tick eradication program that the cattle fever tick and the southern cattle tick are vectors of the causal agent of babesiosis, a severe and often fatal disease of cattle; and that implementing a national strategic plan for the cattle fever tick eradication program is a high priority, among other things.	Same as the House bill. *[Sec. 10302]*	Adopts House and Senate provision. *[Sec. 11008]*

(Continued)

Prior Law/Policy	House-Passed Bill (H.R. 2419)	Senate-Passed Substitute Amendment (H.R. 2419)	Enacted 2008 Farm Bill (P.L. 110-246)
of the southern cattle tick in Mexico.	*[Sec. 11106]*		
Sec. 10407(d)(2) of APHA specifies compensation amounts for seizure, quarantine, and disposal of animals that may carry or have been infected with or exposed to pests or diseases, and are moved through interstate commerce or are imported. *[7 U.S.C. 8306(d)(2)]*	Sense of Congress regarding the voluntary control program for low pathogenic avian influenza program; and that USDA should continue to provide compensation payments to poultry owners and cooperating state agencies at 100% of eligible costs. *[Sec. 11105]*	Amends AHPA to compensate any poultry contract grower or owner participating in the voluntary control program for low pathogenic avian influenza under the National Poultry Improvement Plan. Specifies payments to cooperating state agencies to be 100% of the eligible costs. *[Sec. 10306]*	Amends the AHPA to require the Secretary to compensate industry participants and state agencies that cooperate in voluntary detection and control of diseases with low pathogenicity at 100% of eligible costs. *[Sec. 11011]*
No comparable provision.	No comparable provision.	Sense of Senate that USDA should work with the private insurers to implement an expedited approach for indemnification of livestock producers in cases of catastrophic disease outbreaks. *[Sec. 10308]*	Deletes provision.
Sec. 10411 of AHPA authorizes USDA cooperative agreements with eligible entities, including other governments and associations, to conduct animal health activities. *[7 U.S.C. 8310]*	No comparable provision.	Establishes an advisory committee on national aquatic animal health; details committee membership; requires USDA regulations establishing a national aquatic animal health improvement program under AHPA authority; authorizes appropriations of $15 million for FY2008 and FY2009 for a new producer indemnification fund and for implementation of an animal health task force plan. *[Sec. 11086]*	Permits USDA to enter into cooperative agreements to carry out a project under a national aquatic animal health plan under Sec. 10411 of the AHPA. Requires USDA to determine the nonfederal share of costs (to be either cash or in-kind) on a case-by-case basis. Authorizes such sums as necessary in each fiscal year, FY2008-FY2012. *[Sec. 11013]*
National Animal Identification System			
No comparable provision in AHPA. Under this authority, in 2004, USDA accelerated work on a voluntary National Animal Identification System (NAIS) to trace animals from slaughter through all premises within 48 hours of an animal disease outbreak.	No comparable provision.	Requires USDA to issue regulations, subject to public comment, addressing "the protection of trade secrets and other proprietary and/or confidential business information that farmers and ranchers disclose in the course of participation" in an animal ID system. *[Sec. 10305]*	Deletes provision.

(Continued)

PRIOR LAW/POLICY	HOUSE-PASSED BILL (H.R. 2419)	SENATE-PASSED SUBSTITUTE AMENDMENT (H.R. 2419)	ENACTED 2008 FARM BILL (P.L. 110-246)
Livestock Provision Related to Bioenergy Programs — see section on Energy Programs (below)			
Other Commodity-Specific Provisions			
Sec. 375 of the Consolidated Farm and Rural Development Act (Con Act), as amended, established the National Sheep Industry Improvement Center to provide financial assistance for the enhancement and marketing of U.S. sheep or goat products, focusing on infrastructure development. Funding includes mandatory funds of $28 million for a revolving fund, and appropriations authorized at $30 million. ***[7 U.S.C. 2008j]***	Reauthorizes appropriations of $10 million annually (FY2008-12). Eliminates statutory requirement to eventually privatize the revolving fund. ***[Sec. 6015]***	Also eliminates statutory requirement to eventually privatize the revolving fund. Renames the program as the National Sheep and Goat Industry Improvement Center, and provides for new mandatory funding of $1 million for FY2008, to be available until expended. Authorizes appropriations of $10 million annually for FY2008-12. ***[Sec. 10303]***	Similar to Senate provision, but does not rename the program. ***[Sec. 11009]***
No comparable provision.	No comparable provision.	Requires USDA study and report on the potential economic issues (including costs) associated with animal manure used in normal agricultural operations and as a bioenergy feedstock. ***[Sec. 10307]***	Requires USDA study and report on animal manure use as agricultural fertilizer, potential impact on consumers and agriculture from limitations on its utilization, and effects on agriculture of increasing its use for bioenergy production. ***[Sec. 11014]***
The 2002 farm bill does not include a separate title for animal agriculture.	No new title; includes most animal agriculture provisions as part of the Miscellaneous Title XI.	Creates new farm bill title, Livestock Marketing, Regulatory, and Related Programs (Title X).	Creates new farm bill title, Livestock (Title XI).
TITLE XII: CROP INSURANCE & DISASTER ASSISTANCE PROGRAMS			
Timing of Crop Insurance Payments and Receipts			
The federal government provides three levels of subsidies to the crop insurance program: (1) subsidizing a portion of the farmer-paid premium, (2) reimbursing the private crop insurance companies for most administrative and operating expenses, and (3) absorbing most of the program losses. ***[7 USC 1501 et seq.]***	Changes the timing of crop insurance receipts (premium collections) and the timing of payments to the insurance companies, beginning with the 2012 reinsurance year (which starts July 1, 2011). Two insurance years of program receipts will be received in the same fiscal year (FY2012) and payments will be delayed until the next fiscal year, thus scoring budget savings in FY2012. ***[Secs. 11001(c), 11001(e), and 11010]***	Similar, but not identical, language as the House bill, which effectively requires premiums to be collected from producers slightly earlier, and payments to the insurance companies to be made slightly later, beginning in the 2012 crop year, so that savings can be scored in the last year of the 5-year farm bill (FY2012). ***[Secs. 1906 and 1914]***	Adopts the House provision that changes the premium billing date to August 15 ***[Sec. 12007]***, the Senate provisions that change the timing of reimbursements to the private companies for operating expenses to between October 1 and 31 ***[Sec. 12015]*** and underwriting gains to October 1 ***[Sec. 12018]***, thus allowing the scoring of budget savings in FY2012.

(Continued)

PRIOR LAW/POLICY	HOUSE-PASSED BILL (H.R. 2419)	SENATE-PASSED SUBSTITUTE AMENDMENT (H.R. 2419)	ENACTED 2008 FARM BILL (P.L. 110-246)
Reimbursement of Administrative and Operating Expenses			
Prior law prohibited companies from receiving a reimbursement greater than 24.5% of total premiums. The Standard Reinsurance Agreement (SRA) established the reimbursement rate below the statutory maximum for all insurance plans, ranging from 18.1% to 24.2%. *[7 USC 1508(k)(4)(A)]*	Beginning in the 2009 reinsurance year, the reimbursement rate to the insurance companies for their administrative and operating expenses for all policies declines by 2.9 percentage points from the current rate. The range of reimbursement rates declines to between 15.2% to a maximum of 21.3%. *[Sec. 11001(d)(1)]*	Beginning in the 2009 reinsurance year, the reimbursement rate for additional coverage policies falls by 2 percentage points. An exception is any reinsurance year in any state that has a loss ratio greater than 1.2 (i.e., when indemnity payments exceed total premiums by more than 20%). The reimbursement rate for policies based on area-wide losses is reduced to 17% of total premiums. *[Sec. 1912]*	Beginning in the 2009 reinsurance year, the reimbursement rate for additional coverage policies falls by 2.3 percentage points. Restores one-half of the reduction in states with a loss ratio greater than 1.2. The reimbursement rate for policies based on area-wide losses is reduced to 12% of total premiums. *[Sec. 12016]*
Premiums and Fees			
For catastrophic (CAT) coverage, producers pay no premium, but pay an administrative fee of $100 per crop per county. *[7 USC 1508(b)(5)(A)]* Growers of uninsurable crops are eligible for a separate Noninsured Assistance Program (NAP) and pay a fee of $100 per crop, or $300 per producer per county, not to exceed $900 per producer. [*7 USC 7333(k)(1)]*	Increases the producer-paid fee for catastrophic coverage to $200 per crop per county. For NAP, the fee also is raised to $200 per crop per county, or $600 per producer per county, not to exceed $1800 per producer. *[Secs. 11002 and 11009]*	Similar to the House bill for raising the CAT fee to $200. The NAP fee is increased to $200 per crop per county, or $600 per producer per county, not to exceed $1,500 per producer. *[Secs. 1905 and 1926]*	Raises the CAT fee to $300 per crop per county. *[Sec.12006]* Increases the NAP fee to $250 per crop per county, or $750 per producer per county, not to exceed $1,875 per producer. *[Sec. 12028]*
When permitted by state law, a cooperative or trade association may pay on behalf of its members, any or all of the administrative fee for CAT coverage. *[7 USC 1508(b)(5)(B)]* .	Limits the ability of associations to pay the CAT fee on behalf of a producer. *[Sec. 11001(b)]* Prohibits insurance companies from paying or rebating premiums, or making any inducements to purchase crop insurance. *[Sec. 11001(a)]*	Revises prior law to clarify that the provision applies only to fees for CAT coverage. *[Sec. 1905]*	Adopts the provision in Sec. 1905 of the Senate bill which clarifies that coop-eratives and trade associations can pay only the fees for catastrophic coverage on behalf of their members. *[Sec. 12006]* Adopts the provision in Sec. 11001(a) of the House bill that prohibits the rebating of premiums, with certain exceptions for entities that have already been approved for rebating. *[Sec. 12004]*
Authorizes crop insurance companies to offer customers a discount when the insurance companies adopt efficiencies	Strikes authority for companies to offer a Premium Reduction Plan (PRP) or Premium Rate Reduction Pilot program.	Strikes authority for PRP only, and requires USDA to commission a study on the feasibility of the PRP within 18	Strikes authority for the PRP only. *[Sec. 12010]*

(Continued)

Prior Law/Policy	House-Passed Bill (H.R. 2419)	Senate-Passed Substitute Amendment (H.R. 2419)	Enacted 2008 Farm Bill (P.L. 110-246)
that reduce their administrative and operating costs. *[7USC 1508(b)(5)(A)]*	*[Sec. 1101(f)]*	months of enactment. *[Sec. 1908]*	
No comparable provision.	Reduces the premium subsidy for area risk plans by 4 percentage points. *[Sec. 11013]*	No comparable provision.	Adopts House provision. *[Sec. 12012]*
Requires USDA to set premiums so that the overall program loss ratio is 1.075. *[7 USC 1506(n)]*	No comparable provision	Reduces the statutory loss ratio to 1.0, meaning that total premiums should be established to equal expected total indemnity payments. *[Sec. 1903]*	Adopts Senate provision. *[Sec. 12003*]
Authorizes an Agricultural Management Assistance (AMA) program to in part help certain states make better use of risk management tools. *[7 USC 1524(b)]*	No comparable provision	Allows USDA to use AMA funds to match state funds used to provide additional premium discounts to underserved states. *[Sec. 1923]*	Deletes Senate provision.
Standard Reinsurance Agreement and Risk-Sharing			
The Standard Reinsurance Agreement (SRA) between the federal government and private crop insurance companies determines levels of risk sharing. The agreement requires companies to reinsure 5% of their retained premium with the government.	Requires the private insurance companies to reinsure at least 22% of their retained premiums with the government, and in return the government will provide a ceding commission of 2% to companies, allowing the government to receive some underwriting gains that would otherwise accrue to the companies. *[Sec. 11014]*	No comparable provision.	Deletes House provision.
No comparable provision.	USDA can renegotiate the SRA starting with the 2012-13 reinsurance year, and once every 5 years thereafter. Insurance companies can confer with each other during the process. *[Sec. 11001(d)(2)]*	Similar to the House bill, except that USDA has discretion to renegotiate the SRA more frequently than every 5 years, with congressional notification of such action. Allows crop insurance companies to confer with each other and collectively with USDA during the renegotiation process. *[Sec. 1913]*	Allows USDA to renegotiate the SRA beginning with the 2010-11 reinsurance year and once every 5 years thereafter. Adopts the Senate provision to allow companies to confer with each other and collectively with USDA during renegotiation. SRA can be renegotiated more than once in a 5-year period if one of the changes is required by law, and Congress is notified. *[Sec. 12017]*

(Continued)

PRIOR LAW/POLICY	HOUSE-PASSED BILL (H.R. 2419)	SENATE-PASSED SUBSTITUTE AMENDMENT (H.R. 2419)	ENACTED 2008 FARM BILL (P.L. 110-246)
Program Integrity (Waste, Fraud, and Abuse)			
Annual mandatory funds of $23 million for data mining and program integrity activities expired at the end of FY2005. *[7 USC 1516(k)]* Annual appropriations acts provided $3.6 million in annual discretionary funds (FY2006, FY2007). FY2008 appropriations act authorized mandatory funds of $11.2 million.	Authorizes mandatory funding of $11 million in FY2008, and $7 million in FY2009 and subsequent years for crop insurance program compliance and integrity activities, including data mining. *[Sec. 11008]*	Requires USDA to establish a program whereby crop insurance companies pay USDA a fee for access to its data mining system, and USDA uses proceeds for its data system. *[Sec. 1915]* Prohibits farmers from collecting commissions as agents on policies in which their family has a substantial interest. *[Sec. 1904]*	Provides mandatory funding of up to $4 million per year beginning in FY2009 for data mining, and requires periodic competition for the funds. Adds a new subsection to provide mandatory funds of up to $15 million per year over 4 years (FY2009-13) to upgrade USDA's computer technology for crop insurance. *[Sec. 12021]* Adopts Senate provision that prohibits farmers from collecting commissions as agents on policies in which their family has an interest, modifying the definitions of "family" and "compensation." *[Sec. 12005]*
Risk Management Research and Development			
USDA is required to reimburse an applicant for the R&D costs associated with developing a new plan of crop insurance that is approved by USDA *[7 USC 1522]* and with developing crop insurance education programs. *[7 USC 1524]* Annual mandatory funding is $15 million for R&D reimbursements and $25 million for contracting and partnerships. *[7 USC 1522(e)]*	Authorizes USDA to use no more than $30 million annually in mandatory funds for grants for R&D and education and information programs, of which $5 million is for underserved states. Stipulates criteria for which grants will be awarded. Requires USDA to enter into contracts to improve coverage for organic crops, and to address the needs of beginning and minority farmers. *[Secs. 11003-11006]*	Reduces annual mandatory funding for R&D from $15 million to $7.5 million, and for contracting and partnerships from $25 million to $12.5 million. Prohibits a surcharge on premiums for organic crops, unless greater loss history is confirmed. Establishes an alternative reimbursement grant process. Requires USDA to enter into contracts to expand coverage for organic crops, aquaculture, energy crops such as switchgrass, and to address the needs of beginning and minority farmers. *[Secs. 1917-1919, 1907]*	Adopts Senate provision to reduce mandatory funding for R&D to $7.5 million and for contracting/partnerships to $12.5 million. *[Sec. 12024]* Applicants with approved concept papers for a new policy can receive up to 50% of expenses in advance, and the balance upon approval. *[Sec. 12022]* Adopts House provision to enter into contracts to improve coverage for organic crops, and Senate provisions for energy crops and aquaculture, and other new provisions for poultry, beekeepers, and beginning farmers. *[Sec. 12023]* Requires USDA to develop risk management education programs for beginning, immigrant, socially disadvantaged, retiring, and transitioning farmers. *[Sec. 12026]*

PRIOR LAW/POLICY	HOUSE-PASSED BILL (H.R. 2419)	SENATE-PASSED SUBSTITUTE AMENDMENT (H.R. 2419)	ENACTED 2008 FARM BILL (P.L. 110-246)
Small Business Disaster Loan Program			
The Small Business Administration offers low interest, fixed-rate loans to small businesses to help them recover from economic injury caused by a natural disaster. ***[15 USC 636(b), (c), and (f)]*** *Note*: Farmers generally are not eligible for SBA loans, and instead receive assistance through USDA programs.	No comparable provision.	Makes many changes to SBA disaster loan program authority including raising the loan cap from $1.5 million to $2 million; extending assistance to non-profits; enhancing SBA and FEMA coordination of disaster response, among other provisions. ***[Secs. 11111-11161]***	Adopts Senate provision with changes. ***[Secs. 12051-12086]***
Supplemental Agriculture Disaster Assistance — see section on Trade and Tax Provisions (Title XV)			
"Sodsaver" Provisions Related to Farmland Conservation — see section on Conservation Programs (Title II)			
Other Crop Insurance Provisions			
No comparable provision.	Establishes a National Drought Council within USDA and national drought preparedness plans, including a Drought Assistance Fund to provide technical and financial assistance to states for mitigating drought risk. ***[Sec. 11012]***	No comparable provision.	Deletes House provision.
Ad-hoc emergency disaster payments are available to producers who experienced significant losses to a 2005, 2006, or 2007 crop. (Sec. 9001 of P.L. 110-28, as amended by P.L. 110-161).	Prohibits USDA from using production data from the sweet potato crop insurance pilot program in determining crop disaster payments for 2005 and 2006. ***[Sec. 11016]***	Similar to the House bill. Also requires USDA to extend the disaster application deadline for sweet potato growers, if necessary, to implement this provision. ***[Sec. 1927]***	Adopts Senate provision. ***[Sec. 12029]***
Authorizes USDA to create crop insurance pilot programs. ***[7 USC 1523]***	Mandates a sesame insurance pilot program for Texas. ***[Sec. 11011]***	Creates pilot programs for sesame ***[Sec. 1921]***, camelina ***[Sec. 1920]***, and enterprise/whole farm units ***[Sec. 1909]***.	Authorizes separate insurance pilot programs, for sesame, camelina, grass seed ***[Sec. 12025]*** and for enterprise and whole farm units (with modifications to the Senate provision) ***[Sec. 12011]***
No comparable provision.	No comparable provision.	Makes contract livestock producers eligible for crop insurance, if not covered by other policies. ***[Sec. 1916]*** Requires a USDA report within 180 days of enactment on issues regarding declining crop insurance yields, especially for perennials. ***[Sec. 1928]***	No provision addressing the eligibility of contract livestock producers. Adopts the requirement that USDA report on declining yield issues. ***[Sec. 12030]***

(Continued)

PRIOR LAW/POLICY	HOUSE-PASSED BILL (H.R. 2419)	SENATE-PASSED SUBSTITUTE AMENDMENT (H.R. 2419)	ENACTED 2008 FARM BILL (P.L. 110-246)
No comparable provision.	No comparable provision.	Defines an organic crop as any agricultural commodity organically produced consistent with Section 2103 of the Organic Foods Production Act of 1990 (7 U.S.C. 6502). *[Sec. 1901]*	Adopts Senate provision. *[Sec. 12001]*
TITLE XIII: COMMODITY FUTURES			
Authorization for the Commodity Futures Trading Commission (CFTC), a "sunset" agency established in 1974, expired on Sept. 30, 2005. In the past, Congress has used the reauthorization process to consider amendments to the Commodity Exchange Act, which provides the basis for federal regulation of commodity futures trading. The last reauthorization resulted in the enactment of the Commodity Futures Modernization Act of 2000 (CFMA), the most significant amendments to the CEA since the CFTC was created in 1974. *[7 U.S.C. 2(c)(2))]* Both chambers considered reauthorization bills in the 109th Congress; none was enacted.	No comparable provision.	Reauthorizes the CFTC. *[Sec. 13001-13204]* Clarifies the CFTC's jurisdiction over foreign exchange contracts offered to retail customers. Expands CFTC's authority over non-exchange "electronic trading facilities" where contracts based on metals and energy commodities are traded. If the CFTC determined that trading on such a market played a significant role in price discovery, the facility would have to comply with several core regulatory principles, including maintaining and enforcing rules to prevent price manipulation. Such markets would also have to publish data on prices and trading volume.	Reauthorizes appropriations for the CFTC (FY2008-FY2013). Amends the Commodity Exchange Act to (1) clarify CFTC jurisdiction over retail financial contracts based on foreign currencies, (2) make the CFTC's anti-fraud authority applicable to certain off-exchange or over-the-counter derivatives contracts, (3) increase civil monetary and criminal penalties for violations, (4) permit cross-margining of accounts in security futures and options, and (5) establish CFTC regulation over certain exchange-like trading facilities that are currently exempt from most regulation. *[Sec. 13001-13204]*
TITLE XIV: MISCELLANEOUS			
Section 32 Funding for Nutrition Programs			
Section 32 of the Act of August 24, 1935, authorizes a permanent appropriation equal to 30% of annual U.S. customs receipts, to be used by USDA only for: (1) encouraging the export of farm products through producer payments or other means; (2) encouraging the domestic consumption of farm products by diverting surpluses from normal channels or increasing their use by low-income groups; (3) reestablishing farmers' purchasing power. The Secretary of Agriculture has considerable	No comparable provision.	No comparable provision.	Provides explicit instructions on how each year's Section 32 money must be allocated, in order to fund the fresh fruit and vegetable program (*Sec. 4304* of the nutrition title). Essentially caps Section 32 "unobligated" funds — i.e., the amount the Secretary (through AMS) is permitted to spend after transfers for use in the child nutrition programs and to Commerce for fisheries activities. The program under Sec. 4304 is to be funded with a portion of these unobligated funds. *[Sec. 14222]* To fund a new

(Continued)

Prior Law/Policy	House-Passed Bill (H.R. 2419)	Senate-Passed Substitute Amendment (H.R. 2419)	Enacted 2008 Farm Bill (P.L. 110-246)
discretion in deciding how to achieve these broad objectives. *[7 U.S.C. 612c]*			requirement (in the nutrition title under Sec. 4305) to purchase whole grain products for the school lunch and breakfast programs, the Secretary must make available, in FY2009, $4 million in Section 32 money. *[Sec. 14222]*
Socially Disadvantaged and Limited Resource Producers			
Sec. 2501(a)(1) of the 1990 farm bill established Outreach and Assistance for Socially Disadvantaged Farmers and Ranchers and Limited Resource Farmers and Ranchers, authorizing USDA to carry out outreach and technical assistance to assist socially disadvantaged farmers and ranchers in: owning and operating farms and ranches; and in participating equitably in the full range of agricultural programs offered by USDA *[7 U.S.C. 2279]*	Amends program to specify that technical and outreach assistance program is to be used to enhance the coordination, outreach, technical assistance, and education efforts authorized under USDA programs. Mandates CCC funding of $15 million annually for FY2008-12. No more than 5% of funds are to be used for administrative expenses. *[Sec. 11201]*	Similar to the House bill, with technical differences. Authorizes appropriations of up to $50 million annually for FY2008-12. No more than 5% of funds are to be used for administrative expenses. *[Sec. 11052]*	Allows additional contracting authority and does not require matching funds. *[Sec. 11053]* Adopts Senate provisions with modification to delete language pertaining to renewal of contracts, review of proposals, coordination of with USDA's Office of Outreach, and additional contracting authority. Provides $75 million in mandatory funding for the Outreach and Assistance for Socially Disadvantaged Farmers and Ranchers and Limited Resource Farmers and Ranchers program. [*Sec. 14004*]
Sec. 2501(g)(1) of the 1990 farm bill directs USDA to improve service delivery on Indian reservations. *[7 U.S.C. 2279(g)(1)]*	Reauthorizes improved program delivery by USDA on Indian reservations. *[Sec. 11202*]	Similar to the House bill, with technical differences. *[Sec. 11054]*	Adopts House provision. *[Sec. 14001]*
No comparable provision in statute, but USDA guidance prohibits loan foreclosure when a pending claim of discrimination against USDA exists.	No comparable provision.	Places a moratorium on USDA farm loan foreclosures when there exists a pending claim of discrimination against USDA.	Adopts the Senate provision, with modification; places provision in different U.S. Code section. *[Sec. 14002]*
Sec. 2501A(c)(1) of the 1990 farm bill directs USDA to annually compute the participation rate of socially disadvantaged farmers and ranchers as a percentage of the total participation of all farmers and ranchers for each USDA program *[7 U.S.C. 2279]*	Directs USDA annually to compile program application and participation rate data on socially disadvantaged farmers and ranchers for each county and state. Data are to be made public. *[Sec. 11203*]	Similar to the House bill. Directs USDA, acting through USDA's Assistant Secretary for Civil Rights, to use the enhanced data collection *[Sec. 11056]*, but also to conduct oversight and evaluation of civil rights compliance. *[Sec. 11064*]	Adopts House provision. *[Sec. 14006]*

PRIOR LAW/POLICY	HOUSE-PASSED BILL (H.R. 2419)	SENATE-PASSED SUBSTITUTE AMENDMENT (H.R. 2419)	ENACTED 2008 FARM BILL (P.L. 110-246)
Sec. 7405 of the 2002 farm bill establishes the Beginning Farmer and Rancher Development Program; provides training, education, outreach/technical assistance initiatives. *[7 U.S.C. 3319f]*	Reauthorizes program, and provides $15 million in mandatory funding annually (FY2008-12). *[Sec. 11204]*	Authorizes $30 million in annual appropriations. Incorporates energy conservation efficiency and transition to organic farming into program, limiting grants to $250,000 each. *[Sec. 7309]*	Adopts Senate provision. Moves the program to Title VII (Research) and deletes several provisions. Adds $15 million in mandatory funding for FY2009 and $20 million annually FY2010-2012. *[Section 7410]*
No comparable provision.	Authorizes USDA to provide a receipt for service or denial of service upon request to any applicant for USDA loans and grants. *[Sec. 11205]*	Specifies that the USDA's Farm Service Agency and USDA's Natural Resources Conservation Service as the agencies subject to this provision; requires USDA receipt upon request. *[Sec. 11057]*	Adopts Senate provision. Adds USDA Rural Development to the agencies subject to the provision. *[Sec. 14003]*
No comparable provision.	Directs USDA to ensure that the Census of Agriculture and USDA's Economic Research Service (ERS) track socially disadvantaged and limited resource farmers and ranchers. *[Sec. 11206]*	Similar to the House bill. *[Sec. 11055]*	Adopts Senate provision. *[Sec. 14005]*
No comparable provision.	Directs USDA to prepare a plan to join the technical and support assistance for socially disadvantaged farmers/ranchers within the Office of Outreach, and to relocate USDA's office. *[Sec. 11208]*	No comparable provision.	Adopts House provision with modifications. *[Sec. 14013]*
No comparable provision.	Directs USDA to establish an Advisory Committee on Minority Farmers under USDA's Office of Outreach, which will be responsible for reviewing civil rights cases and ensuring they are processed in a timely manner. *[Sec. 11209]*	No comparable provision.	Adopts House provision with changes. Deletes components pertaining to review of civil rights enforcement, annual reporting on civil rights, and program review. Revises membership of the committee. *[Sec. 14008]*
No comparable provision.	Directs USDA to establish a Coordinator for Chronically Underserved Rural Areas, to be located in USDA's Office of Outreach and to direct resources to high need, poverty rural areas. *[Sec. 11210]*	No comparable provision.	Adopts House provision with an amendment to locate the Coordinator in USDA Rural Development instead of the Office of Outreach. *[Sec. 14118]*
No comparable provision.	No comparable provision.	Creates an Office of Small Farms and Beginning Farmers and Ranchers to	Adopts Senate provision with an amendment that the provision subsume

(Continued)

PRIOR LAW/POLICY	HOUSE-PASSED BILL (H.R. 2419)	SENATE-PASSED SUBSTITUTE AMENDMENT (H.R. 2419)	ENACTED 2008 FARM BILL (P.L. 110-246)
		coordinate USDA activities across agencies to ensure access to all USDA programs for small, beginning, and socially disadvantaged farmers and ranchers, representation in Agriculture Census data, and development of and access to enhanced outreach programs. *[Sec. 11088]*	this office into USDA's Office of Advocacy and Outreach. *[Sec. 14013]*
Pigford Discrimination Decision			
Commonly known as the "Pigford decision," pertains to a federal district court decision resolving a class action discrimination suit between USDA and black farmers. The suit claimed USDA discriminated against black farmers on the basis of race and failed to investigate or properly respond to complaints from 1983-1997. ***[Pigford v. Glickman, No. 97-1978 and No. 98-1693 (D.D.C. July 14, 2000)]***	Permits any claimant in the Pigford decision who has not previously obtained a determination on the merits of a claim, to petition in civil court to obtain such a determination (limits total amount of payment/debt relief pursuant to this court action to $100 million). Restricts USDA from starting foreclosure of a loan if the borrower is a Pigford claimant who can show that a pending foreclosure is related to a Pigford claim. *[Sec. 11312]*	Similar to the House bill, but with technical differences. Authorizes appropriations of such sums as necessary beyond the $100 million in mandatory funding. *[Sec. 5402]*	Adopts Senate provision with modifications. *[Sec. 14012]*
Agricultural Biosecurity			
No comparable provision.	No comparable provision.	Identifies Subtitle A of Title XI with the heading, "Agricultural Security."	Gives the subtitle the name "Agricultural Security Act of 2008." *[Sec. 14101]*
No comparable provision.	No comparable provision.	Defines the terms for purposes of program: agent, agricultural biosecurity, agricultural countermeasure, agricultural disease, agriculture, agroterrorist act, animal, department, development, plant, and qualified agricultural counter-measure. *[Sec. 11011]*	Adopts Senate provision, with modifications. Adds definition for agricultural disease emergency. *[Sec. 14102]*
No comparable provision, but USDA created a "Homeland Security Staff" after the events of September 11, 2001.	No comparable provision.	No comparable provision.	Creates an "Office of Homeland Security" within USDA to coordinate agroterrorism and agricultural disease efforts and to be a liaison with other agencies. Codifies the functions of the

(Continued)

Prior Law/Policy	House-Passed Bill (H.R. 2419)	Senate-Passed Substitute Amendment (H.R. 2419)	Enacted 2008 Farm Bill (P.L. 110-246)
			former homeland security staff. *[Sec. 14111]*
No comparable provision.	No comparable provision.	No comparable provision.	Creates an "Agricultural Biosecurity Communications Center" to collect and share information. *[Sec. 14112]*
No comparable provision, but Homeland Security Presidential Directive 9 (HSPD-9) instructs USDA and DHS to develop a National Plant Disease Recovery System and a National Veterinary Stockpile. ***[HSPD-9(18)]***	No comparable provision.	Establishes (1) a National Plant Disease Recovery System of countermeasures to respond to an outbreak of plant disease within a single growing season; and (2) a National Veterinary Stockpile of counter-measures, available to state veterinarians within 24 hours to leverage the strategic national stockpile. *[Sec. 11012]*	Deletes Senate provision.
No comparable provision.	No comparable provision.	Establishes a competitive grant program at USDA to stimulate R&D of agricultural countermeasures. Waves competitive process in emergencies. Provides for coordination with DHS grants and countermeasure development. Authorizes annual appropriations of $50 million for FY2008-12. *[Sec. 11013]*	Adopts Senate provision, with modifications to remove specificity, including for coordination with DHS. *[Sec. 14121]*
No comparable provision, but HSPD-9 instructs USDA to support the development and promotion of higher education programs for the protection of animal and plant health. ***[HSPD-9(20)]***	No comparable provision.	Establishes a veterinary workforce grant program at USDA to increase the number of veterinarians trained in biosecurity. Authorizes appropriations (such sums as necessary) for FY2008-12. *[Sec. 11014]*	Adopts Senate provision, with modification. Expands program to agriculture, veterinary medicine, and food science. *[Sec. 14122]*
No comparable provision.	No comparable provision.	Provides grants for biosecurity training programs in planning, preparedness and response for food science professionals/ veterinarians. Authorizes appropriations (such sums as necessary) for FY2008-12. Provide grants and low-interest loans to states to assess response capability for food science and veterinary biosecurity. Authorizes annual appropriations of $25 million for FY2008-12. *[Sec. 11015]*	Adopts Senate provision. *[Sec. 14113]*.

PRIOR LAW/POLICY	HOUSE-PASSED BILL (H.R. 2419)	SENATE-PASSED SUBSTITUTE AMENDMENT (H.R. 2419)	ENACTED 2008 FARM BILL (P.L. 110-246)
Prohibits importation of foot and mouth disease (FMD) virus, and limits research on FMD virus to locations outside of the U.S. mainland to lessen the likelihood that an accidental laboratory release of FMD might reach domestic animals. By statute, USDA must explicitly permit research on FMD virus to be performed on the mainland of the United States, and has not yet done so. ***[21 U.S.C. 113a]***	Allows explicitly for USDA to conduct FMD (and other hazardous virus) research on the U.S. mainland without issuing itself a permit. Prohibits anyone other than USDA from possessing certain viruses on a USDA-prescribed list, unless USDA issues a permit. But the prohibition would not apply to "select agents," and FMD virus is a select agent. ***[Sec. 7108]***	Requires USDA to issue a permit to DHS to possess and work with live foot and FMD virus at the proposed National Bio- and Agro-Defense Facility (NBAF), subject to compliance with USDA rules for handling "select agents." Leaves unchanged the current restrictions on possession of FMD and other dangerous viruses. ***[Sec. 11016]***	Adopts Senate provision, with modification. Refers to successor facility to Plum Island (rather than NBAF specifically), limits issuance of permit to one facility. ***[Sec. 7524]***
Sets civil penalties for violations of the Plant Protection Act: $50,000 in the case of any individual (civil penalty may not exceed $1,000 in the case of an initial violation for moving regulated articles not for monetary gain); $250,000 in the case of any other person for each violation; $500,000 for all violations adjudicated in a single proceeding; or twice the gross gain or gross loss for any violation, forgery, counterfeiting, unauthorized use, etc. ***[7 U.S.C. 7734(b)]***	No comparable provision.	Amends civil penalties as follows: $500,000 for each violation adjudicated in a single proceeding; adds a penalty of $1 million for each violation adjudicated in a single proceeding involving a genetically modified organism. Requires action on a violation no later than 5 years after the date of violation. ***[Sec. 11017]***	Adopts Senate provision with modification. No change in statute of limitations, expand scope of penalties, and clarify subpoena authorities. ***[Sec. 10203]*** Identical changes made to the Animal Health Protection Act. ***[Sec. 11012]***
No comparable provision.	No comparable provision.	Requires a DHS report on regulations for the possession of propane in certain quantities, including number of facilities, alternative security programs, and compliance costs. ***[Sec. 11070]***	Adopts Senate provision with modification. ***[Sec. 14206]***
Food Safety Commission			
Sec. 10807 of the Farm Security and Rural Investment Act of 2002 (P.L. 107-171) established a 15-member Food Safety Commission appointed by the President to make recommendations to enhance the U.S. food safety system. Provision not implemented. ***[21 U.S.C. 341 note]***	No comparable provision.	Establishes a Congressional Bipartisan Food Safety Commission to study and make recommendations to modernize food safety programs, including organizational and resource requirements emphasizing prevention and to be based on risk assessment and best-available science. Specifies membership requirements, meeting procedures and	Deletes Senate provision.

(Continued)

Prior Law/Policy	House-Passed Bill (H.R. 2419)	Senate-Passed Substitute Amendment (H.R. 2419)	Enacted 2008 Farm Bill (P.L. 110-246)
		timetables, and other reporting aspects. *[Sec. 11060]* Requires the President to review the report and submit proposed legislation based on recommendations. Expresses Sense of the Senate on the need for additional resources and direction for federal food safety agencies, for agreements between the U.S. and its trading partners, and for comprehensive food safety legislation. *[Sec. 11072]*	
Foods from Cloned Animals			
FDA had asked companies to voluntarily not introduce meat and milk from cloned animals and offspring until it completes a final risk assessment and guidance on their safety. FDA published the final risk assessment/guidance on 1/15/08; USDA has asked that the moratorium on cloned animals (but not offspring) continue until markets are educated on safety.	No comparable provision.	Prohibits FDA from issuing a final risk assessment and lifting the voluntary moratorium until completion of newly mandated National Academy of Sciences and USDA studies, respectively, on the safety and on the market impacts of introducing products from cloned animals. *[Sec. 7507]*	Deletes Senate provision.
Invasive Species			
A number of federal agencies, including USDA, have statutory responsibilities aimed at preventing the introduction of and controlling invasive species.	No comparable provision.	Sets forth new requirements to require cooperation among federal agencies and specifically Hawaii to prevent/control the spread of invasive species in the state; establishes expedited procedures for Hawaii to seek federal approval to adopt restrictions. Authorizes appropriations of such sums as necessary (FY2008-12). *[Sec. 11063]*	Deletes Senate provision.
USDA is authorized under several authorities to provide financial assistance to eradicate outbreaks of invasive pests and diseases, including on an emergency basis.	No comparable provision.	Clarifies that USDA may provide funds on an emergency basis to assist states in combating invasive pest and disease outbreaks for any appropriate period of years after the date of initial detection, as determined by USDA. *[Sec. 11078]*	Deletes Senate provision.

PRIOR LAW/POLICY	HOUSE-PASSED BILL (H.R. 2419)	SENATE-PASSED SUBSTITUTE AMENDMENT (H.R. 2419)	ENACTED 2008 FARM BILL (P.L. 110-246)
The Plant Protection Act (PPA) *[7 U.S.C. 7701 et seq.]* authorizes USDA's Animal and Plant Health Inspection Service (APHIS) to cooperate with states, localities and others to prevent the spread of and eradicate invasive pests and diseases, some which are affecting trees in urban areas, including the Asian Longhorned Beetle and the Emerald Ash Borer.	No comparable provision.	Establishes a revolving loan fund for eligible local governments to borrow up to $5 million (at 2% interest), to finance purchases of equipment to monitor, remove, dispose of and replace infested trees within their jurisdictions; spells out repayment terms. Authorizes appropriations of such sums as necessary *[Sec. 11090]* Requires USDA, in cooperation with states, to allow states to pass along cost-sharing assistance to local government for activities relating to invasive species infestations. *[Sec. 11091]*	Adopts Senate provision with changes to replace the term "invasive species" with the term "pest and disease." Moves provision to Title X (Horticulture and Organic Agriculture). *[Sec. 10205]* Generally adopts Senate language on cooperative agreements, also moving the provision to Title X. *[Sec. 10206]*
Animal Welfare Act			
No comparable provision under the Animal Welfare Act (AWA) as amended *[7 U.S.C. 2131 et seq.]*, which is intended to ensure the humane treatment of research animals, bred for commercial sale, exhibited to the public, or commercially transported; and to prevent animal fighting activities. Authorizes fines of up to $2,500 per violation; each violation and each day is considered a separate offense.	Amends the AWA to prohibit use of live animals for marketing medical devices. Increases the cap for AWA violations to $10,000 per violation, and specifies that each day, each violation, and each animal subject to a violation be considered a separate offense, among other things. *[Sec. 11316]*	No comparable provision on medical marketing.	Increases maximum fines for AWA violations from $2,500 to $10,000 per violation. Deletes House provision on medical device marketing. *[Sec. 14214]*
Sec. 7 of the AWA prohibits research facilities from buying dogs or cats except from certain persons regulated under the AWA.	Replaces Sec. 7 with new language on the definition of a person regulated under this section, and on permissible sources of dogs and cats for research facilities. Introduces an additional penalty of $1,000 for each violation of this section of the AWA. *[Sec. 11317]*	Same as the House provision with an additional provision directing that use of random source dogs and cats from "Class B dealers" is to be phased out within 5 years of enactment. *[Sec. 11079]*	Directs USDA to review "any independent reviews by a nationally recognized panel of experts" on Class B use by researchers, and to report on any recommendations applying to USDA. *[Sec. 14216]*
Sec. 26 of the AWA spells out a series of prohibited acts related to animal fighting and establishes penalties for violations. *[18 U.S.C. 49]* Enables the federal	No comparable provisions on animal fighting or commercial importation of young dogs.	Amends the AWA to strengthen prohibitions on dog and other animal fighting activities; defines a dog fighting venture; and appears to expand who can	Animal fighting provisions generally reflect Senate language. *[Sec. 14207]* Dog importation provisions generally reflect Senate language, with additional

(Continued)

Prior Law/Policy	House-Passed Bill (H.R. 2419)	Senate-Passed Substitute Amendment (H.R. 2419)	Enacted 2008 Farm Bill (P.L. 110-246)
government to collect costs incurred for caring for seized animals. No comparable AWA provision on importation of young dogs.		collect for costs of care of seized animals. Increases the maximum imprisonment from 3 to 5 years. *[Sec. 11076]* Amends the AWA to require HHS and USDA regulations prohibiting importing, for resale, unless they are at least 6 months of age, in good health, and have all necessary vaccinations (exemptions for research or veterinary treatment). *[Sec. 3205]*	limited exceptions for those imported into Hawaii. *[Sec. 14210]*
Other APHIS-Related Provisions			
FDA since 1975 has banned the sale of pet turtles under four inches long due to health concerns (i.e., the risk of children getting *Salmonella* infections from them). Does not ban sale of other pet reptiles and amphibians. *[21 C.F.R. 1240.62 (b)]*	No comparable provision.	Requires FDA to study the prevalence of *Salmonella* in legally-sold reptiles and amphibians in the U.S. compared with the level in pet turtles. If prevalence is similar, USDA is to take additional specified actions. *[Secs. 11101-11103]*	Deletes Senate provision.
USDA's Animal and Plant Health Inspection Service (APHIS) operates a program (other legislative authority) that provides certification as a service to U.S. exporters desiring to meet foreign sanitary and phytosanitary (SPS) requirements.	Requires USDA to coordinate fruit and vegetable market analyses with its Foreign Agricultural Service and with the private sector; requires USDA to list on the Internet the status of all plant-related export petitions, and provide SPS-related information. *[Sec. 11307]*	No comparable provision.	Adopts House language with changes to include a provision, in the Technical Assistance for Specialty Crops program (in Title III), requiring an annual USDA report on sanitary and phyto-sanitary trade barriers. *[Sec. 3203]*
The 1990 farm bill, as amended, authorizes APHIS to collect agricultural quarantine inspection (AQI) user fees for both APHIS and DHS's Customs and Border Protection services in connection with preclearance or the port-of-entry arrival of international passengers and of commercial vessels, trucks, loaded railroad cars, aircraft. *[21 U.S.C. 136a]*	No comparable provision.	Exempts, from AQI user fees, trucks transiting Canada between Alaska and the lower 48 states (and vice versa), so long as they remain sealed when in Canada. *[Sec. 11080]*	Deletes Senate provision.
The USDA-APHIS Wildlife Services program is charged with helping to alleviate wildlife damage to agricultural,	No comparable provision.	Sense of Senate that USDA's Wildlife Services program should not compete with or condone competition with the	Deletes Senate provision.

(Continued)

Prior Law/Policy	House-Passed Bill (H.R. 2419)	Senate-Passed Substitute Amendment (H.R. 2419)	Enacted 2008 Farm Bill (P.L. 110-246)
urban, and natural resources. *[7 U.S.C. 426]*		private sector in managing nuisance birds in urban areas *[Sec. 11085]*	
Miscellaneous Provision Related to Payments to Deceased Farmers — see section on Commodity Programs (above)			
Miscellaneous Rural Development Provisions			
No comparable provision.	No comparable provision.	Amends the Con Act to authorize a new Northern Border Economic Development Commission (VT, NY, NH, and ME); authorizes appropriations of $40 million annually (FY2008-12). *[Sec. 6034]*	Adopts Senate provision with modifications to authorize the Northern Border Development Commission. Authorizes 3 additional regional development commissions: Southeast Crescent Commission, the Southwest Border Regional Commission. Provides each with $30 million each FY2002-2012. *[Sec. 14217]*
No comparable provision.	Prohibits closure or relocation of USDA's Farm Service Agency (FSA), Natural Resources Conservation Service, and Rural Development Agency county offices until one year after enactment. *[Sec. 11306]*	Prohibits any expenditures to close "critical access county FSA offices" through 2012, unless approved by Congress. *[Sec. 11071]*	Adopts the House provision with modification. Extends prohibition to two years after enactment; allows exceptions and requires notice. *[Sec. 14212]*
No comparable provision.	Authorizes USDA to make grants to reduce the availability of anhydrous ammonia to curtail the production of methamphetamine (up to $15 million for) FY2008-12. *[Sec. 11308]*	Similar to the House bill, except it provides that a grant can be used either for a physical lock or a chemical substance. *[Sec. 11062]*	Adopts Senate provision. *[Sec. 14203]*
No comparable provision.	Authorizes USDA to make available to rural areas any excess and surplus USDA computers. *[Sec. 11303]*	No comparable provision.	Adopts House provision with amendment to ensure that the activities authorized under this section are in addition to, and would not replace, activities conducted under other existing authorities of USDA on property disposal. *[Sec. 14220]*
No comparable provision.	No comparable provision.	Provides emergency grants to community-based agencies to assist low-income migrant and seasonal farmworkers (for use in transportation,	Deletes Senate provision.

(Continued)

PRIOR LAW/POLICY	HOUSE-PASSED BILL (H.R. 2419)	SENATE-PASSED SUBSTITUTE AMENDMENT (H.R. 2419)	ENACTED 2008 FARM BILL (P.L. 110-246)
		food, clothing, housing). *[Sec. 11061]*	
No comparable provision.	Sense of Congress regarding "food deserts" or geographically isolated neighborhoods/communities with limited or no access to major grocery stores. Directs USDA, in coordination with other agencies/nonprofits, to conduct a national assessment. *[Sec. 11311]*	No comparable provision.	Adopts Senate provision with an amendment to move the provision to Title VII (Research), to define the term "food desert," and to include an authorization of appropriations for the national assessment. *[Sec. 7527]*
No comparable provision.	No comparable provision.	Authorizes USDA to make grants to community -based organizations to improve the supply, stability, safety, and training of farmworkers. *[Sec. 11066]*	Adopts Senate provision with an amendment to clarify the eligible services that may be provided with grant funds under the program. Specifies assistance may be provided to farm-workers who are citizens or otherwise legally present in the United States. *[Sec. 14204]*
No comparable provision.	No comparable provision.	Directs GAO, in consultation with other agencies and organizations to issue a report on access to health care for farmers. *[Sec. 11074]*	Deletes Senate provision.
Various sections of the Department of Agriculture Reorganization Act of 1994 pertain to the reorganization of USDA. *[7 U.S.C. 6933; 7 U.S.C. 7014]*	Directs USDA to create a Farmworker Coordinator (working in consultation with other agencies and organizations) to assist and support farmworkers and migrant seasonal workers. *[Sec.11207]*	Similar to the House bill, but amends different section of the Department of Agriculture Reorganization Act of 1994. *[Sec. 11059]*	Adopts Senate provision with an amendment to specify the Coordinator shall have responsibility for assisting farmworkers in becoming producers and landowners. *[Sec. 14013]*
No comparable provision.	No comparable provision.	Directs USDA to conduct a study on the economic impacts of local food systems and commerce. *[Sec. 11089]*	Deletes Senate provision.
Other Miscellaneous Title Provisions			
Provision pertaining to U.S. government claims collected from a person through administrative offset. *[31 U.S.C. 3716(e)]*	Eliminates the statute of limitations applicable to collection of debt by administrative offset. *[Sec. 11314]*	Similar to the House bill. *[Sec. 11069]*	Adopts House provision. *[Sec. 14219]*

(Continued)

PRIOR LAW/POLICY	HOUSE-PASSED BILL (H.R. 2419)	SENATE-PASSED SUBSTITUTE AMENDMENT (H.R. 2419)	ENACTED 2008 FARM BILL (P.L. 110-246)
Allows disclosure of financial information in certain circumstances. *[12 U.S.C. 3413]*	Allow financial institutions to disclose an individual's financial records to the government to prevent fraud. *[Sec. 11310]*	Similar to House bill. *[Sec. 11068]*	Adopts Senate provision. *[Sec. 14205]*
No comparable provision.	No comparable provision.	Requires reporting of USDA conference expenditures over $10,000. *[Sec. 11081]*	Adopts the Senate provision, with modification. *[Sec. 14208]*
No comparable provision.	No comparable provision.	Requires a report on ending childhood hunger in the U.S. by 2013. *[Sec. 11082]*	Deletes Senate provision.
No comparable provision.	No comparable provision.	Facilitates use of emergency funds for effects of Kansas tornado. *[Sec. 11083]*	Deletes Senate provision.
No comparable provision.	No comparable provision.	Requires a report on USDA programs with poor scores in the Program Assessment Rating Tool (PART). *[Sec. 11084]*	Deletes Senate provision. Managers statement encourages progress reports.
No comparable provision.	Permanent debarment from participation in USDA programs for fraud. *[Sec. 11304]*	No comparable provision.	Adopts House provision with changes. Debarment may be reduced to 10 years. Excludes debarment in food assistance programs. *[Sec. 14211]*
No comparable provision.	Prohibits discrimination against use of pesticide in conservation programs. *[Sec. 11305]*	No comparable provision.	Deletes the House provision.
Authorizes the USDA Graduate School as a non-appropriated fund instrumentality. *[7 U.S.C. 2279b]*	Prohibits USDA from operating the USDA Graduate School after Oct. 1, 2008. *[Sec. 11309]*	No comparable provision.	Adopts House provision with changes. Provides for the transition of the USDA graduate school to a non-government entity by Oct. 1, 2009. *[Sec. 14213]*
No comparable provision.	Requires a study of wastewater infrastructure near the U.S.-Mexico border. *[Sec. 11313]*	No comparable provision.	Deletes the House provision.

(Continued)

Prior Law/Policy	House-Passed Bill (H.R. 2419)	Senate-Passed Substitute Amendment (H.R. 2419)	Enacted 2008 Farm Bill (P.L. 110-246)
Title XV: Revenue & Offsetting Cost Provisions			
Supplemental Agriculture Disaster Assistance			
Congress periodically provides ad-hoc emergency disaster payments to crop and livestock growers to supplement income following a natural disaster. Most recently, Congress provided emergency supplemental assistance for 2005, 2006, or 2007 production losses. ***[Sec. 9001 of P.L. 110-28, as amended by P.L. 110-161]***.	No comparable provision.	Authorizes a permanent agricultural disaster trust fund that will fund a series of disaster programs that provide payments to crop and livestock growers who experience significant production losses in a USDA-declared disaster area. For FY2008-12, the program is funded through a transfer of 3.34% of annual customs receipts from the U.S. Treasury. Payments are made under four new programs: 1) Supplemental Revenue Assistance Payments (for crops); 2) Livestock Indemnity Payments; 3) Emergency Assistance for Livestock, Honeybees, and Farm-Raised Catfish; and; 4) Tree Assistance Program. ***[Sec. 12101]***	Adopts a variation of the Senate provision. For FY2008-11, five new disaster programs are authorized and funded through a transfer of 3.08% of annual customs receipts. The five new programs are: 1) Supplemental Revenue Assistance Payments (for crops); 2) Livestock Indemnity Payments; 3) Livestock Forage Disaster Program; 4) Emergency Assistance for Livestock, Honeybees, and Farm-Raised Catfish; and 5) Tree Assistance Program. ***[Sec. 15101]***
Sec. 9012 of the U.S. Troop Readiness, Veterans' Care, Katrina Recovery, and Iraq Accountability Appropriations Act of 2007 (P.L. 110-28) required that "in carrying out crop disaster and livestock assistance in this title, the Secretary shall require forage producers to have participated in a crop insurance pilot program or the Non-Insured Crop Disaster Assistance Program (NAP) during the crop year for which compensation is received." In August 2007, separate legislation was enacted (P.L. 110-80), removing this requirement of forage producers.	Amends Sec. 9012 by stating that "the purchase of a Non-insured Assistance Program (NAP) policy shall not be a requirement to receive any Federal livestock disaster assistance." ***[Sec. 11015]*** *Note*: The House farm bill was passed prior to enactment of P.L. 110-80, which removed the crop insurance or NAP purchase requirement for forage producers.	No comparable provision.	No comparable provision.

PRIOR LAW/POLICY	HOUSE-PASSED BILL (H.R. 2419)	SENATE-PASSED SUBSTITUTE AMENDMENT (H.R. 2419)	ENACTED 2008 FARM BILL (P.L. 110-246)
Customs User Fees			
Extension of custom user fees. Sec. 13031 of the Consolidated Omnibus Budget Reconciliation Act of 1985 ("COBRA") authorizes the Secretary of the Treasury to collect certain customs services fees. Customs user fees include passenger and conveyance processing fees and merchandise processing fees. Congress has authorized collection of the passenger and conveyance processing fees through December 27, 2014. The authorization for the collection of the merchandise processing fees is through December 27, 2014.	No comparable provision.	No comparable provision.	Amends Section 13031 of the Consolidated Omnibus Budget Reconciliation Act of 1985 to extend the passenger and conveyance processing fees through September 30, 2017, and extend the merchandise processing fees through November 14, 2017. ***[Sec. 15201]***
Other Revenue and Tax-Related Provisions			
Tax-Treaty Withholding Tax Rates. The U.S. Internal Revenue Code (IRC) applies a 30% withholding tax to interest, dividends, and similar "fixed and determinable" income payments made to foreign persons/firms residing outside the U.S. (nonresident aliens). ***[IRC Sec. 871]*** The tax rate is reciprocally reduced or eliminated in many cases by one of the many bilateral tax treaties the U.S. has signed (acknowledged in IRC Sec. 894).	Provides in IRC Section 894 that if a U.S. subsidiary makes a deductible payment to a foreign corporation that has a common foreign parent, and the withholding tax rate on the payment would be higher if the payment were made directly to the common parent, the higher rate will be applied. ***[Sec. 12001]***	No comparable provision.	Deletes House provision.
Corporate estimated tax payments. Corporations are required to make quarterly tax payments generally equal to a specified percentage of their estimated tax liability for the year. ***[IRC Sec. 6655]*** In 2005, P.L. 109-222 increased the amount of estimated tax payments due in the last quarter of federal FY2012 to 106.25% of the amount otherwise due and reduced firms' next required	Increases the amount of required corporate estimated tax payments falling in the last quarter of FY2012 to 115.75% of the amount otherwise required (a 1.25 percentage-point increase over P.L. 110-28) and reduces firms' next required payments by a corresponding amount. (Note, however, that P.L. 110-142, passed after the House approved H.R. 2419, increased the required payments by	Increases the amount of required corporate estimated tax payments falling in the last quarter of FY2012 to 121.5% of the otherwise required amount and reduces firms' next required payments by a corresponding amount. ***[Sec. 12506]***	Increases the amount of required corporate estimated tax payments falling in the last quarter of FY2012 to 122.25% of the otherwise required amount and reduces firms' next required payments by a corresponding amount. ***[Sec. 15202]***

(Continued)

Prior Law/Policy	House-Passed Bill (H.R. 2419)	Senate-Passed Substitute Amendment (H.R. 2419)	Enacted 2008 Farm Bill (P.L. 110-246)
payments by a corresponding amount. The provision shifted a portion of payments forward from FY2013 to FY2012. In 2007, P.L. 110-28 increased the amount thus shifted to 114.5% of the amount otherwise required.	1.50 percentage points, rendering the House-passed provision inapplicable. *[Sec. 13003]*		
Increase in information return penalties. The tax code requires information returns to be filed that report on a variety of transaction types. *[IRC secs. 6721 — 6723]* The returns are one of the tax system's tool for boosting tax compliance and for monitoring tax shelters. There are penalties for failure to file required information returns.	No comparable provision.	Increases penalties for failure to file information returns. *[Sec. 12508]*	Deletes Senate provision.
Economic substance doctrine. The "economic substance" doctrine is a judicial doctrine that has developed in regard to tax shelter cases. In general terms, it denies the use of tax benefits in transactions not having an economic substance not related to taxes.	No comparable provision.	"Clarifies" the economic substance doctrine by integrating a part of it into the IRC (generally Section 7701, relating to definitions). In cases where a court determines the economic substance doctrine is relevant, applies a two-part ("conjunctive") test to a transaction, requiring that (1) the transaction change the taxpayer's economic position in a meaningful way (an "objective" test); and (2) the taxpayer has a substantial non-federal-tax purpose for engaging in the transaction. *[Sec. 12521]*	Deletes this provision.
No comparable provisions.	No comparable provisions.	Other miscellaneous provisions: — Denies deduction for certain fines and penalties. *[Sec. 12507]* — Modifies penalties for failure to file partnership returns. *[Sec. 12511]* — Denies deduction for interest paid in certain tax-motivated transactions. *[Sec. 12523]*	Other miscellaneous provision: — Deletes Senate provisions.

(Continued)

Prior Law/Policy	House-Passed Bill (H.R. 2419)	Senate-Passed Substitute Amendment (H.R. 2419)	Enacted 2008 Farm Bill (P.L. 110-246)
Tax-Related Conservation Provisions			
Conservation Reserve Tax Credit. USDA provides a conservation reserve program under which farmers receive annual federal payments for contracting to establish resource-conserving covers on farmland (Title XII of the 1985 farm bill). There is no related tax credit, although taxpayers may deduct charitable donations made for conservation purposes under ***IRC Sec. 170***.	No comparable provision.	Creates a new elective conservation reserve tax credit that can be claimed in place of payments received under the conservation reserve program. *[Sec. 12201]*	Deletes Senate provision.
Application of Self-Employment Contributions Act (SECA) social security tax to conservation reserve payments. In December 2006, the Internal Revenue Service (IRS) issued a notice (Notice 2006-108) proposing to apply SECA tax ***[IRC Sec. 1401]*** to conservation reserve program payments.	No comparable provision.	Exempts conservation reserve program payments received by persons receiving social security benefits from SECA tax. *[Sec. 12202]*	Exempts conservation reserve program payments received by persons receiving social security benefits from SECA tax. *[Sec. 15301]*
Charitable contributions for conservation purposes. Deductible contributions of capital gain property are limited by ***IRC Sec. 170*** to generally 30% of an individual's income. Rules were temporarily liberalized (through 2007) for contributions of property for conservation purposes; contributions are deductible to the extent they exceed 50% of what is generally an individual's income over other contributions.	No comparable provision.	Makes permanent the temporary relaxation of the deduction cap for contributions of conservation property. *[Sec. 12203]*	Two year extension of the temporary relaxation of the deduction cap for contributions of conservation property. *[Sec. 15302]*
No comparable provision.	No comparable provision.	Creates a new tax credit for costs incurred or the loss in value to property to carry out a habitat management plan. The credit's rate varies with the length of the qualifying plan. *[Sec. 12204]*	Deletes Senate provision.

(Continued)

PRIOR LAW/POLICY	HOUSE-PASSED BILL (H.R. 2419)	SENATE-PASSED SUBSTITUTE AMENDMENT (H.R. 2419)	ENACTED 2008 FARM BILL (P.L. 110-246)
Deductions for endangered species recovery expenditures. No provision, but under ***IRC Sec. 175***, farmers are permitted to "expense" (deduct immediately) outlays for soil or water conservation rather than treating the outlays as capital expenditures — a treatment that would generally delay the deductions.	No comparable provision.	Permits farmers to treat outlays made pursuant to the Endangered Species Act in the same manner as soil or water conservation outlays (i.e., the outlays can be expensed). ***[Sec. 12205]***	Adopts Senate provision, except that provision is effective for expenditures paid or incurred after December 31, 2008. ***[Sec. 15303]***
Tax-exempt bonds for forest conservation. In general, interest on bonds issued by state and local governments is tax-exempt if used for governmental purposes (IRC Section 103). However, not all state and local bonds used to finance private activities are tax exempt; private activity bonds are subject to certain restrictions and are subject to caps. ***[IRC secs. 141-142]***	No comparable provision.	Establishes (in limited amounts) a new type of tax-exempt private activity bond for bonds whose proceeds are used to finance forest conservation. ***[Sec. 12211]***	Deletes Senate provision.
Deduction for qualified timber gain and timber real estate investment trust (REIT) provisions. For tax years beginning before 2011, individuals' long term capital gain is taxed at reduced rates (generally, 15%; section 1 of the tax code). Taxpayers are permitted to treat the cutting of timber as a sale eligible for capital gains treatment. ***[IRC Sec. 631]*** REITs are specially-defined "pass through entities" that are permitted to deduct dividends they distribute from taxable income, thus eliminating corporate-level tax from distributed income. ***[IRC secs. 856-859]*** REITs are subject to several minimum-distribution requirements and 75% of REIT gross income must consist of certain types of real estate income.	No comparable provision.	Adds new Section 1203 to the Internal Revenue Code, which permits taxpayers to elect to deduct 60% of timber gain from taxable income. (The remaining 40% would be taxed at ordinary-income rates.) Also includes several provisions affecting timber REITs, including clarification that timber gain is qualified real estate income for a REIT. ***[Sec. 12212-12217]***	Redesigns Section 1201 of the Internal Revenue Code, which reduces the rate of gain on qualified timber gains of corporations to 15%. REIT provisions follow the Senate amendment. ***[Sec. 15311-15315]***

Prior Law/Policy	House-Passed Bill (H.R. 2419)	Senate-Passed Substitute Amendment (H.R. 2419)	Enacted 2008 Farm Bill (P.L. 110-246)
No comparable provisions.	No comparable provisions.	Other miscellaneous provisions: — Exclusion for programs and payments for fish, wildlife, forest protection, and pest management. *[Sec. 12206]* — Elective tax credits for payments under conservation programs *[Sec. 12207]*	Deletes Senate provisions.
Tax-Related Energy Provisions			
Credit for cellulosic fuel production. ***IRC Sec. 40*** provides an alcohol fuels 3-part credit consisting of: (1) a 10¢ per gallon small-producer tax credit for ethanol production; (2) a 60¢ per gallon tax credit for alcohol used in a mixture of alcohol and gasoline or other fuels; and (3) a 60¢ per gallon credit for production of alcohol that is not used in a mixture. (The latter two credits are 51¢ per gallon in the case of ethanol.) The credit is scheduled to expire at the end of 2010.	No comparable provision.	Provides a fourth tax credit under section 40 for small producers who produce alcohol fuel from cellulosic materials (e.g., corn stover or switchgrass). The credit is $1.25 per gallon less the amount of small-producer ethanol credit claimed and the alcohol mixture credit claimed for ethanol. *[Sec. 12312]*	Provides a fourth tax credit under section 40, the Cellulosic Biofuel Producer Credit. The credit is $1.01 per gallon less the amount of small-producer ethanol credit claimed and the alcohol mixture credit claimed for ethanol. *[Sec. 15321]*
Alcohol Fuels Credit. For ethanol, the rates for the alcohol mixture credit and the alcohol credit are 51¢ per gallon rather than the 60¢ per gallon rate that otherwise applies.	No comparable provision.	Reduces the tax credit rate for ethanol credits to 46¢ from 51¢ per gallon. *[Sec. 12315]*	Adopts Senate provision with changes to modify the alcohol credit to reduce the ta credit rate for ethanol to 45¢ per gallon. *[Sec. 15331]*
Small-Producer Ethanol Credit. One component of the 3-part credit is the 10¢ per gallon tax credit for small ethanol producers. The credit is scheduled to expire year-end 2010.	No comparable provision.	Extends the small producer ethanol credit for two years through 2012. *[Sec. 12313]*	Deletes Senate provision.
As described in the preceding entries, IRC Sec. 40 provides a three-part tax credit for alcohol fuels. There is no comparable provision for a small producer credit for fossil-free alcohol.	No comparable provision.	Provides a fourth component of the alcohol fuels credit consisting of a 10¢ per gallon credit for fossil-free alcohol fuels. The credit terminates at the end of 2012. *[Sec. 12314]*	Deletes Senate provision.

Prior Law/Policy	House-Passed Bill (H.R. 2419)	Senate-Passed Substitute Amendment (H.R. 2419)	Enacted 2008 Farm Bill (P.L. 110-246)
Alternative Fuels Excise Tax Credits. Provides two alternative fuels tax credits that can be claimed against federal excise taxes: the alternative fuels tax credit and the alternative fuel mixture credit. ***[IRC secs. 6426-6427]*** The credits generally expire at the end of September 2009, but the credit for hydrogen expires at the end of September 2014.	No comparable provision.	Extends the two alternative fuels credits through December 2010 for all fuels other than hydrogen. ***[Sec. 12331]***	Deletes Senate provision.
Tax Credits for Biodiesel Fuels. ***IRC Sec. 40A*** provides income tax credits for biodiesel (scheduled to expire at the end of 2008): (1) biodiesel mixture credit; (2) biodiesel credit; and (3) the small producer agri-biodiesel credit. Other IRC provisions include an excise tax credit for biodiesel fuel mixtures and payments for biodiesel fuel mixtures.	No comparable provision.	Generally extends the biodiesel fuel credits and payment provisions for two years (through 2010). The small producer agri-biodiesel credit is extended through 2012. ***[Sec. 12321]***	Deletes Senate provision.
Renewable diesel fuel incentives. Provides a tax incentive of $1.00 per gallon for qualified renewable diesel fuel. It can be claimed as an excise tax credit ***[IRC secs. 6426-6427]***, an income tax credit ***[IRC Sec. 40A]***, or a government payment. It is scheduled to expire at the end of 2008.	No comparable provision.	Extends the renewable diesel fuel incentives for two years, through 2010. It also places a cap on the credit that can be claimed for co-produced fuel. ***[Sec. 12321]***	Deletes Senate provision.
No comparable provisions.	No comparable provisions.	Other miscellaneous provisions: — Wind credit ***[Sec. 12301]*** — Landowner incentive for electric transmission. ***[Sec. 12302]*** — Modified treatment of energy grants/loans. ***[Sec. 12303]*** — Expansion of depreciation for biofuel plants. ***[Sec. 12311]*** — Modified calculation of fuel credits. ***[Sec. 12316]*** — Extension of temporary duty on ethyl alcohol. ***[Sec. 12317]***	Other miscellaneous provisions: — Adopts Senate provision modifying the calculation of fuel credits. ***[Sec. 15332]*** — Adopts Senate provision extending the temporary duty on ethyl alcohol. ***[Sec. 15333]*** — Adopts Senate provision with changes to limit the drawback of duty on imported ethanol. ***[Sec. 15334]*** — Deletes other Senate provisions.

(Continued)

Prior Law/Policy	House-Passed Bill (H.R. 2419)	Senate-Passed Substitute Amendment (H.R. 2419)	Enacted 2008 Farm Bill (P.L. 110-246)
		— Limits on drawback of duty on imported ethanol. *[Sec. 12318]* — Treatment of qualified fuel mixtures. *[Sec. 12322]* — Extension of credit for installing alternative fuel refueling property. *[Sec. 12332]*	
Agricultural Tax Provisions			
Limitation on farm losses. No comparable provision, although the tax code's passive activity rules ***[IRC Sec. 469]*** may limit farm losses where the taxpayer does not materially participate in the farm business. Section 461 contains special rules governing when deductions can be claimed.	No comparable provision.	Modifies IRC Section 461 to limit deductible farm losses for the year in which losses are incurred to $200,000 in cases where the taxpayer receives Agriculture Program Payments or CCC loans. *[Sec. 12501]*	Adopts Senate provision with changes. Modifies IRS Section 461 to limit, for taxpayer other than C Corporations, deductible farm losses for the year in which losses are incurred to the greater of $300,000 ($150,000 for taxpayers electing married filing separate status) or the taxpayer's total net farm income over the preceding 5 years in cases where the taxpayer receives Agriculture Program Payments or CCC loans. *[Sec. 15351]*
Optional methods for self-employment tax. When farmers have low net income or losses from farming, they may use one of two optional methods for calculating net earnings from self-employment. This can help them earn the quarters of coverage necessary to qualify for Social Security benefits. Over time, the number of quarters of coverage earned under the optional method has been reduced by inflation. The optional methods are: (1) the farm optional method; and (2) the nonfarm optional method. ***[IRC Sec. 1402(a)]***	No comparable provisions.	For the self-employment tax, modifies the farm optional method so that electing taxpayers may be eligible to secure four credits of Social Security benefit coverage each taxable year by increasing and indexing the thresholds. The provision makes a similar modification to the nonfarm optional method. *[Sec. 12502]*	Adopts Senate provision increasing and indexing dollar threshold for computing net earnings from self-employment under the optional methods. *[Sec. 15352]*
Limitation on tax deferral for like-kind exchanges. No gain or loss is recognized if property used in a trade or business is exchanged for property of a like kind	No comparable provision.	Denies non-recognition treatment for like-kind exchanges that include unimproved real estate in cases where the owner receives Agriculture Program	Deletes Senate provision.

(Continued)

PRIOR LAW/POLICY	HOUSE-PASSED BILL (H.R. 2419)	SENATE-PASSED SUBSTITUTE AMENDMENT (H.R. 2419)	ENACTED 2008 FARM BILL (P.L. 110-246)
used in a trade or business. ***[IRC Section 1031]***.		Payments or CCC loans. ***[Sec. 12504]***	
Installment Sale Rules for Farm Property. Taxpayers in some cases are permitted to recognize income from installment sales gradually, as income is actually received. ***[IRC Sec. 453]*** Separately, taxpayers who recognize gain from the sale of depreciable assets are required to "recapture" a portion of the gain in accordance with depreciation deductions that have been claimed on the asset; recaptured gain is taxed at ordinary income tax rates rather than capital gains rates. ***[IRC secs. 1245 and 1250]*** Income that is treated as ordinary income under the recapture rules is taxed in the year of the property's disposition rather than in installments.	No comparable provision.	Repeals the immediate recognition of recapture income for sales of single-purpose agricultural or horticultural structures (i.e., barns and similar farm buildings). ***[Sec. 12402]***	Deletes Senate provision.
Tax-credit bonds for investment in qualified rural infrastructure projects. In general, interest on bonds issued by state/local governments is tax-exempt if used for governmental purposes. ***[IRC Sec. 103]*** Not all state and local bonds used to finance private activities are tax exempt; private activity bonds are subject to certain restrictions and are subject to caps. An additional type of tax-favored bond are "tax credit" bonds, for which the purchaser can claim a tax credit rather than receive a tax exemption. Several different types of tax credit bonds have been created in recent decades, including qualified zone academy bonds ***[IRC Sec. 1397E]***, clean renewable energy bonds ***[IRC Sec. 54]***, and gulf tax credit bonds ***[IRC Sec. 1400N]***.	No comparable provision.	Creates a new type of tax credit bond (rural renaissance bonds) under new Section 54A of the tax code for investment in certain rural projects, generally including utilities programs, distance learning or telemedicine projects, electric programs, rural telephone programs, broadband access programs, and rural community facility programs. The amount of such bonds that can be issued is limited to $400 million. ***[Sec. 12404]***	Deletes Senate provision.

(Continued)

PRIOR LAW/POLICY	HOUSE-PASSED BILL (H.R. 2419)	SENATE-PASSED SUBSTITUTE AMENDMENT (H.R. 2419)	ENACTED 2008 FARM BILL (P.L. 110-246)
No comparable provision.	No comparable provision.	Creates a new tax credit under new IRC Section 45P equal to 50% of a taxpayer's expenditures on the testing of new drugs for "minor" species (e.g., sheep and goats, but not cattle and poultry). *[Sec. 12406]*	Deletes Senate provision.
Reduced depreciation recovery period for certain farm machinery/equipment. Federal tax rules (generally *IRC Sec. 167* and related regulations) specify the rules regarding income tax deductions for depreciation of tangible capital assets, including recovery periods (years over which deductions must be spread) and recovery methods (portion of an assets cost that can be deducted in each year). Rules generally assign a recovery period of 10 years to farm machinery.	No comparable provision.	Shortens the recovery period for farm machinery and equipment to 5 years from 10 years. The shortened recovery period applies only to assets placed in service before 2010. *[Sec. 12407]*	Deletes Senate provision.
Reduced depreciation recovery period for all race horses. Generally assigns a recovery period of three-years to any race horse that is more than two years old at the time it is placed in service and a seven-year recovery period is assigned to any race horse that is two years old or younger at the time it is placed in service. *[IRC Sec. 168]*	No comparable provisions.	Shortens the recovery period for seven-year race horses to three-years. *[Sec. 12509(a)]*	Adopts Senate provision, except that the provision applies to any race horse that is two years old or younger at the time it is placed in service after December 31, 2008 and before January 1, 2014. *[Sec. 15344]*
"Aggie" bonds. Qualified small issue bonds up to $250,000, known as "aggie bonds," are tax-exempt bonds issued by State and local governments to finance private business manufacturing facilities or the acquisition of land and equipment by certain first-time farmers. *[IRC Sec. 144]*	No comparable provisions.	Increase loan limit for aggie bonds from $250,000 to $450,000 and adjust the limit for inflation after 2008. *[Sec. 12401]*	Adopts Senate provision to increase loan limit for aggie bonds. *[Sec. 15341]*
No comparable provision.	No comparable provisions.	Other miscellaneous provisions: — Allow IRC Sec. 1031 like-kind exchange treatment involving	Other miscellaneous provisions: — Adopts Senate provision on the like-kind exchange treatment

Prior Law/Policy	House-Passed Bill (H.R. 2419)	Senate-Passed Substitute Amendment (H.R. 2419)	Enacted 2008 Farm Bill (P.L. 110-246)
		mutual ditch, reservoir, or irrigation stock. *[Sec. 12403]* — Create an agricultural chemicals security tax credit. A 30% tax credit, subject to limits, for qualified chemical security expenses. *[Sec. 12405]* — Expensing of broadband internet access. *[Sec. 12408]* — Tax credit for energy-efficient motors. *[Sec. 12409]*	involving mutual ditch, reservoir, or irrigation stock. *[Sec. 15342]* — Adopts Senate provision creating the agriculture business security tax credit. *[Sec. 15343]* — Deletes other Senate provisions.
Other Provisions			
Key provisions covering income tax relief in times of disaster: casualty loss deductions *[IRC Sec. 165]*; deferral of gain from involuntary conversions *[IRC Sec. 1231]*; delayed filing deadlines; abatement of fines/fees (P.L. 109-73); and tax exemption for certain disaster relief payments. *[IRC Sec. 139]*	No comparable provision.	Provides several tax relief measures for areas in Kansas affected by the tornados of May 2007. Among the proposals are suspension of limits on certain casualty losses and relaxation of time requirements in the case of involuntary conversions. *[Sec. 12701]*	Adopts Senate provision. *[Sec. 15345]*
No comparable provisions.	No comparable provisions.	Other miscellaneous provisions: — Income treatment of Exxon Valdez litigation. *[Sec. 12801]* — Extension of special rule for charitable contributions of food inventory. *[Sec. 12802]* — Increase exclusion amount for mileage reimbursements to volunteers. *[Sec. 12803]* — Technical correction for treatment of stock basis in S corporations making charitable contributions. *[Sec. 12804]* — Payment test for pro-sports facility bonds. *[Sec. 12805]* — Rehabilitation credit and depreciation schedules to low-income housing for the elderly.	Adopts some provisions and deletes others: — Adopts Senate provision, with modifications to the tax treatment for forestry con-servation bonds. *[Sec. 15316]* — Adopts Senate provision modifying the awards authority for advanced coal-based electricity credits and gasification credits. *[Sec. 15346]* — Deletes other Senate provisions.

(Continued)

Prior Law/Policy	House-Passed Bill (H.R. 2419)	Senate-Passed Substitute Amendment (H.R. 2419)	Enacted 2008 Farm Bill (P.L. 110-246)
		[Sec. 12806] — Modifies awards authority for advanced coal-based electricity credits and gasification credits. *[Sec. 12807]* — Tax treatment for forestry conservation bonds. *[Sec. 12808]*	
Address deficiencies in the Haitian Hemispheric Opportunity through Partnership Encouragement Act of 2006 (HOPE I).	No comparable provisions.	No comparable provisions.	Establishes the Haitian Hemispheric Opportunity through Partnership Encouragement Act of 2008 (Hope II) to help Haitian industry attract new investment and create immediate jobs, generate income for workers to cover increased food costs and pay for other necessities, and continue to provide incentives to encourage the use of inputs manufactured by U.S. companies. *[Sec. 15401-15407 and 15410-15411]*
Extend The Caribbean Basin Economic Recovery Act, as amended by the United States-Caribbean Basin Trade Partnership Act (CBTPA).	No comparable provisions.	No comparable provisions.	Extends The Caribbean Basin Economic Recovery Act, as amended by the United States-Caribbean Basin Trade Partnership Act (CBTPA) through September 30, 2010. *[Sec. 15408-15409]*
Define "commercially interchangeable" in Section 313(j)(2) of the Tariff Act of 1930. Sec. 313(j) of the Tariff Act of 1930 provides for unused merchandise drawback. Unused drawback is permitted if imported merchandise is exported or destroyed within 3 years of import without being used in the United States. Pursuant to Sec. 313(j)(2) of the Tariff Act of 1930, domestic or imported merchandise that is commercially interchangeable with the imported merchandise may be substituted for the imported merchandise and drawback granted on the export or	No comparable provisions.	No comparable provisions.	The conference agreement amends Sec. 313(j)(2) of the Tariff Act of 1930, to provide a standard for what is considered to be "commercially interchangeable" for purposes of unused merchandise drawback for wine. *[Sec. 15421]*

(Continued)

PRIOR LAW/POLICY	HOUSE-PASSED BILL (H.R. 2419)	SENATE-PASSED SUBSTITUTE AMENDMENT (H.R. 2419)	ENACTED 2008 FARM BILL (P.L. 110-246)
destruction of the substituted merchandise within the 3-year period beginning on the date of importation. "Commercially interchangeable", however, is not defined.			

End Notes

[1] Concurrent Resolution on the Budget for Fiscal Year 2008, *Deficit-Neutral Reserve Fund for the Farm Bill* (H.Rept. 110-153, conference report, Section 307).

[2] For more information, see CRS Report RL33144, *WTO Doha Round: The Agricultural Negotiations*; CRS Report RL33697, *Potential Challenges to U.S. Farm Subsidies in the WTO*; and CRS Report RL33853, *Canada's WTO Case Against U.S. Agricultural Support.*

[3] USDA, *USDA's 2007 Farm Bill Proposals*, January 31, 2007, at [http://www.usda.gov/documents/07finalfbp.pdf]. Reports and other related materials are at [http://www.usda.gov/wps/portal/!ut/p/_s.7_0_A/7_0_1UH?navid=FARM_BILL_FORUMS]. Also see CRS Report RL339 16, *The USDA 2007 Farm Bill Proposal: Possible Questions.*

[4] USDA, "Johanns Unveils 2007 Farm Bill Proposals," Release No. 0020.07, January 31, 2007, at [http://www.usda.gov/wps/portal/usdahome].

[5] Office of Management and Budget (OMB), *Statement of Administration Policy, H.R. 2419*, at [http://www.whitehouse.gov/omb/legislative/sap/110-1/hr2419sap-r.pdf] (House, May 25, 2007) and [http://www.whitehouse.gov/omb/legislative/sap/110-1/hr2419sap-s.pdf] (Senate, Nov. 6, 2007).

[6] For information about what could have happened if a new farm bill was not enacted and various provisions of the 2002 farm bill were to expire, see CRS Report RL34154, *Possible Expiration of the 2002 Farm Bill.*

[7] Information on House and Senate Agriculture Committee hearings is at [http://agriculture.house.gov/hearings/index.html] and [http://agriculture.senate

[8] March 12 (P.L. 110-196), April 17 (P.L. 110-200), April 24 (P.L. 110-205), May 1 (P.L. 110-208), and May 14 (P.L. 110-231). Information about what could have happened if the new farm bill was not enacted and provisions of the 2002 farm bill had expired is discussed in CRS Report RL34 154, *Possible Expiration (or Extension) of the 2002 Farm Bill.*

[9] Etimates reflect the cost of the bills' mandatory programs only. The bills also include authorization of appropriations for discretionary programs not included in these estimates. The March 2007 baseline is used because the House, the Senate, and the conferees structured their provisions in relation to this baseline. If the March 2008 baseline were used, the bill's cost would be at least $4 billion over FY2008-FY2017.

[10] USDA, "USDA Implementation of the 2008 Farm Bill (as of June 17, 2008)." Information provided to House Agriculture Committee staff.

[11] J. Hagstrom, "USDA Says It Needs More Cash to Implement Farm Bill," National Journal, June 26, 2008.

[12] S. Wyant, "USDA Forges Ahead with Farm Bill Implementation," High Plains Journal, June 5, 2008, at [http://hpj .com/archives/2008/jun08/jun9/USDAforgesaheadwithfarmbill. cfm].

[13] USDA, "Farm Bill Implementation Status report for Congressional Staff," October 3, 2008. Information provided to House Agriculture Committee staff.

[14] Statements during a briefing to House Agriculture Committee staff, October 3, 2008.

[15] USDA, ERS, *The 2008 Farm Bill Side-By-Side Comparison*, at [http://www.ers.usda.gov/ FarmBill/2008/].

[16] Calculated as follows: $40,000 + $65,000 + $75,000 = $180,000 (doubled to $360,000).

[17] Calculated as follows: $40,000 + $65,000 = $105,000 (doubled to $210,000).

[18] A type of loan where farmers or processors pledge a commodity as collateral to obtain a loan from the Commodity Credit Corporation (CCC) at a commodity-specific, per-unit loan rate. The borrower may repay the loan, with interest, within a specified period and regain control of the commodity. Alternatively, the commodity can be forfeited to the CCC with no penalty if market prices fall below the loan rate at the end of the term. The government takes no recourse beyond accepting the commodity as full settlement of the loan.

[19] White House, Office of the Press Secretary, "Fact Sheet: Congress' Farm Bill Is Bad for American Taxpayers," May 9, 2008, available at [http://www.whitehouse.gov/news/releases/ 2008/05/print/20080509.html].

[20] Acreage in CREP — a subprogram within CRP — would be excluded from the CRP county acreage cap in order to encourage greater program participation.

[21] Funding is made available for contract under the former CSP program.

[22] A provision in the Senate bill to authorize artisanal cheese centers was deleted in conference.

[23] *National Institute for Food and Agriculture: A Proposal*, report of the Research, Education, and Economics Task Force of USDA, July 2004. The report is available at [http://www.ars.usda.gov/Research/Research.htm].

[24] A link to the USDA farm bill research proposal is at [http://www.usda.gov/wps/portal/!ut/p/_s.7_0_A/7_0_1UH?contentidonly=true&contentid=2007_Farm_Bill_Title7.xml].

[25] Available at [http://www.create-21 .org/].

[26] A more complete examination of the issues and legislative proposal is in CRS Report RS22693, *Agricultural Research, Education, and Extension in the 2007 Farm Bill.*

[27] In the 2002 farm bill, Congress established this program and provided a one-time transfer of $5 million in mandatory funds to help transition farmers to organic production.

[28] For more information, see CRS Report RL34555, *Speculation and Energy Prices: Legislative Responses,* and CRS Report RS22912, *The Enron Loophole.*

[29] For a more detailed description of the authorized payment formula, see the full page text box in CRS Report RL34207, *Crop Insurance and Disaster Assistance in the 2008 Farm Bill*.

[30] For additional information on the tax provisions of the House and Senate farm bills, see CRS Report RS22759, *Farm Legislation and Taxes in the 110th Congress*.

In: Farm Bill of 2008: Major Provisions and Legislative Action ISBN: 978-1-60741-750-7
Editor: Mary T. Conner

Chapter 3

CONSERVATION PROVISIONS OF THE 2008 FARM BILL

Tadlock Cowan[1], Ren Renée Johnson[2] and Megan Stubbs[3]
[1]Natural Resources and Rural Development
[2]Agricultural Policy
[3]Agricultural Conservation and Natural Resources Policy

SUMMARY

The 2008 enacted farm bill (Food, Conservation, and Energy Act of 2008, P.L. 110-246) reauthorizes almost all existing conservation programs, modifies several programs, and creates various new conservation programs. A new Conservation Stewardship program replaces the existing Conservation Security Program and a new Agricultural Water Enhancement Program under the Environmental Quality Incentives Program is also authorized with mandatory funding. Other new programs include the Chesapeake Bay Watershed Program and a "Sodsaver" provision to help preserve native sod, including virgin prairie in the Prairie Pothole National Priority Area. Significant modifications to existing programs include a reduction of the maximum enrolled acreage under the Conservation Reserve Program to 32 million acres and an increase in the cap for the Wetlands Reserve Program to over 3 million acres.

Other changes in the enacted bill include modifications to address eligibility requirements, program definitions, enrollment and payment limits, contract terms, evaluation and application ranking criteria, among other administrative issues. Eligibility is expanded for many programs and technical assistance under most programs is broadened to cover forested and managed lands, pollinator habitat and protection, and identified natural resource areas. Beginning, limited resource, and socially disadvantaged producers, specialty crop producers, and producers transitioning to organic production are also targeted for special consideration in many existing programs.

Estimated new spending on the conservation title—not including estimated conservation-related revenue and cost-offset provisions in the bill—is projected to increase by $2.7 billion over 5 years and $4.0 billion over 10 years. Total mandatory spending for the conservation title is projected at $24.3 billion over 5 years (FY2008-FY2012) and $55.2 billion over 10 years (FY2008-FY2017).

A comparison of conservation provisions in the enacted 2008 farm bill with existing law and the House and Senate farm bills is provided in the **Appendix**.

This report will not be updated.

The Food, Conservation, and Energy Act of 2008 (P.L. 110-246), the 2008 farm bill, reauthorizes almost all existing conservation programs, modifies several programs, and creates various new conservation programs. These changes address eligibility requirements, program definitions, enrollment and payment limits, contract terms, evaluation and application ranking criteria, among other administrative issues. In general, the conservation title includes specific changes that expand eligibility and delivery of technical assistance under most programs to cover more broadly, for example, forested and managed lands, pollinator habitat and protection, and identified natural resource areas, among other expansions. Producer coverage across most programs is also expanded to include beginning, limited resource, and socially disadvantaged producers; speciality crop producers; and producers transitioning to organic production. The enacted bill also creates new conservation programs to address emerging issues and priority resource areas, and also new subprograms under existing programs.

Estimated new spending on the 2008 farm bill's conservation title—not including estimated conservation-related revenue and cost-offset provisions in the bill—is projected to increase by $2.7 billion over 5 years and $4.0 billion over 10 years. Total mandatory spending for the title is projected at $24.3 billion over 5 years (FY2008-FY2012) and $55.2 billion over 10 years (FY2008-FY2017).

POLICY QUESTIONS AND ISSUES SHAPING THE CONSERVATION TITLE OF THE 2008 FARM BILL

Agricultural conservation became a significant and visible policy focus in the Food Security Act of 1985. Since then, questions and concerns about conservation program funding, policy objectives, individual program effectiveness, comparative geographic emphasis, and the structure of federal assistance have been recurring issues in the debate shaping each successive omnibus farm bill.[1] The 2008 farm bill is no exception. These long-standing issues arguably became even more apparent in this farm bill as they found their way into many individual program changes. This result may be a continuation of the more active profile taken by many conservation groups and their supporters in the 2002 farm bill. Unlike commodity programs, conservation program participation tends to be well represented by small and mid-size farming operations, according to the United States Department of Agriculture's (USDA) Economic Research Service. The programs also enjoy wider public support. In an environment of pronounced domestic budget constraint, however, conservation groups and producers found themselves competing with other farming interests for the necessary resources to expand and continue many conservation programs. Budget concerns aside, several other issues emerged in the debate leading to enactment of the 2008 farm bill: funding priorities and payment structure, geographic targeting, program complexity, the importance of large-scale conservation efforts, and measurement of costs and effectiveness. These general policy issues—in various forms—raised questions central to the deliberations and outcomes in the enacted conservation title of the farm bill.

Program Funding and Producer Payment Limits

Among the questions raised were: Should payment limits be program-specific, or for some combination of conservation programs? Would the imposition of payment limits change patterns of participation and effectiveness of conservation programs? What are the potential differences in saving under different payment limit options? Where should savings be allocated? Should each conservation program have the same payment limit? How will funding levels affect the existing backlog of interest in program participation that cannot be met?

Geographic Emphasis

Because agricultural production is concentrated in specific regions of the United States, agricultural conservation and environmental issues are not randomly distributed. Some areas and some natural resources reveal greater environmental impact from agricultural activities—actual and potential—than others. For example, pressures on farmland development from urban sprawl are more pronounced on the East and West Coasts. Although many areas of the United States have water quality/quantity concerns or air quality issues, the deterioration of certain watersheds and certain air sheds is more advanced in some places than in others. How is it determined where federal conservation spending would be most effective? Should certain locations (states, regions, or watersheds) be targeted? Does targeting particular producer groups provide an effective strategy in resource management? Should some types of agricultural production or resource concerns receive higher priority in evaluating applications?

Program Complexity

Members of Congress, conservation groups, the Administration, and individual producers raised questions about the desirability of proliferating conservation programs and the complexity of adhering to various regulations that govern the programs. Some producers might have some acreage under one program and other acreage under another program. On-farm wetland management may be regulated by one program, while the environmental management plan of retired acreage is under another program. The question arises whether there is avoidable duplication or other inefficiencies that significantly limit conservation and resource management effectiveness. Could existing programs be combined in certain ways? Would a consolidated effort targeting a few specific resource issues reduce the complexity?

Delivery Effectiveness

Along with the perceived complexity of agricultural conservation programs is the related issue of the effectiveness of program delivery. Does the conservation delivery capacity of USDA agencies need to be further supplemented through partnerships, relationships with other organizations, or expansion of the technical assistance provider system? What opportunities and problems would result if a large portion of staff in the responsible agencies retired in a short time period? Does USDA currently have the staff needed to administer conservation programs if they were all fully funded? How might more market-based solutions to resource conservation improve delivery efficiency and program effectiveness?

Regional and Multistate Conservation Efforts

Local planning and implementation of conservation programs through locally constituted councils is an important aspect of conservation management and a long-standing characteristic of U.S. local-federal relations. A significant question raised by conservation groups, however, is the extent to which conservation efforts would be more effective and more efficient if they took place at larger scales, for example, through regional or multistate resource planning and multistate conservation planning. Because the conservation programs are voluntary, concern was voiced that participation could decline if producers felt that conservation programs were more remote from local planning input. Using watersheds or river basin drainage areas as policy targets, for example, could reduce the planning control of any single area jurisdiction. The trade-off is seen in the efficiencies that potentially stem from pooling financial and planning resources and targeting particular resource issues affecting most if not all producers. The Chesapeake Bay region, the Great Lakes, and specific river basins with significant agricultural impacts are examples where larger scale planning may hold particular advantages.

Program Evaluation

How effective are federal conservation programs in improving environmental management of natural resources? Evaluation is an essential part of effective conservation management. Yet, developing practical and reliable indicators that can be used across programs and with different producers and different resources has been a challenge. Policymakers and stakeholders generally agree that conservation measures have been effective in reducing the environmental impact of agricultural activities, but at what cost, in what locations, and under what specific circumstances are much more difficult questions to answer. The recognized importance of establishing valid indicators has been an ongoing issue in environmental management. Research has advanced considerably in the area, to the point that integrating those findings within the structure of current programs is viewed by many to be more feasible than it was in the past.

Program Changes in the Enacted 2008 Farm Bill (P.L. 110-246)

The 2008 farm bill reauthorizes almost all current conservation programs, modifies several current programs, and creates various new conservation programs. The various conservation programs administered by USDA can be broadly grouped into land retirement and easement programs and so-called "working lands" programs. In general, land retirement and easement programs take land out of crop production and provide for program rental payments and cost- sharing to establish longer term conservation coverage, in order to convert the land back into forests, grasslands, or wetlands. Working lands programs provide technical and financial assistance to assist agricultural producers in improving natural resource conservation and management practices on their productive lands. The enacted 2008 farm bill also creates several new conservation programs under the bill's conservation and other titles. The following sections provide brief overviews of the Title II changes to agricultural conservation programs.

Land Retirement and Easement Programs

Major land retirement and easement programs include the Conservation Reserve Program (CRP), the Wetlands Reserve Program (WRP), the Grasslands Reserve Program (GRP), the Farmland Protection Program (FPP), among other programs.

Conservation Reserve Program

The Conservation Reserve Program (CRP) is a voluntary program that allows producers to enter into 10-15 year contracts that provide annual rental payments and financial assistance to install certain conservation practices and maintain vegetative or tree covers. Its purpose is to conserve and improve soil, water, and wildlife resources by converting highly erodible and other environmentally sensitive acreage to a long-term vegetative cover. CRP is the largest conservation program in terms of total annual funding. It is administered by USDA's Farm Service Agency (FSA) and is funded through the Department of Agriculture's (USDA) Commodity Credit Corporation (CCC). CRP also has several subprograms, one of which is the Conservation Reserve Enhancement Program (CREP), that are designated to support state and federal partnerships through incentive payments for installing specific conservation practices that help protect environmentally sensitive land, decrease erosion, restore wildlife habitat, and safeguard ground and surface water.[2]

The enacted 2008 farm bill (Secs. 2101-2111) caps CRP enrollment at 32 million acres, down from its current cap of 39.2 million acres. The managers report on the conference agreement states this reduction is "not ... an indicator of declining or reduced support for CRP"; however, in other sections of the report USDA is encouraged to assist producers who are transitioning from land retirement to working lands conservation. The farm bill makes certain program changes, including allowing USDA to address state, regional, and national conservation initiatives; providing incentives for beginning and socially disadvantaged farmers/ranchers to purchase CRP land from retiring farmers; allowing certain types of managed haying and grazing and installation of wind turbines on enrolled lands (but at reduced rental rates); requiring that program participants manage lands according to a conservation plan; requiring USDA to survey annually the per-acre estimates of county cash rents paid to CRP contract holders; clarifying the status of alfalfa grown as part of a rotation practice; and establishing cost-sharing rates for certain types of conservation structures.

The bill also amends the pilot program for wetland and buffer acres in CRP. Each state can enroll up to 100,000 acres up to a national maximum of one million acres. This maximum may be raised to 200,000 in each state following a review of the program. Eligible lands for the program include (1) wetlands that have been cropped three of the immediately preceding 10 crop years; (2) land on which a constructed wetland is to be developed to manage fertilizer runoff; and (3) land that has been devoted to commercial pond-raised aquaculture.[3]

Conditions under which managed haying and grazing on CRP acreage may occur have been modified. The farm bill allows managed harvesting and grazing in response to drought and routine grazing to control invasive species. Where routine harvesting is permitted, state technical committees are required to coordinate to ensure appropriate environmental management. In addition to managed harvesting, the installation of wind turbines on enrolled land is now permitted activity. Any of these permitted uses on CRP acreage will result in a rental payment reduction commensurate with the economic value of the authorized activity.

The enacted farm bill permits 50% cost share payments on land used for hardwood trees, windbreaks, shelterbelts, and wildlife corridors for contracts entered into after November 1990. Contracts extend from a minimum of two years up to four years. Funding of $100 million also is authorized to cover cost sharing for the thinning of trees to improve the management of natural resources on the land.

The 2008 farm bill modifies the criteria for evaluating CRP contract applications. Evaluation criteria include the extent to which a CRP contract application would improve soil resources, water quality, or wildlife habitat. The bill also allows the Secretary to establish different criteria in various states or regions that lead to improvements in soil quality or wildlife habitat. Preference in new CRP contracts will be given to land owners and operators who are residents of the county or a contiguous county in which the land is located.

The farm bill also establishes incentives to increase the participation of beginning and socially disadvantaged farmers and ranchers. It authorizes CRP contract modifications to assist these producers in leasing or purchasing land under a CRP contract from a retired or retiring farm owner or operator. The provision authorizes $25 million for assistance in making these land transfers.

Other enacted modifications to CRP include redefining the Chesapeake Bay region as a priority area without limiting the region to the states of Pennsylvania, Maryland, and Virginia. While the new program apples to all watersheds draining into the Chesapeake Bay, the Susquehanna, Shenandoah, Potomac, and Patuxent Rivers will get funding priority. A provision in the Trade and Tax Provisions Title (Section 15301) will permit retired or disabled farmers and ranchers to exclude CRP payments from self-employment taxes beginning January 2008.[4]

Wetlands Reserve Program

The Wetlands Reserve Program (WRP) provides long-term technical and financial assistance to landowners with the opportunity to protect, restore, and enhance wetlands on their property, and to establish wildlife practices and protection. It is a voluntary program administered by USDA's Natural Resources Conservation Service (NRCS).

The enacted 2008 farm bill (Secs. 2201-2210) increases the WRP maximum enrollment cap to over 3.014 million acres (up from an existing cap of 2.275 million acres), and expands eligible lands to include certain types of private and tribal wetlands, croplands, and grasslands, as well as lands that meet the habitat needs of specific wildlife species. The farm bill authorizes a new Wetlands Reserve Enhancement Program, to establish agreements with states similar to that for CREP, which includes a Reserved Rights Pilot program to explore whether reserving grazing rights is compatible within WRP. The bill makes certain program changes, including changing the payment schedule for easements; limiting wetland restoration payments; specifying criteria for ranking program applications; requiring that USDA conduct an annual survey of the Prairie Pothole Region starting with FY2008; and requiring USDA to submit a report to Congress on long-term conservation easements under the program.

Grasslands Reserve Program (GRP)

GRP is a voluntary program administered by USDA's Farm Services Agency (FSA) that helps landowners restore and protect grassland, rangeland, pastureland, and shrubland and provides assistance for rehabilitating grasslands.

The enacted 2008 farm bill (Sec. 2403) adopts a new acreage enrollment goal of an additional 1.22 million acres by 2012, with 40% of funds for rental contracts (10-, 15-, and 20-year duration) and 60% for permanent easements. Requirements for cooperative agreements similar to those under the Farmland Protection Program are also authorized for GRP easements. The farm bill modifies the terms and conditions of GRP contracts and easements to permit fire presuppression and the addition of grazing-related activities, such as fencing and livestock watering. Priority for enrollment is also given to certain expiring CRP lands, and tribal lands are made eligible. The bill does not include a Grassland Reserve Enhancement provision, as proposed in the House.

Farmland Protection Program (FPP)

FFP is a voluntary program administered by USDA's NRCS that provides matching funds to help purchase development rights for eligible farmlands to keep productive farm and ranchland in agricultural uses. USDA partners with state, tribal, or local governments and nongovernmental organizations to acquire conservation easements or other interests in land from landowners, and provides up to 50% of the fair market easement value of the conservation easement.

The enacted 2008 farm bill changes the program's purpose from protecting topsoil to protecting the land's agricultural use by limiting nonagricultural uses and including lands that promote state and local farmland protection (Sec. 2401). The federal share of easement costs are capped at 50%, with the land owner contributing 25% of the costs. The program is also restructured to emphasize longer term and renewable cooperative agreements. The bill also makes other technical changes to the program covering the program's administrative requirements, appraisal methodology, and terms and conditions, among other issues. It does not rename the program the Farm and Ranchland Protection Program, as the program is often referred to by USDA.[5] The bill provides additional budget authority for FPP of $743 million.

Working Lands Conservation Programs

Major working lands programs include the Environmental Quality Incentives Program (EQIP), the (renamed) Conservation Stewardship Program (CSP), the Agricultural Management Assistance (AMA) program, and the Wildlife Habitat Incentives Program (WHIP), among others.

EQIP and CSP are the two largest working lands programs, and received additional budget authority over five years under the 2008 farm bill of $3.4 billion for EQIP and $1.1. billion for CSP. The enacted farm bill did not include a Senate proposal that would have closely coordinated CSP and EQIP under the so-called Comprehensive Stewardship Incentives Program.

Environmental Quality Incentives Program (EQIP)

EQIP is administered by USDA's NRCS and provides technical and cost-share assistance to farmers and ranchers for promoting agricultural production and environmental quality by supporting the installation or implementation of structural and management practices on eligible agricultural land. EQIP includes a number of subprograms, including the Colorado River Basin Salinity Control, Conservation Innovation Grants, the Ground and Surface Water Conservation Program, and the Klamath River Basin.

The enacted 2008 farm bill (Secs. 2501-2510) expands the program to cover practices that enhance soil, surface and ground water, air quality, and conserve energy; it also covers grazing land, forestland, wetlands, and other types of land and natural resources that support wildlife. In evaluating applications, cost-effectiveness and comprehensive treatment of resource issues are given priority. The bill sets aside 5% of the EQIP spending for beginning farmers and ranchers and 5% for socially disadvantaged farmers and ranchers, providing up to 90% of the costs of implementing an EQIP plan for these farmers. It also provides payments to assist tribal or native corporation members, and producers transitioning to organic production.

The 2008 farm bill lowers the EQIP payment limit to $300,000 (down from $450,000) in any 6-year period per entity, except in cases of special environmental significance including projects involving methane digesters, as determined by USDA. Projects with organic production benefits are capped at $20,000 annually or $80,000 in any six-year period. The enacted bill retains the requirement that 60% of funds be made available for cost-sharing to livestock producers, including an incentive payments for producers who develop a comprehensive nutrient management plan.

The bill reserves $37.5 million of annual EQIP funds for the Conservation Innovation Grants program and modifies the grants to cover air quality concerns associated with agriculture (including greenhouse gas emissions). It also replaces the Ground and Surface Water Conservation Program within EQIP with a new Agricultural Water Enhancement Program (AWEP) to address water quality and quantity concerns on agricultural land, highlighting certain priority areas and providing additional mandatory funds for the program. The manager's report accompanying the farm bill suggests priority areas under the programs to include the Eastern Snake Plain Aquifer region, Puget Sound, the Ogallala Aquifer, the Sacramento River watershed, the Upper Mississippi River Basin, the Red River of the North

Basin, and the Everglades. AWEP prioritizes assistance to areas experiencing significant drought. The bill provides a total of $280 million through FY2012 for AWEP activities.

Funding for EQIP is authorized at $1.2 billion (FY2008), $1.337 billion (FY2009), $1.45 billion (FY2010), $1.588 billion (FY2011), and $1.75 billion (FY2012).

Conservation Security (Stewardship) Program

The pre-existing Conservation Security Program is a voluntary program administered by NRCS that provides financial and technical assistance to promote the conservation and improvement of soil, water, air, energy, plant and animal life, and other conservation purposes on tribal and private working lands.[6] However, the 2008 farm bill (Sec. 2301) phases this program out (except for existing contracts) and replaces it with a new and renamed Conservation Stewardship Program (CSP). The new CSP, beginning in 2009, will continue to encourage conservation practices on working lands, but will be different from the former program. It eliminates the three-tier approach, establishes 5-year rather than 10-year contracts, and requires direct attribution of payments, among other changes, thus requiring that USDA promulgate new rules for the program. More than $2 billion in funding is made available for existing contracts under the former CSP program.

Rather than the three-tier payment system, payments for new CSP contracts will be based on meeting or exceeding a stewardship threshold—the level of resource conservation and environmental management required to improve and conserve the quality and condition of at least one resource concern. The stewardship threshold also must be met for at least one priority resource concern identified at the state level as a priority for a particular watershed or area of the state. Payments are based on the actual costs of installing conservation measures, any foregone income, and the value of the expected environmental outcomes. The CSP reserves 5% of the funds each for beginning farmers and ranchers and disadvantaged farmers and ranchers (Sec. 2704). Monitoring and evaluation of the stewardship plan to assess the environmental effectiveness is also an element of the new CSP.

The bill sets a target of enrolling 12.8 million acres annually under the new CSP. Individual producer payments are limited to $200,000 in any 5-year period per entity. Rather than annual sign-ups for the program, CSP enrollment will be contracted on a continuous basis. The type of eligible lands is expanded to include priority resource concerns, as identified by states; certain private agricultural and forested lands; and also some nonindustrial private forest lands (limited to not more than 10% of total annual acres under the program). Technical assistance will also be provided to specialty crop and organic producers, along with a pilot testing of producers who engage in innovative new technologies or participate in on-site conservation research. Producers may also receive supplemental payments for resource-conserving crop rotations that provide specific environmental benefits such as improving soil fertility, thus reducing the need for irrigation. Program payments may not be used for the design, construction, or maintenance of animal waste storage or treatment facilities or associated waste transport or transfer devices.

Wildlife Habitat Incentives Program (WHIP)

WHIP is a voluntary program designed for the development and improvement of habitat primarily on private land. Through WHIP, USDA's Natural Resources Conservation Service provides both technical assistance and up to 75% cost-share assistance to establish and improve fish and wildlife habitat. The terms of WHIP agreements between NRCS and the participant generally are from 5 to 10 years from the date the agreement is signed. The 2008 farm bill (Sec. 2602) reauthorizes WHIP at current funding levels, but limits program eligibility to focus on "the development of wildlife habitat on private agricultural land, nonindustrial private forest land, and tribal lands," thus potentially excluding some previously covered areas (i.e., non-agricultural lands). It also allows USDA to provide priority to projects that address issues raised by state, regional, and national conservation initiatives. The manager's report emphasizes that the program address various specific wildlife initiatives at state and local levels. The 2008 farm bill provides \$425 million (FY2008-FY2012) and also increases the limit on cost-share payments to 25% for long-term projects. Payments to an individual entity are limited to \$50,000 per year.

Other Programs

The 2008 farm bill also authorizes a \$15 million increase in funding to \$20 million annually for FY2008-FY2012 for the Grassroots Source Water Protection Program (Sec. 2603). Funding is also increased to \$100 million (FY2009) for the Small Watershed Rehabilitation Program, to remain available until expended (Sec. 2803). The Grassroots Source Water Protection Program, a partnership between the Farm Service Agency (FSA) and the National Rural Water Association, is designed to help keep surface and groundwater water pollution from affecting drinking water. The Small Watershed Rehabilitation Program is administered by NRCS and works to rehabilitate older community dams.

The farm bill (Sec. 2801) also provides additional mandatory funding (\$15 million annually, FY2008-FY2012) for the Agricultural Management Assistance Program (AMA) and includes Hawaii as an eligible state under that program. AMA provides cost share assistance to agricultural producers to voluntarily address issues such as water management, water quality, and erosion control by incorporating conservation into their farming operations.

The enacted bill amends the Resource Conservation and Development Program (RC&D) to emphasize *locally led* planning processes and to provide assistance for implementing area plans (Sec. 2805). The RC&D program designates RC&D areas and assists the capability of elected and civic leaders to plan and carry out projects for resource conservation and community development.

Also reauthorized through FY2012 is the Farm Viability Program (Sec. 2402). The Farm Viability Program, as authorized in the 2002 farm bill, provides authority for USDA to provide grants to eligible entities for the purpose of carrying out farm viability programs. To date, Congress has not appropriated funds to implement the program.

The Great Lakes Basin Program for Soil Erosion and Sediment Control is a federal-state partnership providing demonstration and technical assistance projects throughout the Great Lakes region. The program is coordinated by the Great Lakes Commission in partnership with the NRCS, the Environmental Protection Agency and the U.S. Army Corps of Engineers. The enacted bill modifies the program to implement the recommendations of the Great Lakes Regional Collaboration Strategy (Sec. 2604).

Technical Assistance and Conservation Program Administration

Site-specific technical assistance (TA) is provided for producers and landowners in constructing and installing conservation and natural resource management technologies.

Producers often need TA in designing and implementing appropriate conservation strategies. The enacted 2008 farm bill makes changes in the TA component of various conservation programs to respond to these producer needs and clarifies the purposes of TA. The bill also requires a review of conservation practice standards and includes specific provisions to ensure that speciality crops, organic producers, and precision agricultural producers receive adequate conservation TA. Although NRCS provides TA directly, the enacted bill also authorizes a national certification process for third-party providers, including non-federal providers (Sec. 2706). In addition, the bill creates an Agriculture Conservation Experienced Services (ACES) Program to make use of the talents and skills of older, non-USDA employees (Sec. 2710). The bill further establishes state TA committees composed of various state conservation officials and agricultural producers for each state to assist in the implementation and technical aspects of conservation programs (Sec. 2711).

The bill authorizes a new Cooperative Conservation Partnership Initiative (CCPI) (Section 2707) as a component of the Conservation Technical Assistance program. It is authorized to target technical and financial resources on conservation priorities on agricultural and nonindustrial private forest land on a state, local, multistate, and regional basis. The manager's report especially encourages locally developed projects. Eligible CCPI programs include EQIP, CSP, WHIP, Great Lakes Basin Sediment Control, Conservation of Private Grazing Land, Chesapeake Bay Region, and Grassroots Water Conservation. The provision reserves 6% of the funding for these programs for initiatives under the CCPI and establishes criteria for prioritization of projects, including projects that provide innovative conservation methods.

Authorized Reports

The enacted 2008 farm bill directs USDA to prepare a number of statistical and evaluation reports regarding various conservation programs: (1) an annual report on conservation program enrollments and payments—those greater than $250,000—under the Wetlands Reserve Program, Farmland Protection Program, Grassland Reserve Program, and the Environmental Quality Incentives Program (for land having special environmental significance). The bill also requires a report on the new Agricultural Water Enhancement Program (Sec. 2705). Other required reports include one on the long term implications of conservation easements (Sec. 2210), an annual report on average county cash rental rates (Sec. 2110), and an appraisal of soil and water conservation programs (Sec. 2804).

New Conservation Programs

In addition to the changes made to existing agricultural conservation programs, the enacted 2008 farm bill also expands the range of USDA conservation activities by creating several new programs, including a program expanding conservation activities in the Chesapeake Bay region, a new state grants program, a provision to limit production on native sod, and a provision promoting market-based approaches to conservation.

Chesapeake Bay Watershed Program

This program (Sec. 2605) is targeted at conserving and protecting the Chesapeake Bay and the water sources that make up the watershed. It applies to all tributaries, backwaters, and side channels, including watersheds, draining into the Chesapeake Bay, but gives priority to the Susquehanna, Shenandoah, Potomac, and Patuxent Rivers. The bill authorizes $188 million in mandatory funding (FY2009-FY2012) and $438 million over 10 years (FY2009-FY2018).

Voluntary Public Access and Habitat Incentives Program

Also referred to as the "Open Fields" program, this program authorizes state grants to encourage land-owners to provide public access for wildlife-dependent recreation, subject to a 25% reduction for the total grant amount if the opening dates for migratory bird hunting in the state are not consistent for residents and non-residents. The bill provides $50 million in mandatory funds (FY2009-FY2012) for the program.

Sodsaver Provision

This 2008 farm bill provision (Sec. 12020) makes producers that plant an insurable crop (over 5 acres) on native sod ineligible for crop insurance and the noninsured crop disaster assistance (NAP) program for the first five years of planting. The conference agreement states that this provision may apply to virgin prairie converted to cropland in the Prairie Pothole National Priority Area, if elected by the state.

Environmental Services Markets

This new conservation provision (Sec. 2709) is intended to facilitate the participation of farmers and landowners in emerging environmental services markets, such as water and air quality, habitat protection, and carbon storage. The farm bill directs USDA to establish a framework for developing consistent standards and processes for quantifying environmental services from the agriculture and forestry sectors, but does not authorize funding for this effort.[7]

Miscellaneous Conservation Provisions

The enacted farm bill also includes changes to several additional programs. The Colorado River Basin Salinity Control Act makes funding available to support resource management activities targeting sources of salinity in the Colorado River (e.g., leaking wells, irrigation, industrial sources). The enacted farm bill establishes a Basin States Program for salinity control activities upstream of the Imperial Dam (sec. 2806).

The 2002 farm bill authorized that mandatory spending of $200 million be transferred to the Bureau of Reclamation to provide water to at-risk natural desert terminal lakes. Section 2807 of the enacted bill provides $175 million for desert terminal lakes, but also designates that part of this funding be used for land and water purchases in the Walker River Basin. The basin lies on California's eastern border with Nevada.

A provision in the Credit title of the enacted 2008 bill (Sec. 5002) establishes a new loan and loan guarantee program to assist producers in financing the cost to the producer for applying for needed conservation installations. The manager's report states that the loan program is only a complement to the assistance provided through the various conservation programs. The bill gives priority for these loans to beginning and socially disadvantaged farmers and ranchers, those converting to organic systems, and producers who need conservation assistance to address various compliance requirements.

Response from Conservation Interest Groups

The majority of agriculture and farmland conservation groups have responded favorably to the expanded provisions and increased funding for programs in the Conservation Title of the 2008 farm bill. Since enactment, a few national wildlife groups have expressed concern about changes to some provisions during the conference negotiations, which are perceived as providing fewer benefits to the protection of wildlife and wildlife habitat. Among the concerns expressed by these groups are the reduction in the CRP acreage enrollment cap reduction, easing of the requirements under the "sodsaver" provision, limitations on the types of lands eligible under Wildlife Habitat Incentives Program, and the new permanent disaster fund, which could encourage marginal land plantings, among other concerns.

APPENDIX. AGRICULTURAL CONSERVATION PROVISIONS IN THE ENACTED 2008 FARM BILL (P.L. 110-246) COMPARED WITH PREVIOUS LAW AND THE HOUSE- AND SENATE-PASSED BILLS (H.R. 2419)

Prior Law/Policy (P.L. 107-171) (unless otherwise Indicated)	House Bill (H.R. 2419)	Senate Amendment (H.R. 2419)	Enacted 2008 Farm Bill (P.L. 110-246)
Program Definitions and Funding			
Sec. 1201 of the Food Security Act of 1985 (FSA) (P.L. 99-198, or the 1985 farm bill), as amended, defines 18 terms. *[16 U.S.C. 3801]*	No provisions. (*Note*: some terms added by the Senate bill in this section are defined for specific conservation programs, as noted below.)	Adds definitions of beginning farmer or rancher, Indian tribe, nonindustrial private forest land, socially disadvantaged farmer or rancher, and technical assistance. Authorizes USDA to employ a test of net worth or other measure to qualify. *[Sec. 2001]*	Adopts Senate provision with changes. Removes the test of net worth. Adopts the 1990 farm bill definition of a socially disadvantaged farmer or rancher, with changes to define farm, integrated pest management, person and legal entity, and livestock. *[Sec. 2001]*
Sec. 1241(a) of the FSA, as amended, authorizes mandatory funding through FY2007 to carry out various conservation programs. *[16 U.S.C. 3841]* *Note*: Authorized funding levels for various programs are provided in individual program sections below.	Extends reauthorization through FY2012 with funding specified for CSP, FPP, EQIP, and WHIP. *[Sec. 2401(a)]*	Extends reauthorization through FY2012 with funding specified for CSP, FPP, EQIP, WHIP, GRP, and the Voluntary Public Access and Habitat Incentives Program. *[Sec. 2401(a)]*	Extends reauthorization through FY2012 with the following in additional new budget authority: CSP ($1.1 billion); EQIP ($3,393 million); and FPP ($773 million). *[Sec. 2701]*
Highly Erodible and Wetland Conservation			
Secs. 1211-1212 of the FSA, as amended, make violators of the conservation compliance program ineligible for certain program benefits, with some exceptions from full loss of eligibility. *[16 U.S.C. 3811-3812a, 3812f]*	No comparable provision.	Adds a second level of review by the state or district director, with technical concurrence from USDA's Natural Resources Conservation Service (NRCS) if the Secretary has determined that this exception should apply. *[Sec. 2101]*	Adopts Senate provision, providing for review of good faith determinations related to highly erodible land conservation. *[Sec. 2002]*
Secs. 1221-1222 of the FSA, as amended, makes swampbuster ineligible for certain program benefits, with some exceptions from full loss of eligibility. *[16 U.S.C. 3821-3824, 3822h]*	No comparable provision.	Add a second level of review by the state or district director, with technical concurrence from NRCS if the Secretary has determined that this exemption should apply. *[Sec. 2201]*	Adopts Senate provision, providing for review of good faith determinations related to wetland conservation. *[Sec. 2003]*

(Continued)

Prior Law/Policy (P.L. 107-171) (unless otherwise Indicated)	House Bill (H.R. 2419)	Senate Amendment (H.R. 2419)	Enacted 2008 Farm Bill (P.L. 110-246)
Comprehensive Conservation Enhancement Program			
The 1990 farm bill amended Sec. 1230 of the FSA to establish a program later renamed the Comprehensive Conservation Enhancement Program (CCEP). The CCEP, which includes the Conservation Reserve Program (CRP), Wetlands Reserve Program (WRP), and the Environmental Quality Incentives Program (EQIP), promotes long-term protection for environmentally sensitive lands through easements and technical/ financial assistance. *[16 U.S.C. 3830]* *Note:* Administration of CCEP, the subject of sec. 1243, is described below.	No comparable provision. (*Note:* Amendments to Sec. 1243 described below in the "other conservation programs" subsection.)	Deletes Section 1243 in current law, and moves some provisions, amended, into this (and other) sections. Extends CCEP through FY2012. Makes changes that reduce administrative burdens, streamline the application process, and promote partnerships. Deletes EQIP from CCEP and adds the Healthy Forests Reserve Program. Adds a new exception whereby USDA may exceed the enrollment limitation when a state or local regulation prohibits agricultural water use, requiring USDA to enroll the land within 180 days of receiving a request and pay a rental rate that reflects the rate prior to implementing the regulation. *[Sec. 2301]*	Does not reauthorize the CCEP. The Healthy Forest Reserve Program is retained in the forestry title *[Sec. 8205]*; the county acreage cap is addressed elsewhere *[Sec. 2708]*. The agreement adopts a provision to exclude CREP acreage and continuous CRP acreage from the 25% cap if the county government concurs, and further specifies this provision is separate and distinct from the existing waiver authority. Additional guidance is provided in the Managers statement.
The 1990 farm bill amended Sec. 1243 of the FSA to authorize administration of CCEP. Provisions include avoiding duplication of required conservation plans, limiting enrollment under CRP and WRP to 25% of the cropland in a county, protecting the interests of share croppers and tenants, allowing approved sources to provide technical assistance, and using up to 5% of the funds from the mandatory funded conservation programs to foster cooperation through partnerships. *[16 U.S.C. 3844]*	Amends administration provisions by moving sections on acreage enrollment limits, tenant protection, and obtaining technical assistance. Establishes a new Cooperative Conservation Partnership Initiative to carry out projects/initiatives using competitive (2-5 years) grants. Specifies 14 criteria to be used in reviewing applications and 9 project priorities. Specifies duties of participant and USDA. Specifies program will be funded with 10% of funds for CSP, EQIP, and WHIP. The federal share for each project will be at least 75% of costs; 90% of the funds will be allocated at the state level (incentives and bonus payments may be used for specified purposes). Limits administrative costs to 5% of any grant. *[Sec. 2403]*	Amends to streamline application process, add new endangered species provisions, and establish new partnerships and cooperation projects for special projects (up to 5 years) with multiple producers and eligible partners to address state conservation recommendations. Specifies five project purposes, lists application contents, and identifies USDA's duties and priorities when selecting projects (including 14 priority water project areas); also requires monitoring and evaluation. Specifies funding of 10% of the mandatory funds allocated to each state (except CRP, CSP, WRP, and the new Conservation Stewardship Program), with 75% of funds for intra-state and 25% for multi-state projects. *[Sec. 2405]*	Adopts House provision with changes and names the initiative the Cooperative Conservation Partnership Initiative (CCPI). *[Sec. 2707]* Allows USDA to make consider local circumstances, goals, and objectives, and provides for adjustments to provide producers preferential enrollment in the applicable program as part of the special project. Applies to all USDA conservation programs except CRP, WRP, FPP and GRP. The stated intent is to provide for applications that include innovative combinations of covered initiative programs, and applications that might work in tandem with the enhancement programs under CRP or WRP. Additional guidance is provided in the Managers statement.

(Continued)

Prior Law/Policy (P.L. 107-171) (unless otherwise Indicated)	House Bill (H.R. 2419)	Senate Amendment (H.R. 2419)	Enacted 2008 Farm Bill (P.L. 110-246)
Conservation Reserve Program			
Sec. 1231(a-d) of the 1985 farm bill (FSA) authorizes the CRP; the program is currently authorized through CY2007 at 39.2 million acres. *[16 U.S.C. 3831(a-d)]*	Extends authorization through CY2012; retains current acreage enrollment limit (39.2 million acres). *[Sec. 2101(a-b)]*	Extends authorization through CY2012. Retains current acreage enrollment limit. Adds pollinator habitat to the general purposes. Expands eligible land to include some types of marginal pasture-land and land enrolled in a new flooded farmland program. *[Sec. 2311(a-c)]*	Extends authorization through CY2012, and allows USDA to address issues under State, regional, and national conservation initiatives. Caps enrollment at 32 million acres *[Sec. 2103]*. Clarifies that alfalfa grown as part of a rotation practice is a commodity for crop history purposes. *[Secs. 2101-2102, 2105]* Provides for pollinator habitats. *[Secs. 2706, 2708]*
Sec. 1231(f) of the FSA, as amended, lists priority areas as the Chesapeake Bay region (PA, MD, VA) , the Great Lakes Region, and Long Island Sound. *[16 U.S.C. 3831f]*	Deletes states but retains Chesapeake Bay region. *[Sec. 2101(b)]*	Similar to the House bill; also adds to the list the prairie pothole region, the Grand Lake St Mary's Watershed, and the Eastern Snake Plain Aquifer. *[Sec. 2311(d)]*	Adopts House provision to include all States that make up the Chesapeake Bay Region as the Conservation Priority Area. *[Sec. 2104]*
Sec. 1231(h) of the FSA, as amended, authorizes a one million acre pilot program within the CRP for wetlands and buffer areas. *[16 U.S.C. 3831h]*	Extends program through CY2012. *[Sec. 2101(e)]*	Extends program through CY2012; expands eligibility to include certain shallow water areas and certain agricultural drainage water treatment collection areas, and expands the eligible buffer acreage. Directs USDA to establish the maximum size of the buffer acreage to be enrolled along with eligible lands. Increases the maximum wetland size to 40 contiguous acres and makes all acres eligible for payment. *[Sec. 2311(e)]*	Adopts Senate provision with changes. Caps enrollment at 100,000 acres in any State and 1 million acres total. Adds conforming changes to the Emergency Forestry Conservation Reserve Program. Expands enrollment of wetland and buffer acreage to include land that had been cropped during 3 of 10 crop years prior to 2002 and after 1990 and is subject to a natural overflow of a prairie wetland. *[Sec. 2106]*
Sec. 1232(a)(7) of the FSA, as amended, specifies a duty of participants is limiting commercial uses, including haying and grazing on enrolled lands; allows managed haying/grazing under certain circumstances. *[16 U.S.C. 3832a(7)]*	Allows managed haying and grazing to control invasive species, and adds detail on allowed uses, enrolled lands, and adjustments to annual contract payments. *[Sec. 2101(f)]*	Allows managed haying and grazing to control invasive species and permits managed haying and grazing that is a part of a conservation plan. *[Sec. 2311(h)]*	Adopts House provision with changes, allowing for routine grazing, including grazing to control invasive species. Additional guidance is provided in the Managers statement. *[Sec. 2108]* Grants "management on land" should not result in a reduced payment, if done in accordance with the contract. *[Sec. 2107]*
Sec. 1234(c) of the FSA, as amended, establishes a framework for calculating	Requires USDA to conduct and make available an annual survey of dryland and	Similar to the House bill; also requires USDA to give preference to local owners or	Adopts Senate provision with changes, adding a new requirement that USDA

(Continued)

Prior Law/Policy (P.L. 107-171) (unless otherwise Indicated)	House Bill (H.R. 2419)	Senate Amendment (H.R. 2419)	Enacted 2008 Farm Bill (P.L. 110-246)
annual rental payments. *[16 U.S.C. 3834c]*	irrigated cropland cash rental rates in all counties with more than 20,000 acres of crop and pasture land. *[Sec. 2101(g)]*	operators when considering competing offers providing equivalent benefits. *[Sec. 2311(j)]*	give priority to offers from local residents if conservation benefits are equivalent among offers. *[Sec. 2110]*
Sec. 1235(e) of the FSA, as amended, allows USDA to terminate CRP contracts after 5 years if contract was in effect before 1/1/95. *[16 U.S.C. 3835e]*	Allows USDA to terminate any contract after 5 years, but prohibits terminating contracts for land enrolled under a continuous signup. *[Sec. 2101(l) and (j)]*	Allow USDA to terminate a contract if a retired or disabled producer has endured financial hardship because of taxes on rental payments. *[Sec. 2311(k)]*	Adopts House provision. *[Sec. 2111]*
No comparable provision.	No comparable provision.	No comparable provision.	Specifies a 50% federal share of cost sharing payments relating to trees, windbreaks, shelterbelts, and wildlife corridors. *[Sec. 2109]*
The 2002 farm bill amended Sec. 1244(a) of the FSA to authorize USDA to provide incentives to beginning farmers/ranchers and Indian tribes to participate in con-servation programs. *[16 U.S.C. 3844(a)]*	Facilitates the transfer of CRP land from a retiring owner to a beginning / socially-disadvantaged producer to return land to production, and allows new owner to begin land improvements or start organic certification process one year before CRP contract expires. *[Sec. 2101(h)]*	No comparable provision. (*Note:* Support for socially disadvantaged and limited resource farmers/ranchers are in other bill sections.)	Adopts House provision with modifications. *[Sec. 2111]*
No comparable provision.	No comparable provision.	Creates new Flooded Farmland Program for the Prairie Pothole region within the CRP. Allows continuous enrollment. Eligible land parcels must exceed 5 acres, been incapable of production preceding 3 crop years, have a cropping history, and have no natural outlet. *[Sec. 2312]*	Deletes this section and modifies CRP and WRP to accomplish the intent of the Senate amendment, including expanding eligible lands under the CRP pilot program for wetlands and buffer areas *[Sec. 2106]* and expands eligible lands under WRP. *[Sec. 2201]*
No comparable provision.	No comparable provision.	Creates new Wildlife Habitat Program for CRP participants with established softwood pine stands using management practices that benefit wildlife (contracts up to 5 years). Program ends September 30, 2011. *[Sec. 2313]*	Deletes this section and modifies CRP to accommodate the intent of the Senate amendment. Additional guidance is provided in the Managers statement.

(Continued)

Prior Law/Policy (P.L. 107-171) (unless otherwise Indicated)	House Bill (H.R. 2419)	Senate Amendment (H.R. 2419)	Enacted 2008 Farm Bill (P.L. 110-246)
Wetlands Reserve Program			
The 1996 farm bill amended Sec. 1237(a) of the FSA to authorize WRP, stating its purpose to restore and protect wetlands. *[16 U.S.C. 3837a]*	Adds to the purposes to create and to enhance wetlands, and to purchase floodplain easements. *[Sec. 2102(a)]*	No comparable provision.	Amends purposes to restore, protect, or enhance wetlands on private or tribal lands. *[Sec. 2201]*
Sec. 1237(b) of the FSA, as amended, sets maximum enrollment at 2.275 million acres, with an annual calendar year enrollment goal of 250,000 acres. *[16 U.S.C. 3837b]*	Sets maximum enrollment at 3.605 million acres. Sets an annual fiscal year enrollment goal of 250,000 acres, of which not more than 10,000 acres may be flood plain easements. *[Sec. 2102(b)]* New section adds language authorizing WRP from FY2008-12. *[Sec. 2402(h)]*	Sets annual fiscal year enrollment goal of 250,000 acres, with no enrollment after FY2012. *[Sec. 2321]*	Sets maximum enrollment at 3.041 million acres. Sets annual fiscal year enrollment goal of 250,000 acres through FY2012. *[Sec. 2202-2203]*
Sec. 1237(c) of the FSA, as amended, establishes requirements for eligible lands through 2007. *[16 U.S.C. 3837c]*	Adds riparian areas to eligible wetlands, and makes eligible floodplain land flooded in the past calendar year or at least twice in the past 10 years, and land that contributes to flood water storage, flow, or erosion control. *[Sec. 2102(c)]*	Reauthorizes through FY2012. *[Sec. 2321(2)]*	Expands eligible lands under WRP to include cropland or grassland that was used for agricultural production prior to flooding from the natural overflow of a closed basin lake or pothole. *[Sec. 2203]* Adds terms to "meet habitat needs of specific wildlife species." *[Sec. 2204]*
Sec. 1237(e) of the FSA, as amended, identifies ineligible land to include lands already planted to timber in the CRP. *[16 U.S.C. 3837e]*	Expands ineligible lands to include floodplains where restoration practices would not be productive or the land is already protected. *[Sec. 2102(d)]*	No comparable provision.	Expands ineligible lands to include farmed wetland or converted wetland, together with the adjacent land that is functionally dependent on the wetlands, except wetlands converted before December 23, 1985. *[Sec. 2203]*
Sec. 1237A(f) of the FSA, as amended, states compensation to be paid in cash (in 5 to 30 payments) and not to exceed the fair market value, as reduced by the easement. *[16 U.S.C. 3837a(f)]*	Limits compensation to lowest of 4 options: percentage of the fair market value; percentage of market value determined by a survey; a geographic cap; or a landowner's offer. Allows USDA to use non-federal contributions to administer program *[Sec. 2102(e)]*	Limits compensation to lowest of 3 options: an amount necessary to encourage enrollment; a limit for a geographic area; or a landowner's offer. Compensation may be in 1 to 30 payments. *[Sec. 2322(b)(3-4)]*	Adopts House provision with changes, revising the process for determining the value of easements and contracts by requiring USDA to provide the lowest amount of compensation based on a comparison of the fair market value of the land, a geographic cap, or an offer made by the landowner. Provides that easements with values less than $500,000 be paid out over 1-30 years; easements with values greater than $500,000 are to

(Continued)

Prior Law/Policy (P.L. 107-171) (unless otherwise Indicated)	House Bill (H.R. 2419)	Senate Amendment (H.R. 2419)	Enacted 2008 Farm Bill (P.L. 110-246)
			be paid out over 5-30 years. *[Sec. 2208]* Provides for the repeal of payment limitations (exception for State agreements for new Wetlands Reserve Enhancement Program. *[Sec. 2209]*
Sec.1237C(c) of the FSA, as amended, lists three considerations USDA is to use when considering offers for WRP contracts. *[16 U.S.C. 3837c(c)]*	Adds new additional criteria for ranking offers (conservation benefits; cost-effectiveness; and offer of a financial contribution) and conservation benefits of floodplain lands. *[Sec. 2102(f)]*	No comparable provision.	Adopts House provision. *[Sec. 2207]*
Sec. 1237D(c)(4) of the FSA, as amended, waives limits for public entities receiving payments through the wetland and environmental enhancement programs. *[16 U.S.C. d(c)(4)]*	Replaces provision with a new language on Wetland Reserve Enhancement program, where states contribute funds so as to increase payments. *[Sec. 2102(g)]*	Authorizes a Wetlands Reserve Enhancement Program (WREP). Makes a conforming change to allow payments for 30-year contracts. *[Sec. 2322(c)]*	Adopts Senate provision authorizing WREP for agreements with States similar to CREP. Authorizes a Reserved Rights Pilot program. *[Sec. 2205-2206]*
No comparable provision.	No comparable provision.	Requires a report to House and Senate Agriculture Committees by 1/1/2010 on the implications of long-term easements on USDA resources. *[Sec. 2322(d)]*	Adopts Senate provision. *[Sec. 2210]*
Conservation Security Program			
The 2002 farm bill amended the FSA to establish the Conservation Security Program (CSP) for FY2003-11. Defines eligible producers and eligible lands and excluded lands (land enrolled in multiyear land retirement programs and land not in crop production at least 4 of the preceding 6 years). Specifies terms for 3 tiers of conservation contracts. Identifies topics that may be addressed in contracts. Contracts are 5 years under tier 1, and 5 to 10 years under tiers 2 and 3. Specifies circumstances and requirements for modifying, terminating, and renewing contracts. Contracts may be renewed for 5 to 10	Establishes a new CSP for FY2012-2017. Eligible producers must submit an offer addressing "at least one priority resource of concern to a minimum level of management intensity." Eligible land excludes incidental forest land. Limits program to one type of contract of 5 years; describes five elements to be in all contracts, but eliminates list of topics to be addressed. Prohibits termination of contracts, without penalty, by a producer who is required to modify a contract. Allows contracts to be renewed for one additional 5 year period. Adds new provisions on evaluating offers from organic	Authorizes through FY2012 a new CSP as a conforming amendment, and reauthorizes current CSP for existing contracts only. Future CSP contracts would be replaced by a Comprehensive Stewardship Incentives Program consisting of a Conservation Stewardship Program with similar provisions of the existing CSP and EQIP. Eligible land must have been planted to crops 4 of preceding 6 years. Specifies contract are for 5 years, with renewal under certain conditions. Allows for terminating and changing contracts. Specifies how to evaluate contract offers, producer duties, enhancement terms, and supple-mental payments. Defines 15 terms.	Defines program terms for the new CSP. Adopts elements of the Senate provision. Expands eligible lands to include nonindustrial private forest land (limited to not more than 10% of total annual acres under the program). Permits 5-year extension of contracts. Excludes under the program, land used for cropland that had not been planted, considered to be planted, or devoted to crop production for 4 of the 6 years prior to the date of enactment of the act (unless the land had previously been enrolled in CRP; had been maintained in a long term crop rotation; or was incidental land needed for efficient operation). *[Sec. 2301]*

(Continued)

Prior Law/Policy (P.L. 107-171) (unless otherwise Indicated)	House Bill (H.R. 2419)	Senate Amendment (H.R. 2419)	Enacted 2008 Farm Bill (P.L. 110-246)
years. Defines 15 terms pertaining to the program. *[16 U.S.C. 3838]*	producers. Defines twelve terms that are new terms or differ from current law. *[Sec. 2103(a)]*	*[Secs. 2391 and 2341]*	
Secs 1238A(d-3) of the FSA, as amended, specifies terms for 3 tiers of conservation contracts. Identifies topics that may be addressed in contracts. Contracts are 5 years under tier 1, and 5 to 10 years under tiers 2 and 3. Specifies circumstances and requirements for modifying, terminating, and renewing contracts. Contracts may be renewed for 5 to 10 years *[16 U.S.C. 3838a(d-e)]*	Limits program to 1 type of contract of 5 years and describes 5 elements to be in all contracts, but eliminates list of topics to be addressed. Contracts could no longer be terminated, without penalty, by a producer who is required to modify a contract. Contracts may be renewed for 1 additional 5 year period. New provisions on evaluation of offers and coordination with organic certification are added. *[Sec. 2103(a)]*	Defines eligible land and eligible producers; land must have been planted to crops 4 of preceding 6 years. Contents of contracts are specified and are for 5 years, with renewal if certain conditions are met. Specifies considerations in evaluating contract offers, producer duties, enhancement terms, and supplemental payments. Adds provisions on terminating and changing contracts. *[Sec. 2341]*	Adopts Senate provision with modifications. Authorizes supplemental payments for producers who adopt a beneficial crop rotation that provide significant conservation benefits and are not limited to a particular crop, cropping system, or region of the country. Allows for on-farm conservation research and demonstration activities and pilot-testing as part of contract offers. Allows for contract modification. *[Sec 2301]*
Sec. 1238C of the FSA, as amended, specifies that duties of the Secretary include making payments early in each fiscal year, the components of payments for each tier, annual payment limits for each tier ($20,000 for tier 1, $35,000 for tier 2, and $45,000 for tier 3), minimum requirements for practices, and requirements for implementing regulations *[16 U.S.C. 3838c]*	Alters duties of the Secretary to include identification of priority resources of concern at the state level (limited to 5 concerns in any geographic area of a state). Limits total payments under a contract to $150,000 (5 years); allows for the environmental needs associated with agriculture to be considered in state allocations; requires USDA to compile data of specified program contract and payment topics. *[Sec. 2103 (a)]*	Alters duties of the Secretary to allow for continuous enrollment (allowing a producer to apply at any time during the year), providing assistance to producers, and maintaining contract and payment information that will support program monitoring and evaluation, and enabling specialty crop producers to participate. Specifies an acreage allocation, limiting payments to $240,000 (6-year). *[Secs. 2391 and 2341]*	Provides that state acreage allocations be based on each state's proportion of eligible acres to the total eligible acres nationwide (available to all producers, not only specific watersheds/geographic regions), allowing for input from USDA. Directs USDA to adopt continuous enrollment, but allows for USDA to determine when to rank applications. Directs USDA to provide technical assistance to specialty crop and organic producers. Directs USDA to encourage producers who are transitioning from land retirement programs to enroll in CSP and other working lands programs. Limits payments to $200,000 in any 5-year period. *[Sec. 2301]*
No comparable provision.	No comparable provision.	Provides for enrollment of up to 79.628 million acres and attempted annual enrollment of 13.273 million acres, nationwide and at a average annual cost of $19 per acre. Provides for small farm participation, and allocates to each state each year the lesser of 20,000 acres or 2.2%	Deletes Senate provision. Enrolls an additional 12.8 million acres annually FY2008-2017. *[Sec. 2301]*

(Continued)

Prior Law/Policy (P.L. 107-171) (unless otherwise Indicated)	House Bill (H.R. 2419)	Senate Amendment (H.R. 2419)	Enacted 2008 Farm Bill (P.L. 110-246)
		of the eligible land. *[Sec. 2341]*	
No comparable provision.	No comparable provision.	Requires regulations to be issued within 180 days of enactment. *[Sec. 2341]*	Deletes Senate provision. Directs USDA to promulgate regulations. *[Sec. 2301]*
Sec. 1241(a)(3) of the FSA, as amended, authorizes mandatory funding for the CSP at $1.954 billion for FY2006-10 and $5.65 billion from FY2006-15. *[16 U.S.C. 3841(a)(3)]*	Prohibits any new contracts under the terms of the 2002 CSP, such that no new CSP contracts may be entered into after October 1, 2007 (although payments may be continued until contracts expire). For contracts signed before 10/1/07, provides a total of $1.5 billion in mandatory funding for FY2007-12, and $1.9 billion for FY2012-17. For contract signed after 10/1/11, provides $0.5 billion for FY2012 and $4.6 billion for FY2012-17. *[Sec. 2401(b)]*	Prohibits any new contracts under the terms of the 2002 CSP, such that no new CSP contracts may be entered into after October 1, 2007 (although payments may be continued until contracts expire). Authorizes $2.3 billion in mandatory funding for contracts entered into before the date of enactment, (available until spent) and an unspecified amount for the new CSP (enrollment in the new program is measured in acres rather than dollars). *[Sec. 2401(a)(3-4)]*	Adopts Senate provision. The Manager's report states that the bill provides for a $1.1 billion increase in budget authority above current baseline (FY2008-2017). Provides such sums as necessary to carry out existing CSP contracts. *[Sec. 2701]*
Environmental Quality Incentives Program			
The 1996 farm bill amended Sec. 1240 of the FSA to authorize EQIP, stating its purpose as promoting production and environmental quality as compatible goals, and optimizing environmental benefits by working in 5 specified areas. *[16 U.S.C. 3839aa]* Defines 6 terms: beginning farmer or rancher, eligible land, land management practice, livestock, practice, and structural practice.	Adds forest management and organic transition to the program purposes. Revises the descriptions of 2 of the 5 purposes to recognize energy conservation and conservation on forest lands. *[Sec. 2105(a)]* Adds forestry, forest management practices, and coordinated implementation to the "land management practice" definition. Adds alpacas and bison to the "livestock" definition. Adds definitions of "integrated pest management" and "socially disadvantaged farmer or rancher." *[Sec. 2105(b)]*	Adds forest management to the statement of program purposes, and recognizes pollinators and fuels management in the amplifying statements. *[Secs. 2351, 2352, and 2354]* Adds aquaculture to the "eligible land" definition; forestry is added to the "land management practice" definition; adds conservation planning practices to "practices;" defines "producer" to include custom feeding businesses and contract growers; and adds firebreaks to "structural practice." *[Sec. 2352]*	Adds forest management to purposes, and adds language regarding forest lands on EQIP program plan and duplication. *[Secs. 2505-2506]* Adopts the Senate provisions with an amendment to modify eligible land. *[Sec. 2502]* Allows for technical assistance to farmers that promote pollinator habitats, and farmers transitioning to organic farming, among activities. *[Sec. 2501]* Further clarifies duties of the Secretary. *[Sec. 2507]*
Sec. 1240B(a-c) of the FSA, as amended, authorizes EQIP through FY2010. Eligible practices are defined. Contracts are 1 to 10 years. *[16 U.S.C. 3839aa-2(a-c)]*	Reauthorizes through FY2012. Expands types of eligible practices to include organic certification using technical services from approved providers, and improved energy efficiency, renewable energy systems. *[Sec. 2105(c)]*	Reauthorizes through FY2012. Expands permitted practices to include conservation planning. Limits contracts to a maximum of 5 years. Removes prohibitions on bidding down. *[Sec. 2353(a-c)]*	Reauthorizes through FY2012. Limits payments to 75% of costs. *[Sec. 2503]* Lowers individual payment limits from $450,000 to $300,000 during any 6-year period , except in cases of special environmental significance (including

(Continued)

Prior Law/Policy (P.L. 107-171) (unless otherwise Indicated)	House Bill (H.R. 2419)	Senate Amendment (H.R. 2419)	Enacted 2008 Farm Bill (P.L. 110-246)
			projects involving methane digesters), allowing USDA to raise the limit to not more than $450,000. *[Sec. 2508]* Directs USDA to develop criteria for evaluating applications to address national, State, and local conservation priorities, allowing for prioritization and grouping of applications based on cost-effectiveness, how address the designated resource concern(s), how best fulfills the purpose of EQIP, and if improves conservation practices or systems. *[Sec. 2504]*
Sec. 1240B(d)(2) of the FSA, as amended, allows limited resource and beginning producers to receive not more than a 90% federal cost share. *[16 U.S.C. 3839aa-2(d)]*	Sets the federal cost share at 90% for socially disadvantaged producers. Provides increased federal cost-share of 90% for using gasifier technology for certain purposes. *[Sec. 2105(d-e)]*	Sets the federal cost share at 90% for socially disadvantaged producers. Allows for advanced payments to purchase materials and contracting. *[Sec. 2353(c)]*	Adopts Senate proposal. *[Sec. 2503]*
Sec. 1240B(e) of the FSA, as amended, provides incentive payments to perform land management practices, with special emphasis given to practices that promote specified goals. *[16 U.S.C. 3839aa-2(e)]*	Expands purposes for incentive payments: (1) receiving technical services from approved third party providers, (2) developing a comprehensive nutrient management plan, and (3) implementing energy efficiency and renewable energy projects. Pollinator habitats will receive special emphasis. *[Sec. 2105(f)]*	Expands purposes receiving special emphasis to include predator species protected under the Endangered Species Act, gray wolves, grizzly bears, and black bears. *[Sec. 2353(c)(3)]*	Adopts House provision. *[Secs. 2503, 2706 and 2708]* Includes special rule that USDA may accord significance to practices promoting residue, nutrient, air quality, pest, and invasive species management; pollinator habitat; and animal carcass management technology. The conference report recognizes as consistent with the purposes of EQIP options to deter predators protected by the Endangered Species Act, and also delisted populations.
Sec. 1240B(g) of the FSA, as amended, requires that 60% of payments go to practices related to livestock production requirement. *[16 U.S.C. 3839aa-2(g)]*	Extends through FY2012 the 60% of payments to livestock production requirement. *[Sec. 2105(g)(2)]*	Similar to the House bill. *[Sec. 2353]*	Extends through FY2012 the 60% of payments to livestock production requirement. *[Sec. 2503]*
No comparable provision.	Requires USDA to reserve at least 5% of program funds for beginning and socially disadvantaged producers for at least 90 days after the program funds	Amends the cost-share rate exception for beginning and socially disadvantaged farmers or ranchers to allow variable payment, not to exceed 90%, and authority to provide	Adopts Senate provision for advance payments for beginning, socially disadvantaged and limited resource farmers or ranchers and deletes Senate

(Continued)

Prior Law/Policy (P.L. 107-171) (unless otherwise Indicated)	House Bill (H.R. 2419)	Senate Amendment (H.R. 2419)	Enacted 2008 Farm Bill (P.L. 110-246)
	have been made available. *[Sec. 2105(g)]*	advance payments up to 30% for the purchase of materials or contracting. *[Sec. 2353]*	provision for guaranteed loan eligibility. Adopts Senate provision with an amendment for cost-share rates and advance payments for beginning, socially disadvantaged, and limited resource farmers or ranchers. *[Sec. 2503]*
No comparable provision.	No comparable provision.	Gives priority to improving water conservation and air quality, under certain conditions. Requires participants to have/expect at least $15,000 in gross sales from farming. *[Sec. 2353(c)(6)]*	Deletes Senate provision.
No comparable provision.	Expands eligibility to market agencies and custom feeders. *[Sec. 2105(h)]*	Expands 'producer' eligibility to include a custom feeding business and a contract grower or finisher. *[Sec. 2352]*	Adopts Senate provision with an amendment to modify eligible land. *[Sec. 2502]*
Sec. 1240C of the FSA, as amended, gives higher priority for participation in EQIP to producers using cost-effective conservation practices and practices that address national conservation priorities. *[16 U.S.C. 3839aa-3]*	Identifies 5 priorities for program applications. Specifies a streamlined evaluation process for operations with substantial and sound environmental management systems involving a limited number of practices. *[Sec. 2105(l)]*	Adds a higher priority for improving conservation practices or systems in place at the time of the contract offer. *[Sec. 2354]*	Adopts House provision with changes to prioritize State, regional, or local resource concerns, and to allow for the grouping of applications of similar agriculture operations. *[Sec. 2502]*
Sec. 1240E of the FSA, as amended, defines the general contents of a producer's EQIP plan, and calls on USDA to avoid duplication with other conservation plans. *[16 U.S.C. 3839aa-5]*	Adds to the planning requirements the need to be consistent with forest plans, and allows as an acceptable plan consideration of an air or water quality permit that meet regulatory requirements as an acceptable plan. *[Sec. 2105(k)]*	Includes forestry language similar to House bill, but also allows a producer organization to act on behalf of its membership in submitting applications or conducting similar activities to facilitate program participation. *[Sec. 2356]* Establishes a Chesapeake Bay Watershed Conservation Program under EQIP to assist producers in applying conservation practices on agricultural/nonindustrial private forestland in the Bay watershed to address natural resource concerns related to agriculture, funded at $165 million for FY2008-12. *[Sec. 2361]*	Adopts House forestry provision *[Sec. 2502]*, but strikes Senate provision on producer organizations.
Sec. 1240F of the FSA, as amended, provides for USDA funding, information, and training to develop and	Lists 3 criteria that must be met before USDA can provide assistance for practices with a primary purpose of	No comparable provision.	Deletes House provision.

(Continued)

Prior Law/Policy (P.L. 107-171) (unless otherwise Indicated)	House Bill (H.R. 2419)	Senate Amendment (H.R. 2419)	Enacted 2008 Farm Bill (P.L. 110-246)
implement conservation plans. *[16 U.S.C. 3839aa-6]*	water conservation. *[Sec. 2105 (l)]*		
Sec. 1240H of the FSA, as amended by the 2002 farm bill, provides for a competitive grants program within EQIP, on a matching basis, to implement innovative conservation practices; examples listed are using market systems in pollution reduction, and using innovative practices, such as storing carbon in soil; no funding is specified. *[16 U.S.C. 3839aa-8]*	Adds to the Conservation Innovation Grants provisions under EQIP. Adds detail on qualities of eligible projects, establishes a pilot program for conservation planning in the Chesapeake Bay watershed, and adds a new subsection to assist producers who are meeting state and local regulatory air quality requirements. Provides funding from EQIP: $30 million (FY2008) rising to $75 million (FY2012), with specified funds for air quality and for organic and specialty crop producers. *[Sec. 2105(m)]*	Adds to the Conservation Innovation Grants provisions under EQIP. Adds nonindustrial private forest lands to the list of potential recipients of innovative technologies. Adds two items to the list of examples: (1) transfers of innovative technologies to nonindustrial private forest land in production, and (2) assistance for specialty crop production. *[Sec. 2358]*	Adopts House provisions related to forest resource management and air quality. Provides $37.5 million annually (FY2009-2012) to implement air quality plans. *[Sec. 2509]* The conference report states conservation programs should recognize the use of innovative technology such as enhanced efficiency fertilizers.
Sec. 1240I of the FSA, as amended by the 2002 farm bill, creates a Ground and Surface Water Conservation Program (GSWCP) within EQIP for activities that will result in a net savings of ground or surface water; lists 6 types of eligible activities (improve irrigation systems, for example), and provides mandatory funding of $25 million in FY2002, growing to $60 million annually in FY2004-07. *[16 U.S.C. 3839-aa-9]*	Replaces GSWCP with a Regional Water Enhancement Program to address water quality, make eligible governmental entities (including irrigation and water districts) and Indian tribes, and to implement program on a regional scale through cooperative agreements. Expands the list of eligible activities and requires the Secretary to identify priority areas. Lists 5 priority areas, which together may receive no more than 50% of the available funds. Establishes a process for soliciting/selecting proposals and developing implementation agreements. Provides mandatory funds of $60 million annually through FY2012 (limits administrative expenses to no more than 3% of the total). *[Sec. 2106]*	Maintains GSWCP and provides an increase in funding from $60million to $65 million annually. Provides funding for each state that received funding under the program in previous years (simple average of funds provided for FY2002- 2007 or the amount provided in 2007, whichever is greater), except for states over the Ogallala Aquifer, which will receive not less than the greater of $3 million or the average of funds provided for FY2002-2007. Provides at least $20 million for the Eastern Snake Plain Aquifer. *[Sec. 2359]*	Replaces GSWCP with the Agricultural Water Enhancement Program (AWEP) under EQIP. Provides mandatory funding: $73 million annually (FY2009-2010), $74 million (FY2011), and $60 million annually (FY2012 and each year thereafter). Recognizes the purpose as addressing water quality/quantity concerns on agricultural land, with the role of AWEP partners as leveraging federal funds and encouraging producers to address these concerns. The Managers report emphasizes the importance of addressing groundwater management in the Ogallala region, promoting water use efficiency projects, converting irrigated farming operations to a dryland farming; and providing assistance to construct on farm reservoirs/irrigation ponds in drought-stricken areas. Identifies six priority areas: The Eastern Snake Plain Aquifer region, Puget Sound, the Ogallala Aquifer, the Sacramento River watershed, Upper

(Continued)

Prior Law/Policy (P.L. 107-171) (unless otherwise Indicated)	House Bill (H.R. 2419)	Senate Amendment (H.R. 2419)	Enacted 2008 Farm Bill (P.L. 110-246)
			Mississippi River Basin, the Red River of the North Basin, and the Everglades. *[Sec. 2510]* Follows same payment limits as under EQIP. *[Sec. 2508]* Provides for a transition period for the existing GSWCP through September 30, 2008. *[Sec. 2903]*
Sec. 1240I(c)(2) of the FSA, as amended by the 2002 farm bill, provides $50 million to carry out water conservation activities in the Klamath River basin (OR, CA) *[16 U.S.C. 3839aa-9(c)(2)]*	Lists the Klamath River basin as one of the 5 listed priority areas under the Regional Water Enhancement Program. *[Sec. 2106 (b)(2)]*	No comparable provision. (*Note*: The Klamath Basin is listed as 1 of 14 priority areas in the Partnerships and Cooperation Program; see above.)	Deletes House provision.
No comparable provision.	No comparable provision.	Adds program at end of EQIP to assist farmers who are converting to organic production (with contracts of 3-4 years). Payments are limited to $20,000 per year. *[Sec. 2360]*	Adopts Senate provision. Includes organic practices as eligible management systems and limits payment to an aggregate of $80,000 in any 6-year period. *[Secs. 2501 and 2503]*
Sec. 1241(a)(6) of the FSA, as amended, authorizes EQIP funding, rising from $0.4 billion in FY2002 to $1.3 billion in FY2010. *[16 U.S.C. 3841(a)(6)]*	Authorizes EQIP funding: $1.25 billion (FY2008), $1.6 billion (FY2009), $1.7 billion (FY2010), $1.8 billion (FY2011), and $2 billion (FY2012). *[Sec. 2401(d)]*	Authorizes EQIP funding: $1.27 billion annually (FY2008-09), $1.3 billion each (FY2010-FY2012). *[Sec. 2401(a)(7)]*	Provides additional budget authority. Authorizes EQIP funding: $1.2 billion (FY2008); $1.337 billion (FY2009); $1.45 billion (FY2012); $1.588 billion (FY2011); and $1.75 billion (FY2012). *[Sec. 2701]*
Farmland Protection Program			
The 1996 farm bill amended Sec. 1238H of the FSA to authorize the Farmland Protection Program (FPP), defining eligible entity, land, Indian tribes, and programs. *[16 U.S.C. 3838h-I]* The program, as amended, provides for the purchase of conservation easements to protect topsoil by limiting the land's nonagricultural uses subject to a pending offer. The federal cost may not exceed 50% of the value of the easement; the value of a charitable donation by the seller may not exceed 25% of the value of the easement. If	Reauthorizes program, and renames to Farm and Ranchland Protection Program (FRPP). Expands eligible land definition to include historic and archaeological resources. States will be certified (reviewed every 3 years) to participate and receive program funds based on 4 listed requirements. States may spend up to 10% of funds for administrative costs. Lists terms and conditions for agreements with eligible entities (reviewed every 3 years). Provides that USDA may require a contingent right to enforce easement,	Reauthorizes the program. Modifies definition of eligible forest land, and makes eligible other land that is needed for efficient administration of an easement. Changes purpose of program from protecting topsoil to "protecting agricultural use and related conservation values." Adds new requirements for cooperative agreements with participants, cost-sharing, and protection of the federal investment. Allows USDA to enter into cooperative agreements with eligible entities under certain circumstances, and requires the protection of federal investment through an executory	Reauthorizes the program through 2012, but does not rename program. Changes administrative requirements, appraisal methodology, and terms and conditions of cooperative agreements. Adopts terms/conditions for cooperative agreements similar to Senate provision. Clarifies the purpose of the program as protecting land for agricultural use by limiting nonagricultural uses. Adopts Senate provision to modify the definition of eligible land to include forestland and other land that contributes to the economic viability of an operation. Establishes a

(Continued)

Prior Law/Policy (P.L. 107-171) (unless otherwise Indicated)	House Bill (H.R. 2419)	Senate Amendment (H.R. 2419)	Enacted 2008 Farm Bill (P.L. 110-246)
multiple applications are comparable, USDA may not use cost alone to determine which ones will be funded.	and requires the use of a conservation plan for highly erodible cropland. Retains a federal contingent right of enforcement or executory limitation in an easement to ensure its enforcement. Provides cost-share assistance for purchasing an easement, but assistance may not exceed 50% of the appraised fair market value of the easement. The fair market value is determined by an appraisal using an industry-approved method. *[Sec. 2110]*	limitation. Limits the amount USDA can share in the costs of purchasing the easement to 50% of the appraised fair market value and establishes minimum amounts entities pay based on the amount of landowner contributions. Requires appraisals based on uniform standards of professional appraisal practice or any other industry-approved standard. *[Sec. 2371]*	certification process similar to the House bill for all eligible entities. To become certified, entities must have the authority and resources to enforce easements, polices in place that are consistent with the purposes of the program, and clear procedures to protect the integrity of the program. Includes a limit on impervious surfaces consistent with agricultural activities, and clarifies agreement terms for certified and non-certified entities. *[Sec. 2401-2402]*
Sec. 1241(a)(4) of the FSA, as amended, authorizes mandatory funding for the FPP at: $50 million in FY2002, $100 million in FY2003, $125 million in FY2004 and FY2005, $100 million in FY2006, and $97 million in FY2007. *[16 U.S.C. 3841(a)(4)]*	Mandatory funding for the renamed Farm and Ranchland Protection Program is authorized at: $125 million in FY2008, $150 million in FY2009, $200 million in FY2010, $240 million in FY2011, and $280 million in FY2012. *[Sec. 2401(c)]*	Mandatory funding for the FPP is authorized at $97 million annually from FY2008 through FY2012. *[Sec. 2401(a)(5)]*	Provides additional budget authority. Authorizes FPP funding: $97 million (FY2008); $121 million (FY2009); $150 million (FY2010); $175 million (FY2011); and $200 million (FY2012). *[Sec. 2701]*
Grassland Reserve Program			
The 2002 farm bill amended Sec. 1238N of the FSA to authorize the Grasslands Reserve Program (GRP), setting maximum enrollment for at 2.0 million acres (all enrolled parcels in at least 40 contiguous acres). Requires 40% of land enrolled in 10-20 year, and 60% in 30 year agreements. *[16 U.S.C. 3838n]*	Sets the GRP enrollment ceiling at an additional 1.34 million acres, with at least 60% of these acres to be enrolled using 30 year rental agreements and easements. *[Sec. 2104(a) and (b)]*	Adds definitions: eligible entity, eligible land, and permanent conservation easement. Eligible entity and authority would allow for USDA to enter cooperative agreements with entities to purchase easements. Provides for GRP enrollment options through a 30 year contract, 30 year easement, and permanent easement. *[Sec. 2381]*	Adopts an acreage enrollment goal of an additional 1.22 million acres by 2012. Includes 10-, 15-, and 20-year rental contracts and permanent easements. Deletes House priority for 60% of acreage in long term contracts; retains current law that 60% of funds would be dedicated to easements, while 40% of funds would be dedicated to short term contracts. Adopts Senate definition of eligible entity, authority, and eligible land (with technical corrections). Adopts a priority for enrollment of CRP land with a modification to clarify that the priority applies upon expiration of the CRP contract. *[Sec. 2403]*
No comparable provision.	Allows USDA to transfer certain land currently in the CRP into the GRP, but	Allows USDA to transfer certain land currently in the CRP to be transferred to a	Adopts House provision regarding the method for determining fair market value

(Continued)

Prior Law/Policy (P.L. 107-171) (unless otherwise Indicated)	House Bill (H.R. 2419)	Senate Amendment (H.R. 2419)	Enacted 2008 Farm Bill (P.L. 110-246)
	limits the total in any calendar year to no more than 10% of GRP acres enrolled. Requires USDA pay the lowest of four specified ways to calculate fair market value. *[Sec. 2104(c)]*	permanent easement under GRP, but limits the total transferred in any calendar year to 10% of the total funding available for the GRP in that year. *[Sec. 2381]*	with a technical correction. *[Sec. 2403]*
Sec. 1238O of the FSA, as amended, specifies the duties and requirements of landowners in the GRP, terms of easements and agreements, and how applications are to be evaluated. *[16 U.S.C. 3838o]*	No comparable provision.	Specifies landowner duties and USDA considerations in evaluating offers. Specifies how to determine compensation levels and technical assistance. Specifies terms/conditions that apply to GRP contracts/easements, such as permitted and prohibited uses, minimum require-ments for cooperative agreements, and other considerations. *[Sec. 2381]*	Adopts Senate provisions with changes. *[Sec. 2403]*
No comparable provision.	Authorizes a Grasslands Reserve Enhancement Program. *[Sec. 2104 (d)]*	No comparable provision.	Deletes House provision.
Sec. 1238Q(a) of the FSA, as amended, allows USDA to transfer the title of an easement in the GRP to a state or private organization. *[16 U.S.C. 3838q(a)]*	Requires USDA to transfer the title of an easement to a private organizations or a state. *[Sec. 2104 (e)]*	Provides authority for USDA to enter into cooperative agreements with eligible entities for those entities to purchase, own, enforce, and monitor easements. *[Sec. 2381]*	Adopts the Senate amendment provision for cooperative agreements between USDA and eligible entities with a modification to the language specifying that eligible entities shall assume costs of administering and enforcing easements. Adopts a requirement for a contingent right of enforcement. *[Sec. 2403]*
Sec. 1241(a)(5) of the FSA, as amended, limits funding for the GRP to a total of $254 million from FY2003-07. *[16 U.S.C. 3841(a)(5)]*	No comparable provision. (*Note:* Sets acreage enrollment limit in GRP provisions, but no funding limit.)	Total GRP funding limited to $240 million for FY2008-12, with no acreage enrollment limit. *[Sec. 2401(a)(6]*	Deletes Senate provision.
Wildlife Habitat Incentives Program			
The 1996 farm bill amended Sec. 1240N of the FSA to authorize Wildlife Habitat Incentives Program (WHIP), providing cost-sharing to landowners who improve habitat, with up to 15% of the total made available in any years for agreements that are longer than 15 years.	Reauthorizes WHIP through FY2012; allows additional funds to be used to meet regulatory requirements that "reduces the economic scope of the producer's operation;" increases portion of funds for long-term agreements from 15% to 25% of funding.	Reauthorizes WHIP through FY2012; increases portion of funds for long-term agreements from 15% to 25% of funding; requires USDA to give priority to projects that foster the goals of state, regional, and national fish and wildlife conservation plans. *[Sec. 2393]*	Limits program eligibility to focus on lands "for the development of wildlife habitat on private agricultural land, nonindustrial private forest land, and tribal lands." Raises limit on cost-share payments for long-term projects to 25% and limits total payments to $50,000 per year. Allows USDA to provide priority to

(Continued)

Prior Law/Policy (P.L. 107-171) (unless otherwise Indicated)	House Bill (H.R. 2419)	Senate Amendment (H.R. 2419)	Enacted 2008 Farm Bill (P.L. 110-246)
[16 U.S.C. 3839bb-1]	*[Sec. 2112]*		projects that address issues raised by State, regional, and national conservation initiatives. *[Sec. 2602]*
Sec. 1241(a)(7) of the FSA, as amended, authorizes mandatory funding raising from $15 million to $60 million between FY2002-04, and $85 million annually (FY2005-07). *[16 U.S.C. 3841(a)(7)]*	Mandatory funding for WHIP is authorized at $85 million annually through FY2012. *[Sec. 2401(e)]*	Similar to the House bill. *[Sec. 2401(a)(8)]*	Reauthorizes program through FY2012 at current levels. *[Sec. 2701]*
Other Conservation Programs			
The 2002 farm bill amended the FSA to provide grants to implement a Farm Viability Program. Authorizes appropriations "such sums as are necessary" through FY2007. *[16 U.S.C. 3838j]*	Reauthorizes discretionary funding for program through FY2012. *[Sec. 2111]*	Reauthorizes discretionary funding for program through FY2012. *[Sec. 2396]*	Reauthorizes the program through 2012. *[Sec. 2402]*
The 1996 farm bill amended Sec. 1240M(e) of the FSA to authorize the Conservation of Private Grazing Land Program. Authorizes appropriations of $60 million annually through FY2007. *[16 U.S.C, 3839bb(e)]*	Extends authorization of appropriations through FY2012. *[Sec. 2108]*	Extends authorization of appropriations through FY2012. *[Sec. 2392]*	Extends authorization of appropriations through FY2012. *[Sec. 2601]*
The 2002 farm bill amended Sec. 1240O of the FSA to authorize a Grassroots Source Water Protection Program to assist state rural water associations that operate wellhead and groundwater protection programs. Authorizes appropriations of $5 million annually through FY2007. *[16 U.S.C. 3839bb-2]*	Authorizes $20 million annually in discretionary funds (FY2008-12) and one-time funding of $10 million in mandatory funding to remain available until spent. *[Sec. 2107]*	Authorizes $20 million annually in discretionary funding (FY2008-12). *[Sec. 2394]*	Adopts Senate provision. *[Sec. 2603]*
The 2002 farm bill amended Sec. 1240P of the FSA to authorize a Great Lakes Program for Soil Erosion and Sediment Control, and provides $5 million annually through FY2007. *[16 U.S.C. 3839bb-3]*	Extends authorization of appropriations through FY2012. *[Sec. 2109]*	Extends authorization of appropriations through FY2012; specifies program will help implement recommendations of a collaborative restoration strategy, giving priority to certain projects. *[Sec. 2395]*.	Reauthorizes program, and authorizes $5 million annually FY2008-2012. *[Sec. 2604]*

(Continued)

Prior Law/Policy (P.L. 107-171) (unless otherwise Indicated)	House Bill (H.R. 2419)	Senate Amendment (H.R. 2419)	Enacted 2008 Farm Bill (P.L. 110-246)
Sec. 524(b)(1) of the Federal Crop Insurance Act authorizes an Agricultural Management Assistance (AMA) program for listed states that have historic low participation rates in the Federal Crop Insurance Program. *[7 U.S.C. 1524(b)]*	Adds Hawaii and Virginia to the list of eligible states. Allocates 50% of funds to USDA's NRCS; 10% to Agricultural Marketing Service (for organic certification assistance); and 40% to the Risk Management Agency. *[Sec. 2201]*	Reauthorizes the program through FY2012; adds Idaho to the list of eligible states. *[Sec. 2601]*	Adopts House provision with changes to include Hawaii as an eligible State. Provides an additional $25million in mandatory funding (FY2008-2012). *[Sec. 2801]*
Secs. 1528-1537 of the 1981 farm bill (Agriculture and Food Act of 1981, P.L. 108-7) authorizes the Resource Conservation and Development Program (RC&D) to develop and implement a regional plan to address conservation, water/land management, or community development. *[16 U.S.C. 1528-1527]*	Amends RC&D program to provide a designated coordinator to assist each approved area. Eliminates requirement to submit a program evaluation to the House and Senate Agriculture Committees before June 30, 2005. *[Sec. 2202]*	Similar to House provision, clarifying that an area plan must be developed through a locally led process, and that the planning process, and that the planning process must be conducted by a local council. Provides for a coordinator to improve technical assistance to councils, as designated by USDA. *[Sec. 2605]*	Adopts Senate provision. *[Sec. 2805]*
Sec. 14(h) of the Watershed Protection and Flood Prevention Act (P.L. 106-472) authorizes discretionary and mandatory funding for a Small Watershed Rehabilitation Program. *[16 U.S.C. 1012]*	Authorizes $50 million annually in mandatory funding (FY2009-12); extends FY2007 discretionary funding level through FY2012. *[Sec. 2203]*	Authorizes such sums as necessary in discretionary funding annually (FY2008-12). *[Sec. 2604]*	Adopts House provision and provides $100 million in mandatory funding for FY2009 to remain available until expended. *[Sec. 2803]*
The 2002 farm bill amended Sec. 1241(d) of the FSA to authorize a program to promote regional equity, giving each state a total of at least $12 million annually from certain mandatory programs. *[16 U.S.C. 3841d]*	Annual funding for regional equity is raised to at least $15 million *[Sec. 2404]*	Annual funding for regional equity is raised to at least $15 million, and crop insurance payments are added to this calculation. Directs USDA to update state allocation formulas. *[Sec. 2402]*	Adopts Senate provision with changes. *[Sec. 2703]*
The 2002 farm bill amended Sec. 1242 of the FSA to authorize delivery of technical assistance directly or using a third party provider and specifies how providers are to be approved by USDA. Authorizes cooperative agreements with non-federal entities to provide technical assistance. *[16 U.S.C. 3842]*	Expands use of third party providers using contracts. Specifies providers should get at least prevailing market rates, calls for a review/update of all technical assistance specifications, including the needs of specialty crop producers. *[Sec. 2402]*	Expands use of third party providers using contracts. Directs USDA to develop national certification criteria and approve established state standards. Provides funding through each conservation program, specifies minimum and maximum contract terms, among other considerations. Includes similar provisions for specialty crop producers as	Adopts Senate provision with changes regarding the delivery of technical assistance. *[Sec. 2706]*

(Continued)

Prior Law/Policy (P.L. 107-171) (unless otherwise Indicated)	House Bill (H.R. 2419)	Senate Amendment (H.R. 2419)	Enacted 2008 Farm Bill (P.L. 110-246)
		the House bill. *[Sec. 2404]*	
The 2002 farm bill amended Sec. 1244(a) of the FSA to authorize USDA to provide incentives to beginning farmers/ranchers and Indian tribes to participate in conservation programs. *[16 U.S.C. 3844(a)]*	Expands access to program incentives to include socially disadvantaged and limited resource farmers and ranchers. Requires USDA to develop a streamlined application process. *[Sec. 2405]*	Requires USDA to develop a streamlined application process for conservation programs. Provides for Safe Harbor assurances to the landowner under the Endangered Species Act. Allows producers to apply for programs through a producer organization. *[Sec. 2405]*	Adopts House provision with changes to include certain acreage limitations and exemptions, and also a pollinator protection provision. Requires USDA report to Congress on the completion of the requirements not later than 1 year after enactment. *[Sec. 2708]*
The 1990 farm bill amended Sec. 1261 of the FSA to authorize state technical committees (STC), including members and interests to be represented, outlining duties, and specifying that committees are advisory with no implementation or enforcement authority. *[16 U.S.C. 3861-3862]*	Specifies STC have at least 12 producers representing agriculture; removes requirement for persons knowledgeable about conservation; adds new provisions creating subcommittees and lists potential topics; describes responsibilities in more general terms. *[Sec. 2408]*	Adds non-industrial private forest land owners to the list of groups represented on the STC. Requires USDA to develop standard operating procedures to be used by the State technical committee in the development of technical guidelines for the implementation of the conservation provisions of this title. Makes local work groups subcommittees of the State technical committee. *[Sec. 2501]*	Adopts House provision with changes to require USDA to develop standard committee operating procedures, updates the names of participating agencies, and deletes the requirement for establishing specific issue-area subcommittees. Requires that public notice be given for meetings of the State technical committee and adds local working groups as subcommittees. *[Sec. 2711]*
The 1996 farm bill amended Sec. 351 of the FSA to authorize a National Natural Resources Conservation Foundation to raise private funds that will be used to promote conservation. Program has never been implemented. *[16 U.S.C. 5801-5809]*	No comparable provision.	Amends numerous provisions authorizing the Foundation. *[Sec. 2606]*	Deletes Senate provision.
The 2002 farm bill amended Sec. 2507 of the FSA to authorize USDA to transfer $200 million of CCC funds to the Bureau of Reclamation for water to at-risk natural desert terminal lakes. *[43 U.S.C. 2211note]*	No comparable provision.	Amends the desert terminal lakes provision to allow funds to be used to lease water or to purchase land and related interests in the Walker River Basin. *[2607]*	Adopts Senate provision with changes. Provides $175 million in mandatory funding. *[Sec. 2807]*
The 1990 farm bill amended Sec. 1261 of the FSA to authorize state technical committees (STC), including members and interests to be represented, outlining duties, and specifying that	Specifies STC have at least 12 producers representing agriculture. Removes requirement for persons knowledgeable about conservation. Adds new provisions creating subcommittees issue	Adds non-industrial private forest land owners to the list of groups represented on the STC. *[Sec. 2501]*	Adopts House provision with changes to require USDA to develop committee standard operating procedures, and to update the names of participating agencies. Deletes the requirement for

(Continued)

Prior Law/Policy (P.L. 107-171) (unless otherwise Indicated)	House Bill (H.R. 2419)	Senate Amendment (H.R. 2419)	Enacted 2008 Farm Bill (P.L. 110-246)
committees are advisory with no implementation or enforcement authority. *[16 U.S.C. 3861-3862]*	areas. Describes responsibilities in more general terms. *[Sec. 2408]*		establishing specific issue-area subcommittees. *[Sec. 2711]*
New Conservation Programs			
No comparable provision.	Authorizes a new Chesapeake Bay Program for Nutrient Reduction and Sediment Control to carry out restoration, enhancement, and preservation projects. Identifies four specified watersheds. The non-federal cost share for each project will be at least 35%, but will not exceed $5 million. Sets mandatory funding at $10 million (FY2008), rising to $55 million (FY2012). *[Sec. 2301]*	No comparable provision. (*Note*: The Chesapeake Bay program is authorized as a part of EQIP.)	Adopts House provision with changes. Renames program Chesapeake Bay Watershed Program. Applies to all tributaries, backwaters, and side channels, including watersheds, draining into the Chesapeake Bay, but gives priority to the Susquehanna, Shenandoah, Potomac, and Patuxent Rivers. Provides mandatory funds of $23 million (FY2009); $43 million (FY2010); $72 million (FY2011); and $50 million (FY2012). *[Sec. 2605]*
No comparable provision.	The so-called "Open Fields" provision authorizes state grants through a new Voluntary Public Access and Habitat Incentive Program to encourage land-owners to provide public access for wildlife-dependent recreation. Sets application contents and award priorities. Authorizes discretionary funding of $20 million annually through FY2012. *[Sec. 2303]*	Similar to the House bill, and includes a priority to States where the location of participating lands would be available to the public and provides $20 million per year in mandatory funding annually (FY2008-2012). *[Sec. 2399 and Sec. 2401(a)(9)]*	Adopts Senate provision with an amendment, providing $50 million in mandatory funds for the period FY2009-2012. Includes a 25% reduction for the total grant amount if the opening dates for migratory bird hunting in the State are not consistent for residents and non-residents. *[Sec. 2606]*
No comparable provision.	No comparable provision.	Creates a new Conservation Access program, requiring 10% of the funds (or acres in the cases of WRP and CRP) be used to assist beginning and socially disadvantaged farmers and ranchers with annual gross sales of $15,000 or more. *[Sec. 2403]*	Adopts Senate provision with changes. Provides that 5% of CSP acres and 5% of EQIP funds be used to assist beginning farmers or ranchers, and an additional 5% of each to assist socially disadvantaged farmers or ranchers. *[Sec. 2704]*
No comparable provision.	Authorizes a new Muck Soils Conser-vation Program for eligible land, defined by five characteristics. Authorizes	No comparable provision.	Deletes provision.

(Continued)

Prior Law/Policy (P.L. 107-171) (unless otherwise Indicated)	House Bill (H.R. 2419)	Senate Amendment (H.R. 2419)	Enacted 2008 Farm Bill (P.L. 110-246)
	appropriations of $50 million annually through FY2012, with payments between $300-$500 per acre. *[Sec. 2303]*		
No comparable provision. (*Note:* Current law limits CRP payments to $50,000 per year, sets payment limits for each of 3 tiers in the CSP, and limits EQIP payments to $450,000 for all contracts in any 6-year period. *[16 U.S.C. 3834, 3838c, and 3839aa-7]*	Authorizes new payment limits, deleting existing conservation payment limit language. Limits annual payments to $60,000 for any single program; limits total payments to $125,000 under all conservation (except WRP, FRPP, GRP). Defines how payments should be attributed to individuals. *[Sec. 2409]*	No comparable provision.	Deletes provision. Places payment limits within each section of the bill and deletes this section.
No comparable provision.	Requires USDA to submit an annual report on specialty crop producer participation in conservation programs, including how to improve producer program access. *[Sec. 2406]*	No comparable provision.	Deletes provision. The managers report states it has modified the compliance and performance provisions of Section 1244 of FSA to accommodate the intent of the House bill.
No comparable provision.	Authorizes a new provision to develop agriculture and forestry based environmental services to promote market-based conservation. Specifies use of USDA-funded research, contracts, and award grants. Establishes a USDA-led Environmental Services Standards Board of senior federal officials to facilitate the development of credit markets and disseminate performance standards to federal agencies. Authorizes $50 million in discretionary funding, with appropriated amounts to remain available until spent. *[Sec. 2407]*	Authorizes a new provision to develop agriculture and forestry based environmental service markets, giving priority to developing carbon storage. Directs USDA to use a collaborative process with specified government and non-government interests to develop a framework and identifies relevant framework components (quantification, accounting, and verification). Requires three reports to Congress. Authorizes discretionary funding of "such sums as are necessary." *[Sec. 2406]*	Adopts House provision with changes. Directs USDA to work in consultation with other federal and state government agencies, nongovernmental interests and other interested persons, as determined by USDA, to establish technical guide-lines for measuring environmental services and to establish a verification process (allowing for consideration of third party verifiers). Directs USDA to focus initially on carbon markets. Does not authorize funds, expecting USDA to use available resources. *[Sec. 2709]*
No comparable provision.	Adds income from affiliated packing and handling operations to definition of farm income when calculating adjusted gross income limitation to determine eligibility for conservation programs. *[Sec. 2501]*	No comparable provision.	Deletes provision.

(Continued)

Prior Law/Policy (P.L. 107-171) (unless otherwise Indicated)	House Bill (H.R. 2419)	Senate Amendment (H.R. 2419)	Enacted 2008 Farm Bill (P.L. 110-246)
No comparable provision.	Allows USDA to encourage development of voluntary sustainable practices for specialty crops. *[Sec. 2502]*	No comparable provision.	Deletes provision.
No comparable provision.	Requires USDA to develop information on the importance of productive farmland and designate at least one farmland information center to distribute this and related information. Specifies federal matching funds of at least $400,000 (not exceeding 0.5% of the amount provided to implement the FRPP). *[Sec. 2503]*	No comparable provision.	Deletes provision.
No comparable provision.	Requires USDA to contract with a peanut producer for a 4 year crop rotation; authorizes appropriations up to $10 million annually (FY2008-12). *Sec. 2504]*	Similar to House provision, but provided within CSP, directing USDA to provide additional payments to producers who agree to adopt resource-conserving crop rotations to achieve optimal crop rotations. *[Sec. 2341]*	Adopts Senate provision. *[Sec. 2301]*
No comparable provision. (*Note*: See the Conservation Security Program, above, in existing programs.)	See section 2103 on the Conservation Security Program (described above), for some related changes. For example, the House bill defines "priority resources of concern;" however, the House bill does not create a new program.	Authorizes a new Comprehensive Stewardship Incentives Program to coordinate administration of a new Conservation Stewardship Program (see above) and EQIP. Addresses defined resource concerns, meets regulatory demands, encourages conservation, and promotes conservation and production as compatible goals. *[Sec. 2341]*	Deletes Senate provision. Renames CSP the Conservation Stewardship Program. *[Sec. 2301]*
No comparable provision.	No comparable provision.	Authorizes a Discovery Watershed Demonstration Program to reduce loss of nutrients into surface waters in 30 small watersheds in the Upper Mississippi River basin. Authorizes discretionary funds as are necessary. *[Sec. 2397]*	Deletes provision.
No comparable provision.	No comparable provision.	Authorizes an Emergency Landscape Restoration Program to repair landscapes damaged by natural events. Replaces two others emergency conservation and	Deletes provision.

(Continued)

Prior Law/Policy (P.L. 107-171) (unless otherwise Indicated)	House Bill (H.R. 2419)	Senate Amendment (H.R. 2419)	Enacted 2008 Farm Bill (P.L. 110-246)
		watershed programs. *[Sec. 2398]*	
No comparable provision.	No comparable provision.	Directs USDA to assist producers who apply for programs indirectly through certain organizations, if this will increase participation and program benefits; payment limits apply to each producer, not the organization. *[Sec. 2405]*	Deletes provision. Places payment limits within each section of the bill and deletes this section.
No comparable provision.	No comparable provision.	Authorizes a new Agriculture Conservation Experienced Service Program, such that USDA can enter into agreements with organizations to provide technical assistance using qualified individuals 55 years or older. *[Sec. 2602]*	Adopts Senate provision, but limits individuals employed under this authority to providing only technical assistance (excluding administrative tasks). *[Sec. 2710]*
No comparable provision.	No comparable provision.	Amends the Soil Conservation and Domestic Allotment Act of 1935 by providing definitions and creates new technical assistance provisions. Reauthorizes the Soil and Water Resources Conservation Act of 1977 through 2028; requires a national appraisal of soil, water and related resources to be issued every 10 years. *[Sec. 2603]*	Adopts Senate provision. *[Sec. 2802]*
No comparable provision.	The "sodsaver" provision noncropland (including native grassland and pastureland) planted to an insurable crop ineligible for crop insurance for the first 4 years of planting. *[Sec. 11007]*	Makes native sods planted to an insurable crop (over 5 acres) ineligible for crop insurance and the noninsured crop disaster assistance program. Directs USDA to report within 180 days of enactment, and annually thereafter, on changes in cropland acreage, by county, since 1995. *[Sec. 2608]*	Makes native sods planted to an insurable crop (over 5 acres) ineligible for crop insurance and the noninsured crop disaster assistance program for the first 5 years of planting. May apply to virgin prairie converted to cropland in the Prairie Pothole National Priority Area, if elected by the state. *[Sec. 12020]*
No comparable provision.	No comparable provision.	Requires that no producers in Texas lose program benefits as a result of participating in a study of the Ogallala Aquifer's recharge potential. *[Sec. 2609]*	Adopts Senate provision. *[Sec. 2901]*
No comparable provision.	No comparable provision.	Amends the Federal Insecticide, Fungicide, and Rodenticide Act (FIFRA) *[7 U.S.C. 136o(d)]* to require the State Department to	Adopts the Senate provision on payment of expenses. *[Sec. 14209]* Deletes Senate provision making technical corrections to

(Continued)

Prior Law/Policy (P.L. 107-171) (unless otherwise Indicated)	House Bill (H.R. 2419)	Senate Amendment (H.R. 2419)	Enacted 2008 Farm Bill (P.L. 110-246)
		pay expenses incurred by EPA employees associated with certain international activities. *[Sec. 2610]* Amends sec. 33 of FIFRA *[7 U.S.C. 136w-8]* to allow the EPA Administrator to waive a portion of the pesticide registration service fee under certain circumstances. *[Sec. 2612]*	pesticide registration, but includes a container recycling provision. *[Sec. 14109]*
Sec. 202(a) of the Colorado River Salinity Control Act of 1974 authorizes DOI to construct, operate, and maintain the specific salinity control units as the initial stage of the Colorado River Basin salinity control program. *[43 U.S.C. 1592(a)]*	No comparable provision.	Amends the act to create a basin states program implementing specified salinity control activities. Requires DOI to consult with the Colorado River Basin Salinity Control Advisory Council related to assistance in the form of grants, grant commitments, or the advancement of funds to federal or non-federal entities. Requires a planning report to Congress describing the proposed program implementation; stipulates that no funds may be expended until 30 days after the report is submitted. *[Sec. 2611]*	Adopts Senate provision. The Mangers report states this provision to be fiscally neutral both as to appropriations and as to draws on the basin funds. States there are no changes to the cost share ratios already established in the act; the percentage split between the two funds; or the 15% cap requirement on the basin states cost share derived from the Upper Colorado River Basin Fund. *[Sec. 2806]*
No comparable provision.	No comparable provision.	No comparable provision.	Authorizes USDA to accept contributions to support conservation programs to establish a sub-account for each USDA conservation program to accept contributions of non-Federal funds. Provides that contributions of non-Federal funds received for a conservation program be deposited and shall be available to USDA, without further appropriation and until expended. *[Sec. 2702]*
No comparable provision.	No comparable provision.	No comparable provision.	Direct USDA to submit to Congress an annual report regarding enrollments and assistance under conservation programs, including (1) WRP, FPP, and GRP payments valued at $250,000 or greater; (2) EQIP payments for land determined to have special environmental significance; and (3) AWEP payments

(Continued)

Prior Law/Policy (P.L. 107-171) (unless otherwise Indicated)	House Bill (H.R. 2419)	Senate Amendment (H.R. 2419)	Enacted 2008 Farm Bill (P.L. 110-246)
			subject to the waiver of adjusted gross income limitations. Allows for waivers granted by USDA to protect environmentally sensitive land of special significance. *[Sec. 2705]*
No comparable provision.	No comparable provision.	No comparable provision.	Amends the Soil and Water Resources Conservation Act of 1977 to require USDA to conduct two comprehensive appraisals and inventory of soil, water, and related natural resource conservation (completed by year-end 2010 and 2015). Requires a report in early 2012 on the types of improvements to appraisals and programs. *[Sec. 2804]*
No comparable provision.	No comparable provision.	No comparable provision.	Other miscellaneous provisions: —Names the National Plant Materials Center at Beltsville, MD, in honor of Norman A. Berg. *[Sec. 2902]* —Directs USDA, in consultation with the CCC, to promulgate regulations not later than 90 days after the date of enactment, necessary to implement title II. *[Sec. 2904]*

Author Contact Information

Tadlock Cowan
Analyst in Natural Resources and Rural Development
tcowan@crs.loc.gov, 7-7600

Megan Stubbs
Analyst in Agricultural Conservation and Natural Resources Policy
mstubbs@crs.loc.gov, 7-8707

Renée Johnson
Specialist in Agricultural Policy
rjohnson@crs.loc.gov, 7-9588

End Notes

[1] Farmland conservation programs and initiatives initially were enacted as part of the 1985 farm bill (P.L. 99-198, Food Security Act of 1985), which also included for the first time a conservation title. These programs have been reauthorized, modified, and expanded, and several new programs have been created, particularly in subsequent omnibus farm bills.

[2] For more detailed information on CRP, see CRS Report RS21613, *Conservation Reserve Program: Status and Current Issues*, by Tadlock Cowan.

[3] Acreage in CREP—a subprogram within CRP—and other continuous enrolled land would be excluded from the CRP county acreage cap (25%) in order to encourage greater program participation.

[4] For additional information on CRP's tax status, see CRS Report, CRS Report RL34457, *Conservation Reserve Program Payments: Self-Employment Income, Rental Income, or Something Else?*, by Carol A. Pettit.

[5] The program was authorized as FPP in the 2002 farm bill, and the name has not been changed in statute.

[6] For more detailed information on CRP, see CRS Report RS21740, *Conservation Security Program: Implementation and Termination*, by Tadlock Cowan.

[7] For more information on environmental services and the role of markets in environmental conservation planning, see CRS Report RL34042, *Environmental Services Markets in the 2008 Farm Bill*, by Renée Johnson.

In: Farm Bill of 2008: Major Provisions and Legislative Action ISBN: 978-1-60741-750-7
Editor: Mary T. Conner

Chapter 4

RENEWABLE ENERGY POLICY IN THE 2008 FARM BILL

Tom Capehart
Agricultural Economics

SUMMARY

The Food, Conservation, and Energy Act of 2008 (P.L. 110-246, the 2008 farm bill) extends and expands many of the renewable energy programs originally authorized in the 2002 farm bill. The bill also continues the emphasis on the research and development of advanced and cellulosic bioenergy authorized in the 2007 Energy Independence and Security Act (P.L. 110-140).

Farm bill debate over U.S. biomass-based renewable energy production policy focused mainly on the continuation of subsidies for ethanol blenders, continuation of the import tariff for ethanol, and the impact of corn-based ethanol on agriculture. The enacted bill requires reports on the economic impacts of ethanol production, reflecting concerns that the increasing share of corn production being used for ethanol is contributing to high commodity prices and food price inflation.

Title VII, the research title of the 2008 farm bill, contains numerous renewable energy related provisions that promote research, development, and demonstration of biomass-based renewable energy and biofuels. The Sun Grant Initiative coordinates and funds research at land grant institutions on biobased energy technologies. The Agricultural Bioenergy Feedstock and Energy Efficiency Research and Extension Initiative provides support for on-farm biomass energy crop production research and demonstration.

Title IX, the energy title of the farm bill, authorizes mandatory funds of $1.1 billion and discretionary funds (subject to appropriations) totaling $1.0 billion for the FY2008-FY2012 period. Energy grants and loans provided through initiatives such as the Bioenergy Program for Advanced Biofuels promote the development of cellulosic biorefinery capacity. The Repowering Assistance Program supports increasing efficiencies in existing refineries. Programs such as the Rural Energy for America Program (REAP) assist rural communities and businesses in becoming more energy-efficient and self-sufficient, with an emphasis on small operations. The Biomass Crop Assistance and Development Program, the Biorefinery Assistance Program, and the Forest Biomass for Energy Program provide support to develop alternative feedstock resources and the infrastructure to support the production, harvest, storage, and processing of cellulosic biomass feedstocks. Cellulosic feedstocks—for example, switchgrass and woody biomass—are given high priority both in research and funding.

Title XV of the 2008 farm bill contains tax and trade provisions. It continues current biofuels tax incentives, reducing those for corn-based ethanol but expanding tax credits for cellulosic ethanol. The tariff on ethanol imports is extended.

BACKGROUND

Renewable energy policy in the 2008 farm bill builds on earlier programs first established in the 2002 farm bill. The Farm Security and Rural Investment Act of 2002 (P.L. 107-171, FSRIA) was the first omnibus farm bill to explicitly include an energy title (Title IX). The energy title authorized grants, loans, and loan guarantees to foster research on agriculture-based renewable energy, to share development risk, and to promote the adoption of renewable energy systems.[1] For example, the U.S. Department of Agriculture's (USDA's) Bioenergy Program (Sec. 9006 of P.L. 107-171)—whose funding expired in FY2006—provided direct incentives to expand actual production of bioenergy. See **Table 2** for a list of provisions in the 2002 farm bill's energy title and their funding levels.

Since enactment of the 2002 farm bill, interest in renewable energy has grown rapidly, due in large part to a strong rise in domestic and international petroleum prices and a dramatic acceleration in domestic biofuels production (primarily corn-based ethanol).[2] Many policymakers view agriculture-based biofuels as both a catalyst for rural economic development and a response to growing energy import dependence. Ethanol and biodiesel, the two most widely used biofuels, receive significant federal support in the form of tax incentives, loans and grants, and regulatory programs.[3]

The 2008 farm bill became law six months after the enactment of the Energy Independence and Security Act of 2007 (EISA, P.L. 110-140), and many of its provisions also build on the goals of EISA.[4] The emphasis on facilitating production of biofuels derived from cellulosic feedstocks reflects the goals of the renewable fuels standard (RFS) in EISA. It includes a significant expansion of the RFS to 36 billion gallons by 2022, with carve-outs for biodiesel (1 billion gallons by 2012) and cellulosic ethanol (16 billion gallons by 2022) and an implicit cap on cornstarch ethanol (15 billion gallons by 2015). This RFS expansion is likely to have a significant impact on the U.S. agriculture sector.[5]

The emphasis on cellulosic ethanol also reflects increasing concerns about the economic and environmental issues associated with corn-starch-based ethanol. Record high commodity prices in 2007 and mid-2008, combined with high energy costs, have resulted in sharp increases in livestock feed costs, export prices, and domestic food price inflation. For the first time, an agricultural commodity is directly competing with petroleum in the marketplace. Ethanol production, the profitability of which depends directly on both petroleum and corn prices, accounts for about a third of U.S. corn production. The increase in corn used for U.S. ethanol production exceeds the increase in corn produced during the past three years. Today, as petroleum prices rise so has demand for ethanol as a substitute, which in turn increases both the demand for and price of corn. The "food versus fuel" debate intensified during the farm bill debate as food price inflation accelerated both in the U.S. and globally—highlighting some of the potential problems associated with replacing even a small share of the nation's gasoline consumption with corn-based ethanol. Competition for limited corn supplies between livestock producers, ethanol refiners, exporters, and other domestic users have resulted in calls for at least a partial waiver of the RFS in 2009.[6]

Many of the federal programs that currently support renewable energy production in general, and agriculture-based energy production in particular, are outside the purview of USDA and have legislative origins outside of the farm bill.[7] For example, the RFS mandates the inclusion of an increasing volume of biofuels in the national fuel supply. This originated with the Energy Policy Act of 2005 (P.L. 109-58) and was more recently expanded in EISA. Similarly, the federal tax credits available to biofuels blenders and wind energy producers were initially contained in the American Jobs Creation Act of 2004 (P.L. 108-357) but have now been incorporated in the farm bill.

MAJOR ENERGY PROVISIONS IN THE 2008 FARM BILL

The Food, Conservation, and Energy Act of 2008 (P.L. 110-246, H.R. 6124), enacted in June 2008,[8] significantly expands existing programs to promote biofuels. Like the previous farm bill, it contains a distinct energy title (Title IX) that covers a wide range of energy and agricultural topics with extensive attention to biofuels, including corn-starch based ethanol, cellulosic ethanol, and biodiesel. Research provisions relating to renewable energy are found in Title VII and tax and trade provisions are found in Title XV.

The final 2008 farm bill keeps the structure of Title IX as it was in the Senate-passed version of the farm bill. Title IX serves as a substitute amendment to the 2002 farm bill Title IX and consists of 3 sections. The first section, 9001, contains 13 new provisions which effectively replace the provisions of the 2002 bill. (See the **Appendix** for a side-by-side comparison of previous law with the energy titles from the enacted farm bill and the House- and Senate-passed versions of the 2008 farm bill.)

Key biofuels-related provisions in the enacted 2008 farm bill include:

- emphasis on cellulosic ethanol production through new blender's tax credits, promotion of cellulosic feedstocks production, feedstocks infrastructure and refinery development;
- grants and loan guarantees for biofuels (especially cellulosic) research, development, deployment, and production;
- studies of the market and environmental impacts of increased biofuel use;
- expansion of biofuel feedstock availability;
- expansion of the existing biobased marketing program to encourage federal procurement of biobased products;
- support for rural energy efficiency and self-sufficiency;
- reauthorization of biofuels research programs within the U.S. Department of Agriculture and Environmental Protection Agency;

- an education program to promote the use and understanding of biodiesel;
- reduction of the blenders' tax credit for corn-based ethanol;
- continuation and expansion of the federal bio-products certification program;
- environmental safeguards through greenhouse gas emission requirements on new biofuel production, and
- continuation of the import duty on ethanol.

ENERGY POLICY ISSUES IN THE 2008 FARM BILL

Cellulosic Ethanol

The 2008 farm bill energy title provides $1 billion in financial incentives and support to encourage the production of advanced (mainly cellulosic) biofuels.[9] Grants and loan guarantees leverage industry investments in new technologies and the production of cellulosic feedstocks. For instance, the Biomass Crop Assistance Program supports the production of dedicated crop and forest cellulosic feedstocks and provides incentives for harvest and post production storage and transport. Advanced biofuels refinery capacity construction is assisted under the Biorefinery Program for Advanced Biofuels through grants and loans for the development, construction, and retrofitting of commercial-scale refineries to produce advanced biofuels. These programs are supported by increased funding for advanced biofuels research under the Agricultural Bioenergy Feedstock and Energy Efficiency Research and Extension Initiative, and the Sun Grant Program which support and coordinate advanced biofuels research, extension, and development between government agencies, universities, and research institutions.

Cellulosic ethanol is produced from cellulose, hemicellulose, or lignin which is derived from the structural material that provides much of the mass of plants. Cellulosic ethanol is produced from non-food crops and trees that can be grown on marginal land with low inputs, and appears to offer environmental benefits over corn-based ethanol.[10] Common sources are grasses, corn stover, and wood chips.

Currently, cellulosic ethanol is not produced on a commercial scale. As of July 2008, there was one commercial scale refinery under construction, two demonstration- scale plants and three pilot-scale plants completed. Industry sources expect commercial production of cellulosic ethanol to begin in 2010 or 201. In January 2009, USDA announced funding for a cellulosic biofuels plant under the Biorefinery Program with projected output of 20 million gallons annually, beginning in 2010. The RFS mandates cellulosic ethanol production of 100 million gallons in 2010. For more information on cellulosic ethanol see CRS Report RL34738, *Cellulosic Biofuels: Analysis of Policy Issues for Congress*, by Tom Capehart.

Table 1. 2008 Farm Bill (P.L. 110-246): Authorized Funding for Energy Provisions, FY2008-FY2012

Title IX Provisions	Section and Provision Name	Funding Status	FY2008	FY2009	FY2010	FY2011	FY2012	Sum: FY2008-FY2012
			$ millions					
Title VII Research								
Sec. 7205	Nutrient Management Research and Extension Initiative	Discretionary	SSAN	SSAN	SSAN	SSAN	SSAN	SSAN
Sec. 7207	Bioenergy Feedstock and Energy Efficiency Research and Extension Initiative	Discretionary	50	50	50	50	50	250
Sec. 7526	Sun Grant Program	Discretionary	75	75	75	75	75	375
Title IX Energy								
Sec. 9002	Biobased Markets Program	Mandatory	1	2	2	2	2	9
	Additional funding authorization	Discretionary	0	2	2	2	2	8
Sec. 9003	Biorefinery Assistance	Mandatory	0	75	24	5 0	0	320
	Additional funding authorization	Discretionary	0	150	150	150	150	600
Sec. 9004	Repowering Assistance	Mandatory	0	35	0	0	0	35
	Additional funding authorization	Discretionary	0	15	15	15	15	60
Sec. 9005	Bioenergy Program for Advanced Biofuels	Mandatory	0	55	55	85	105	300
	Additional funding authorization	Discretionary	0	25	25	25	25	100
Sec. 9006	Biodiesel Fuel Education Program	Mandatory	1	1	1	1	1	5

Table 1. (Continued)

Title IX Provisions	Section and Provision Name	Funding Status	FY2008	FY2009	FY2010	FY2011	FY2012	Sum: FY2008-FY2012
			$ millions					
Sec. 9007	Rural Energy for America Program	Mandatory	0	55	60	70	70	255
	Additional funding authorization	Discretionary	0	25	25	25	25	100
Sec. 9008	Biomass Research and Development Act	Mandatory	0	20	28	30	40	118
	Additional funding authorization	Discretionary	0	35	35	35	35	140
Sec. 9009	Rural Energy Self-Sufficiency Initiative	Mandatory	0	0	0	0	0	0
	Additional funding authorization	Discretionary	0	5	5	5	5	20
Sec. 9010	Feedstock Flexibility Program for Bioenergy Producers	Mandatory	0	0	0	0	0	0
	Additional funding authorization	Discretionary	SSAN	SSAN	SSAN	SSAN	SSAN	SSAN
Sec. 9011	Biomass Crop Assistance Program	Discretionary	SSAN	SSAN	SSAN	SSAN	SSAN	SSAN
Sec. 9012	Forest Biomass for Energy	Discretionary	0	15	15	15	15	60
Sec. 9013	Community Wood Energy Program	Discretionary	0	5	5	5	5	20
Sec. 9003	Renewable Fertilizer Study	Discretionary	0	1	0	0	0	1
	Total Discretionary Funding	**Discretionary**	**$125**	**$403**	**$402**	**$402**	**$402**	**$1,734**
	Total Mandatory Funding	**Mandatory**	**$2**	**$243**	**$391**	**$188**	**$218**	**$1,042**

Source: P.L. 1 1 0-246 (Food and Energy Security Act of 2008.

Notes: “SSAN” = Such sums as necessary.

Table 2. 2002 Farm Bill Title IX Energy Funding (Presidential Request, Authorization, and Enactment), by Provision, FY2002 to FY2007

Current law provision		FY2002			FY2003			FY2004			FY2005			FY2006			FY2007 [b]		
		Pres. Req.	Auth.	Enact.[a]	Pres. Req.	Auth.	Enact.[a]	Pres. Req.	Auth.	Enact.[a]	Pres. Req.	Auth.	Enact.[a]	Pres. Req.	Auth.	Enact.[a]	Pres. Req.	Auth.	Enact.[a]
		$ millions			$ millions			$ millions			$ millions			$ millions			$ millions		
Sec. 9002	Federal Biobased Product Procurement	1	1	1	1	1	1	1	1	1	1	1	1	1	1	1	1	1	0
Sec. 9003[e]	Biorefinery Development Grants	0	SSAN	0	0	SSAN	0	0	SSAN	0	0	SSAN	0	0	SSAN	0	0	SSAN	0
Sec. 9004	Biodiesel Fuel Education Program	1	1	1	1	1	1	1	1	1	1	1	1	1	1	1	1	1	0
.																			
Sec. 9005[e]	Energy Audit and Renewable Energy Program	0	SSAN	0	0	SSAN	0	0	SSAN	0	0	SSAN	0	0	SSAN	0	0	SSAN	0
Sec. 9006	Renewable Energy Systems[c]	23	23	23	18	23	23	3	23	23	11	23	23	10	23	23	10	23	0
Sec. 9007[e]	Hydrogen and Fuel Cell Technologies	0	SSAN	0	0	SSAN	0	0	SSAN	0	0	SSAN	0	0	SSAN	0	0	SSAN	0
Sec. 9008[e]	Biomass Research and Development[d]	5	54	5	14	63	14	14	63	14	14	63	14	12	214	12	12	214	0

Table 2. (Continued)

Current law provision		FY2002			FY2003			FY2004			FY2005			FY2006			FY2007 b		
		Pres. Req.	Auth.	Enact.[a]	Pres. Req.	Auth.	Enact.[a]	Pres. Req.	Auth.	Enact.[a]	Pres. Req.	Auth.	Enact.[a]	Pres. Req.	Auth.	Enact.[a]	Pres. Req.	Auth.	Enact.[a]
		$ millions			$ millions			$ millions			$ millions			$ millions			$ millions		
Sec. 9009	Cooperative Research and Extension Projects	0	0	0	0	0	0	0	0	0	0	0	0	0	0	0	0	0	0
Sec. 9010[e]	Continuation of Bioenergy Program	0	0	0	150	150	116	150	150	150	100	150	100	60	150	60	0	0	0
Total	Energy Programs	30	79	30	184	238	155	169	238	189	127	238	139	84	389	97	24	239	0

Note: "Pres. Req." = Presidential request; "Auth." = Farm bill authorized level; "Enact." = Enacted; "SSAN" = Such sums as necessary. Sec. 9003 and Sec. 9005 were not implemented.

a. Represents a combination of the amount authorized by the 2002 Farm Bill, less any reductions in annual appropriations acts, plu s discretionary funding provided where applicable.

b. Complete data on appropriations by program is not yet available.

c. The Deficit Reduction Act of 2005 (P.L. 109-171) reduced Sec. 9006 mandatory funding by $20 million in FY2007, leaving $3 milli on as the authorized level for FY2007.

d. Section 310 (7 U.S.C. 8609) of the Biomass Research and Development Act of 2000 (P.L. 106-224) provides discretionary authorization of $200 million for each of FY2006 through FY2015.

e. Joint program with the Department of Energy.

Tax Credits and Tariffs

Title XV of the farm bill contains provisions which extend and modify tax credits and tariffs on ethanol. In keeping with the promotion of cellulosic ethanol, a blenders' credit of $1.01 per gallon applies to ethanol produced from qualifying cellulosic feedstocks. This tax credit is intended to spur investment in cellulosic ethanol production. The ethanol blender's tax credit of $0.51 per gallon (which applies to all ethanol blended, including imports) was reduced to $0.45 per gallon in January 2009. Section 15331 of the farm bill requires the reduction starting the first year following that year in which U.S. ethanol production and imports exceed 7.5 billion gallons. Production and imports in 2008 are expected to have exceeded 9 billion gallons.

The import tariff for ethanol benefits the U.S. ethanol industry by protecting U.S. ethanol from lower-cost imports. The tariff also keeps imported ethanol from benefitting from the blenders' tax credit when it is blended into gasoline in the United States. The tariff of $0.54 per gallon that was set to expire at the end of 2008 is now extended to the end of 2010.

Economic Impacts of Ethanol Production

The impact of increased ethanol production on agricultural and rural economies was a subject of debate during the farm bill process. As a result, the farm bill includes provisions requiring a series of reports assessing how ethanol production may be impacting the farm economy, the environment, and consumer food prices. Among these are the Comprehensive Study of Biofuels (to be conducted by the U.S. Department of Agriculture (USDA), the Environmental Protection Agency (EPA), the Department of Energy (DOE), and the National Academy of Sciences) and the Biofuels Infrastructure Study by USDA, DOE, Department of Transportation (DOT), and EPA. The Biomass Crop Assistance Program requires an assessment of the economic impacts of expanded cellulosic biomass production on local economies and infrastructures. Likewise, the Biomass Research and Development Program requires a report on the economic impacts of rural economies of biorefinery expansion and conversion by USDA.

Funding for Energy Programs

Table 1 illustrates mandatory and discretionary spending levels for renewable energy programs in the 2008 farm bill. Title IX authorizes $1.1 billion in mandatory funding for FY2008 through FY2012, compared with $800 million in the 2002 farm bill (FY2002-FY2007).[11] Of the mandatory funding, $320 million was allocated to the Biorefinery Assistance Program, $300 million to the Bioenergy Program for Advanced Biofuels and $255 million to the Rural Energy for America Program (REAP). Authorizations for appropriations in the 2008 farm bill total $1 billion, four times the $245 million in the 2002 farm bill. Most of the increase is for the Biorefinery Assistance Program, which has an authorization $600 million higher than in the 2002 farm bill. **Table 2** provides a list of provisions in the 2002 farm bill's energy title for FY2002 through FY2007 and their funding levels (as requested by the President, and as authorized and provided by Congress).

Appendix. Comparison of the Enacted 2008 Farm Bill (P.L. 110-246) with Previous Law and the House- and Senate-Passed 2008 Farm Bill (H.R. 2419)

2002 Farm Bill (FSRIA, P.L. 107-171) or Other Law (as indicated)	House-Passed Farm Bill (H.R. 2419)	Senate-Passed Farm Bill (H.R. 2419)	Enacted Farm Bill (P.L. 110-246)
TITLE VII: Agricultural Research and Extension			
Nutrient Management Research and Extension Initiative			
Section 1673(h) of the Food, Agriculture, Conservation, and Trade Act of 1990 (P.L. 101-624) authorized matching grants under the farm bill nutrient management research and extension initiative for finding innovative methods and technologies for economic use or disposal of animal waste. Extended through 2007 in section .7120 of the 2002 farm bill. Such sums as necessary were appropriated annually for FY1999-FY2007. [7 U.S.C. 5925a]	Extends the nutrient management research and extension initiative through 2012 and adds dairy cattle waste as a type of waste to be studied. The authorization of appropriations for such sums as necessary is extended annually for FY2008-12. [Sec. 7307]	Extends program through FY2012. [Sec. 7103]	The Conference substitute adopts the House provision with an amendment to include the production of renewable energy from animal waste as an eligible activity to receive grants under this section. Authorizes such sums as necessary annually for FY2007-12. [Sec. 7205]
Agricultural Bioenergy Feedstock and Energy Efficiency Research and Extension Initiative			
No Provision.	Initiates multi-regional biomass crop experiments with a network of laboratories to provide a sound knowledge base on the production of biomass energy crops, including crop species, nutrient requirements, management practices, environmental impacts, greenhouse gas implications, and economics. Competitive grants to be awarded to a network of laboratories. Discretionary appropriations of $50 million authorized annually FY2008-FY2012. [Sec 7410]	New Section 9010 of FSRIA. Regional Biomass Crop Experiments. Establishes a program of regional biomass crop experiments at 10 geographically dispersed and competitively selected land-grant universities. Individual grants total $4 million for FY2008-FY2010. Crop experiments are to include all appropriate biomass species, including perennials, annuals, and woody biomass species in coordination with the Biomass Research and Development Board and with the Sun Grant Centers. Establishes a "best practices" database on all aspects of biomass crop production. Mandatory CCC funding of $10 million in FY2008, $20 million in FY2009, and $10 million in	Establishes the Agricultural Bioenergy Feedstock and Energy Efficiency Research and Extension Initiative, a program to award competitive matching (up to 50%) grants for projects with a focus on supporting on-farm biomass crop research and the dissemination of results to enhance the production of biomass energy crops and the integration of such production with the production of bioenergy. Discretionary appropriations of $50 million annually are authorized for FY2008-12. [Sec. 7207]

(Continued)

2002 Farm Bill (FSRIA, P.L. 107-171) or Other Law (as indicated)	House-Passed Farm Bill (H.R. 2419)	Senate-Passed Farm Bill (H.R. 2419)	Enacted Farm Bill (P.L. 110-246)
		FY2010. [Sec. 9001]	
New Century Farm Project			
No provision.	No provision.	New section 9011 of FSRIA. Establishes the New Century Farm Project to serve as a model farm. Authorizes appropriations of $15 million for FY2008 through FY2012, to remain available until expended.	The purpose of this section Is incorporated in section 7207.
Biochar Research, Development, and Demonstration			
No provision.	No provision.	New section 9012 of FSRIA. Creates a program of competitive grants for research and demonstration of the production and use of biochar in the agricultural sector. Provides discretionary funding of $3 million for each of FY2008 through FY2012. [Sec. 9001]	The purpose of this section Is incorporated in the as a research priority in section 7203.
Farm Energy Production Pilot Program			
No provision.	Farm Energy Production Pilot Program. Provides grants to farmers to demonstrate the feasibility of making a farm energy neutral using existing technologies. Discretionary funds of $5 million are authorized to be appropriated for FY2008-12. [Sec. 9010]	No provision.	The purpose of this section Is incorporated in the Agricultural Bioenergy and Biobased Products Research Initiative of section 7207.
Farm Energy Production Pilot Program			
No provision.	Provides grants of not more than $1 million to no more than 20 universities to develop a 3-year demonstration program of supplementing corn as an ethanol feedstock with sweet sorghum and switchgrass. Discretionary funds of $20 million are authorized for FY2008-FY2012. [Sec. 9020]	No provision.	The purpose of this section Is incorporated in the Agricultural Bioenergy and Biobased Products Research Initiative of section 7207.

(Continued)

2002 Farm Bill (FSRIA, P.L. 107-171) or Other Law (as indicated)	House-Passed Farm Bill (H.R. 2419)	Senate-Passed Farm Bill (H.R. 2419)	Enacted Farm Bill (P.L. 110-246)
Research and Development of Renewable Energy			
No provision.	No provision.	New section 9022 of FSRIA. USDA will conduct research, with the Colorado Renewable Energy Collaboratory, relating to various aspects of renewable energy including biomass crops adapted to arid and semi-arid regions, and storage and conversion technologies for wind and solar energy. Authorizes discretionary funding of $5 million for each of FY2008 through FY2012. Additional discretionary funding of $110 million to USDA Under Secretary for Research, Education, and Economics for cellulosic biofuel research for each of FY2008 through FY2012; and additional discretionary funding of $110 million to USDA/DOE for development of small-scale biorefineries for each of FY2008 through FY2012.[Sec. 9001]	The purpose of this section is incorporated in the Agricultural Bioenergy and Biobased Products Research Initiative of section 7207.
Northeast Dairy Nutrient Management and Energy Development Program			
No provision.	No provision.	New section 9023 of FSRIA. Establishes USDA program to provide grants to a consortium of Northeast U.S. land-grant colleges and universities to conduct research, extension, and demonstration projects for dairy nutrient management and energy development . Authorizes discretionary funding of such sums as are necessary. [Sec. 9001]	The purpose of this section Is incorporated in section 7204.
Future Farmsteads Program			
No provision.	Section 9015. Establishes a program to equip a farm entity (house and lands), in each of 5 regions of the country, with technologies to improve farm energy production and use efficiencies. Each designated entity would serve as a	New section 9025 of FSRIA. Same as House bill. [Sec. 9001]	The purpose of this section Is incorporated in the Agricultural Bioenergy and Biobased Products Research Initiative of section 7207.

(Continued)

2002 Farm Bill (FSRIA, P.L. 107-171) or Other Law (as indicated)	House-Passed Farm Bill (H.R. 2419)	Senate-Passed Farm Bill (H.R. 2419)	Enacted Farm Bill (P.L. 110-246)
	working example to farmers and as an educational, research, and demonstration facility for research related to renewable energy or energy conservation technologies. Authorizes appropriations of such sums as are necessary to carry out the program.		
Sun Grant Program			
Section 9011. This provision was added subsequent to the 2002 farm bill under the Sun Grant Research Initiative Act of 2003. Establishes 5 national Sun Grant research centers based at land-grant universities, each covering a different national region, to enhance coordination and collaboration between USDA, DOE, and land-grant universities in the development, distribution, and implementation of biobased energy technologies. Competitive grants are available to land-grant schools within each region. Authorized appropriations of $25 million in FY2005, $50 million in FY2006, and $75 million for each of FY2006 through FY2010 for total discretionary funding of $375 million during FY2005-10. [7 U.S.C. 8109]	Adds a Sun Grant center at the University of Hawaii. Reauthorizes the Sun Grant program through FY2012 with authorized discretionary appropriations of $75 million for each of FY2008-12, for a total in discretionary funding of $375 million. [Sec. 9008]	New section 9009 of FSRIA. Reauthorizes the Sun Grant Program through FY2012 and establishes a 6th regional center—Western Insular Pacific Sub-Center—at the University of Hawaii. Provides, for the first time, mandatory funding for the Sun Grant program of $5 million in FY2008, $10 million in each of FY2009 and FY2010 (available until expended) for a total of $25 million for FY2008-12. Also authorizes appropriations of $70 million in each of FY2008-12. [Sec. 9001]	Adopts the Senate provision with amendments and moves the provision in statute to the research title of this act. No mandatory funding is provided. The conference substitute authorizes $75 million per year for FY2008-FY2010. [Sec. 7526]
TITLE IX: Energy			
Definitions			
Sec. 9001. Defines Administrator, Biomass, Biobased Product, Procuring Agency, Renewable Energy, Rural Small Business, and Secretary. [7 U.S.C. 8101]	No changes to definitions. [Sec. 9001]	Sec. 9001. Adds several definitions including "Advanced Biofuels," which excludes any fuel derived from corn starch, but includes ethanol derived from other plant starches (e.g., sorghum), sugar, as well as cellulosic biomass or organic waste; it also includes organically-derived biogas, butanol or other alcohols;	New section 9001 of FSRIA. Adopts the Senate definitions with amendments. Advanced biofuels includes aviation, jet, and heating fuels made from cellulosic biomass. [Sec. 9001]

(Continued)

2002 Farm Bill (FSRIA, P.L. 107-171) or Other Law (as indicated)	House-Passed Farm Bill (H.R. 2419)	Senate-Passed Farm Bill (H.R. 2419)	Enacted Farm Bill (P.L. 110-246)
		and, notably, biodiesel. Other definitions are biobased product; biomass conversion facility, biorefinery, intermediate ingredient or feedstock; renewable biomass, and renewable energy. [Section 9001]	
Biobased Markets program			
Sec. 9002. Requires federal agencies to purchase biobased products under certain conditions and authorize a voluntary biobased labeling program. USDA regulations define biobased products, identify biobased product categories, and specify the criteria for qualifying those products for preferred procurement. Mandatory Commodity Credit Corporation (CCC) funding of $1 million is authorized for each of FY2002 through FY2007 for testing biobased products. [7 U.S.C. 8101]	Continues federal commitment to biobased product preference in its purchases. Requires annual reports from agencies to Administrator of Federal Procurement Policy. Also requires annual reports from USDA to Congress on implementation status. Clarifies that products with at least 5% of intermediate ingredients and feedstock that are biobased should be considered under the preference. Requires USDA to complete rulemaking on labeling regulation. Increases mandatory CCC funding to $2 million for each of FY2008 through FY2012 for bio-product testing, labeling, and procurement research, promotion, and awareness initiatives. [Sec. 9002]	New section 9002 of FSRIA. Continues the Federal Biobased Products Procurement Program, clarifying that biobased intermediate ingredients and feedstocks qualify under the program. It also requires USDA to establish a voluntary labeling program for biobased products—"USDA Certified Biobased Product." Additionally, this section provides grants for education and awareness of bioenergy and biobased products. Provides $3 million per year in mandatory funding for FY2008 through FY2012. [Sec. 9001]	New section 9002 of FSRIA. Renames program as the Biobased Markets Program. Requires procuring agencies to establish a program and specifications for procuring biobased products (excluding motor vehicle fuels, heating oil, or electricity). Establishes the voluntary labeling program: "USDA Certified Biobased Product." Requires USDA to establish a national registry of biobased testing centers and a report on implementation. Mandatory CCC funding of $1 million in 2008, and $2 million annually for FY2009-FY2012. Discretionary funding of $2 million annually is authorized for FY2009-FY2012. [Sec. 9001]
Biorefinery Assistance			
No provision.	Included in section 9003, below.	Included in new section 9005 of FSRIA, below.	New section 9003 of FSRIA. Biorefinery Assistance Program. Assists in the development of new and emerging technologies for the development of advanced biofuels. Provides competitive grants and loan guarantees for construction and retrofitting of biorefineries for the production of advanced biofuels. Biorefinery grants provided for up to 30% of total cost. Each loan guarantee is limited to $250 million or 80% of project cost. Mandatory

(Continued)

2002 Farm Bill (FSRIA, P.L. 107-171) or Other Law (as indicated)	House-Passed Farm Bill (H.R. 2419)	Senate-Passed Farm Bill (H.R. 2419)	Enacted Farm Bill (P.L. 110-246)
			funding of $75 million in FY2009 and $245 million in FY2010, available until expended for loan guarantees. Discretionary funding of $150 million annually is authorized for FY2009-12. [Sec. 9001]
Repowering Assistance			
Section 9003. Established a grant program to help finance the cost of developing and constructing biorefineries and biofuel production plants to carry out projects to demonstrate the commercial viability of converting biomass to fuels or chemicals. No mandatory funding was authorized and no discretionary funding has been appropriated for the program. Therefore, no implementation regulations have been developed. [7 U.S.C. 8103]	Renamed as the Biorefinery Development Program and extended through FY2012. It replaces grants with new loan guarantee authority of $2 billion for biorefineries—one-half for loans less than $100 million, and the other half for loans between $100 and $250 million. The loan guarantee would cover 90% of an eligible loan. Requires that construction contractors and subcontractors on federally assisted loan guarantee projects pay their employees not less than the prevailing wage in the same locality under the Davis-Bacon Act. Authorizes mandatory CCC funding of $75 million in FY2008; $100 million in FY2009; $125 million in FY2010; $200 million in FY2011; and $300 million in FY2012 for a total in new mandatory funding of $800 million during FY2008-FY2012. [Sec. 9003]	New section 9005 of FSRIA: Renamed the Biorefinery and Repowering Assistance Program, it refocuses emphasis on "advanced biofuel" production in renewable-fuel-powered biorefineries. Provides competitive grants for up to 50% of project costs for pilot- and demonstration-scale biorefineries; matching funds for feasibility studies; competitive grants for up to 20% of total costs for the repowering of fossil-fueled biomass conversion facilities with renewable resources; and loan guarantees for up to 80% of total eligible project costs for the development and construction of commercial-scale biorefineries and the repowering of biomass conversion facilities with renewable energy. Any loan guaranteed for commercial scale biorefineries cannot exceed $250 million and a loan guaranteed for repowering cannot exceed $70 million. Provides $300 million in mandatory CCC funding during FY2008, to remain available until expended. [Sec. 9001]	New section 9004 of FSRIA. Provides payments to encourage biorefineries in existence on the date of enactment to convert from fossil fuel to renewable energy power sources. Encourages new production of energy for refineries from renewable biomass. Mandatory funding of $35 million for FY2009, to remain available until expended. Discretionary funding of $15 million annually for FY2009-12 is authorized. [Sec. 9001]

(Continued)

2002 Farm Bill (FSRIA, P.L. 107-171) or Other Law (as indicated)	House-Passed Farm Bill (H.R. 2419)	Senate-Passed Farm Bill (H.R. 2419)	Enacted Farm Bill (P.L. 110-246)
Bioenergy Program for Advanced Biofuels			
Section 9010. Originally created by a 1999 Executive Order during the Clinton Administration, the bioenergy program provides CCC incentive payments to biofuels producers based on year-to-year increases in the quantity of biofuel produced. Mandatory CCC funding of $150 million was made available for each of FY2002 through FY2006. No funding was authorized for FY2007. [7 U.S.C. 8108]	.Renews and extends the bioenergy program through FY2012. Ethanol produced from corn starch is excluded. Expands eligibility for the production incentive to the production of heat and power at a biofuels plant, biomass gasification, hydrogen for fuel cells, and renewable diesel. Provides mandatory CCC funding of $225 million in FY2008; $250 million in FY2009; $275 million in FY2010; $300 million in FY2011; and $350 million in FY2012 for a total of $1.4 billion during FY2008-12. [Sec. 9007]	New section 9006 of FSRIA. Renews and extends the Bioenergy Program through FY2012. Bases the payment rate on: (1) advanced biofuel production, (2) feedstock prices, and (3) net non-renewable energy content of the biofuel. This will benefit those purchasing feedstocks for cellulosic biofuels and biodiesel. The program is not available to those claiming a biofuel blenders' tax credit (see Sec. 12312) or those with biofuel production capacity greater than 150 million gallons per year. Provides $245 million in mandatory funding for FY2008 through FY2012. [Sec. 9001]	New section 9005 of FSRIA. Establishes the Bioenergy Program for Advanced Biofuels to encourage production of advanced biofuels. Not more than 5% of the funds can go to facilities with total refining capacity exceeding 150 million gallons per year. Producers of advanced biofuels contracts with USDA to receive payments based on the quantity and duration of production of advanced biofuels, the net renewable energy content of the biofuel, and other factors. Payments limited to ensure equitable distribution. Mandatory funding of $55 million for 2009, $55 million for FY2010, $85 million for FY2011, and $105 million for FY2012. Discretionary funding of $25 million annually is authorized for FY2009-12. [Sec. 9001]
Biodiesel Fuel Education Program			
Section 9004. Administered by USDA's Cooperative State Research, Education, and Extension Service, awards competitive grants to nonprofit organizations that educate governmental and private entities operating vehicle fleets, and educates the public about the benefits of biodiesel fuel use. Mandatory CCC funding of $1 million was authorized for each of FY2003 through FY2007. [7 U.S.C. 8104]	Extends the Biodiesel Fuel Education Program through FY2012 with mandatory funding of $2 million for each of FY2008 through FY2012. [Sec. 9017]	New section 9003 of FSRIA. Same as H.R. 2419. [Sec. 9001]	New section 9006 of FSRIA. Extends the Biodiesel Fuel Education Program through 2012. Mandatory CCC funds of $1 million are provided annually for FY2008-12. [Sec. 9001]

(Continued)

2002 Farm Bill (FSRIA, P.L. 107-171) or Other Law (as indicated)	House-Passed Farm Bill (H.R. 2419)	Senate-Passed Farm Bill (H.R. 2419)	Enacted Farm Bill (P.L. 110-246)
Energy Audit and Renewable Energy Development Program			
Section 9005. A competitive grant program for eligible entities to provide energy audits and technical assistance to agricultural producers and rural small businesses to assist them in becoming more energy efficient and in using renewable energy technology and resources. Authorized appropriations of such sums as are necessary to carry out the program for each of FY2002 through FY2007. [7 U.S.C. 8105]	Section 9004. Extends the Energy Audit and Renewable Energy Development Program through FY2012.	New section 9007 of FSRIA. Extends the Energy Audit and Renewable Energy Development Program through FY2012, but folds it into the new Rural Energy for America Program where mandatory funding is available (see the Renewable Energy Systems section for more information). [Sec. 9001]	See new section 9007 of FSRIA below.
Rural Energy for America Program			
The Renewable Energy Systems and Energy Efficiency Program (Section 9006), administered by USDA's Rural Development Agency, authorizes direct loans, loan guarantees, and grants to farmers, ranchers, and rural small businesses to purchase and install renewable energy systems and to make energy efficiency improvements. Grant funds may be used to pay up to 25% of project costs; combined grants and loans or loan guarantees may fund up to 50% of project cost. Eligible projects include those that derive energy from wind, solar, biomass, or geothermal sources. Projects using energy from those sources to produce hydrogen from biomass or water are also eligible. Mandatory CCC funding of $23 million was made available for each of FY2003 through FY2007. Unspent money lapses at the end of each year. [[7 U.S.C. 8106]	Renamed as the Rural Energy for America Program. Includes other agricultural producers as eligible recipients. Raises the loan guarantee level from $10 million to $25 million. Grant fund cost share is retained at up to 25% of project costs; however, combined grants and loans (including guarantees) is expanded to 75% of project cost. 10% of funds are available for feasibility studies, but excludes recipient from other assistance for feasibility studies. Includes sale of electricity from renewable energy systems as eligible activities for grant and loan funding. 15% of funds available reserved for projects of $50,000 or less. Increases mandatory CCC funding to $50 million in FY2008; $75 million in FY2009; $100 million in FY2010; $125 million in FY2011; and $150 million in FY2012 for total new mandatory funding of $500 million during FY2008-FY2012. [Sec. 9005]	New section 9007 of FSRIA. Renamed as the Rural Energy for America Program, this section adds the option to receive a production incentive payment in lieu of a grant, and stipulates that a grant may cover up to 25% of project costs while a loan guarantee may cover up to 75% of the cost. Preference given to state-funded projects. Direct loans are excluded. Adds state agencies and public power entities to eligible participants. Energy from manure: Provides grants and loan guarantees to build and evaluate on-farm and community animal manure- to-energy facilities for which agricultural producers, rural small businesses, and cooperatives are eligible participants. Criteria based on energy quality, efficiency, environmental issues, greenhouse gas emissions, diversity factors and cost. Establishes streamlined review and application process for small-scale projects (under $20,000), and requires that 20% of funds be used for such projects. The Energy Star Program is extended to identify and	New section 9007 of FSRIA. Renamed as the Rural Energy for America Program. Provides grants and loan guarantees to State governments, tribal, or local governments, land-grant institutions, rural electric cooperatives or utilities to provide energy audits and renewable energy assistance, and financial assistance for energy efficiency improvements and renewable energy systems. Grants up to 25% of cost are provided. Loan guarantees up to $25 million. Combined amount of grant and guaranteed loans limited to 75% of cost. 20% of funds made available in this section to be reserved for grants of $20,000 or less until the end of the fiscal year. Mandatory CCC funds of $55 million in FY2009, $60 million in FY2010, $70 million in FY2011, and $70 million in FY2012. Discretionary funding of $25 million annually is authorized to be appropriated for FY2009-12. [Sec. 9001]

(Continued)

2002 Farm Bill (FSRIA, P.L. 107-171) or Other Law (as indicated)	House-Passed Farm Bill (H.R. 2419)	Senate-Passed Farm Bill (H.R. 2419)	Enacted Farm Bill (P.L. 110-246)
		promote energy-efficient equipment and facilities in the agricultural sector. Provides $230 million in mandatory funding for FY2008 to remain available until expended, with not less than 15% of this funding dedicated to the animal manure-to-energy program and not less than 5% for the Energy Audit and Renewable Energy Development Program provision. [Sec. 9001]	
Biomass Research and Development Program			
This program—created originally under the Biomass Research and Development Act of 2000 (BRDA, P.L. 106-224)—provides competitive funding for research, development, and demonstration projects on biofuels and bio-based chemicals and products, under the Biomass Research and Development Initiative, administered jointly by USDA and DOE. Creates Biomass research and Development Board to coordinate government activities in biomass research, and the Biomass Research and Development Technical Advisory Committee to advise on proposal direction and evaluation. Authorized mandatory CCC funding of $5 million in FY2002 and $14 million for each of FY2003 through FY2007 (available until expended). Additional appropriation authority of $200 million for each of FY2006 through FY2015. [7 U.S.C. 8101]	Extends the program through FY2012. Authorizes mandatory CCC funding of $35 million in FY2008; $60 million in FY2009; $75 million in FY2010; $100 million in FY2011; and $150 million in FY2012 for total new mandatory funding of $420 million during FY2008-FY2012. Continues the appropriations Appropriation authority of$200 million in each of FY2008 through FY2015. [Sec. 9006]	New section 9008 of FSRIA. Extends the program and transfers statutory authority to this act. Adds new emphasis on utilization of byproducts such as dried distillers grains and solubles (DDGS) and development of technologies for harvest, storage, pre-processing and transportation of renewable biomass feedstocks. Provides mandatory funding (to remain available until expended) of $15 million in FY2008, $25 million in FY2009, and $35 million in FY2010 for a total in new mandatory funding of $75 million during FY2008-FY2012. Also includes the appropriations authority for $85 million in each of FY2008 through FY2012. [Sec. 9001]	New section 9008 of FSRIA. Moves the Biomass Research and Development Act of 2000 in statute to Title IX of FSRIA of 2002. Defines biobased product. Expands advisory committee. New technical areas for grants include feedstock development, biofuels and biobased products development, and biofuels development analysis with a minimum of 15% of funding going to each area. Minimum cost-share requirement for demonstration projects increased to 50% and research projects to 20%. Provides for coordination of biomass research and development, including life cycle analysis of biofuels, between USDA and DOE. Authorizes mandatory funding of $20 million for FY2009, $28 million for FY2010, $30 million for FY2011, and $40 million for FY2012. Discretionary funding of $35 million is authorized to be appropriated annually for FY2009-12. [Sec. 9001]

(Continued)

2002 Farm Bill (FSRIA, P.L. 107-171) or Other Law (as indicated)	House-Passed Farm Bill (H.R. 2419)	Senate-Passed Farm Bill (H.R. 2419)	Enacted Farm Bill (P.L. 110-246)
Rural Energy Self-Sufficiency Initiative			
No provision.	Establishes the Rural Energy Self-Sufficiency Initiative to provide cost-share grants to enable eligible rural communities (with population less than 25,000) to substantially increase their energy self-sufficiency. Grants may not exceed 75% of costs. Requires a report to Congress on best practices/approaches. Total grants under this section are limited to no more than 5 per year. Authorizes discretionary appropriations for grants of not more than $5 million for FY2008, and such sums as are necessary for FY2009-12. [Sec. 9011 (appended as section 9014 of FSRIA)]	New section 9015 of FSRIA. Establishes a program—the Rural Energy Systems Renewal Program—of competitive cost-shared grants (50% maximum federal share) for rural communities to assess their energy systems and formulate strategies for improvements. Authorizes discretionary funding of $5 million annually for FY2008-12. [Sec. 9001]	New section 9009 of FSRIA. Establishes the Rural Energy Self-Sufficiency Initiative to assist rural communities with community-wide energy systems that reduce conventional energy use and increase the use of energy from renewable sources. Grants are made available to assess energy use in a rural community, evaluate ideas for reducing energy use, and develop and install integrated renewable energy systems. Grants not to exceed 50% of the total cost of the activity. Appropriations of $5 million annually are authorized for FY2009-12. [Sec. 9001]
Feedstock Flexibility Program for Bioenergy Producers			
No provision.	Requires that USDA establish (in FY2008) and administer a sugar-for-ethanol program using sugar intended for food use but deemed to be in surplus. USDA would implement the program only in those years where purchases are determined to be necessary to ensure that the sugar program operates at no cost. The use of such sums as necessary to carry out the program is authorized. [Sec. 9013] (would be appended as Section 9016 of FSRIA).	See Title I, Section 1501, Sugar Program, for a similar sugar-for-ethanol provision entitled the "Feedstock Flexibility Program for Bio-Energy Producers." Identical to the House provision.	New section 9010 of FSRIA. Adopts the House provision with amendments. [Sec. 9001]

(Continued)

2002 Farm Bill (FSRIA, P.L. 107-171) or Other Law (as indicated)	House-Passed Farm Bill (H.R. 2419)	Senate-Passed Farm Bill (H.R. 2419)	Enacted Farm Bill (P.L. 110-246)
Biomass Crop Assistance Program			
No provision.	Establishes a new Biomass Energy Reserve (BER) to encourage the production of dedicated energy crops, establish biomass energy reserve project areas, and to provide financial and technical assistance to owners and operators for harvesting, storing, and transporting cellulosic material to bioenergy facilities. Provides financial and technical assistance (including five-year contracts) for landowners and operators to grow dedicated energy crops as feedstock for cellulosic ethanol and other energy production. BER projects must be within a 50-mile radius of a bioenergy facility. Authorizes annual mandatory funding of such sums as are necessary for FY2008-FY2012. [Sec. 9018]	New section 9004 of FSRIA. Creates the Biomass Crop Transition Assistance Program to stimulate production of biomass crops. USDA provides technical and financial assistance for the production of perennial bioenergy crops used in a biomass conversion facility. Restricted to advanced biofuels, biobased products, or heat or power from a biomass conversion facility. Provides $130 million in mandatory funding for FY2008 through FY2012 to support biomass production. Also provides $10 million in mandatory funding for each of FY2009 through FY2011 (to remain available until expended) for a fixed assistance payment to agricultural producers for production, collection, harvest, storage and transportation of biomass crops for use in a biomass conversion facility. Requires USDA provide public information on the production potential, environmental impacts, and best practices for production, harvesting, transport and storage of biomass crops and requires participants to provide information as needed for that purpose. [Sec. 9001]	New section 9011 of FSRIA. Establishes the Biomass Crop Assistance Program (BCAP) to provide producers committing to biomass production or a biomass conversion facility with contracts which will enable producers in a BCAP project area to receive financial assistance for crop establishment costs and annual payments t for biomass production. Producers must be within economically practicable distance from a biomass facility and adhere to resource conservation requirements. Cost-share payments cover costs of establishing crop, for collection, harvest, storage and transportation to a biomass conversion facility. Annual payments authorized to producers to support biomass production. A report is required no later than 4 years after enactment. Mandatory CCC funds of such sums as necessary are made available for each of FY2008-12. [Sec. 9001]
Forest Biomass for Energy			
No provision.	.Establishes a new competitive research and development program to encourage the use of woody biomass for bioenergy production. Provides mandatory CCC funding of $15 million for each of FY2008-FY2012. [Sec. 9019]	New section 9013 of FSRIA. Same as the House provision, but with discretionary (not mandatory) authorization of $5 million for each of FY2008-12. [Sec. 9001]	New section 9012 of FSRIA. Forest Service competitive research and development program to encourage use of forest biomass for energy. Priority to projects that use low-value forest byproduct biomass for the production of energy, develop processes to integrate bioenergy from forest biomass into existing manufacturing streams and

(Continued)

2002 Farm Bill (FSRIA, P.L. 107-171) or Other Law (as indicated)	House-Passed Farm Bill (H.R. 2419)	Senate-Passed Farm Bill (H.R. 2419)	Enacted Farm Bill (P.L. 110-246)
			develop new transportation fuels and improve the production of trees for renewable energy. Authorized appropriations of $15 annually for FY2009-12. [Sec. 9001]
Community Wood Energy Program			
No provision.	Establishes Community Wood Energy Program to provide matching grants to State and local governments to acquire community wood energy systems for public buildings and to implement a community wood energy plan to meet energy needs with reduced carbon intensity through conservation, reduce costs, and utilize low-value wood sources, and increase awareness of energy consumption. Authorizes discretionary funding of such sums as are necessary. [Sec. 9019]	New section 9014 of FSRIA. Similar to House provision except specifies matching grants of up to $50,000. Provides discretionary funding of $5 million for each of FY2008-12. [Sec. 9001]	New section 9013 of FSRIA. Same as Senate provision. [Sec. 9001]
Biofuels Infrastructure Study			
No provision.	No provision.	New section 9018 of FSRIA. Directs USDA, in collaboration with DOE, DOT and EPA, to conduct a study of the infrastructure needs associated with a significant expansion in biofuel production and use. Specifically includes dedicated ethanol pipeline feasibility studies and examination of water resource needs. Requires a report to Congress on the results of the studies. Authority to appropriate $1 million for each of FY2008 and FY2009. [Sec. 9001]	Requests USDA, DOE, EPA, and DOT to jointly report on the infrastructure needs and development approaches for expanding the domestic production, transport, and distribution of biofuels, and the required infrastructure. Mandatory funding of $1 million for FY2008, and $2 million annually for each of FY2009-12. Discretionary appropriations of $2 million annually are authorized for FY2009-12. [Sec. 9002]

(Continued)

2002 Farm Bill (FSRIA, P.L. 107-171) or Other Law (as indicated)	House-Passed Farm Bill (H.R. 2419)	Senate-Passed Farm Bill (H.R. 2419)	Enacted Farm Bill (P.L. 110-246)
Rural Fertilizer Study			
No provision.	No provision.	New section 9019 of FSRIA. Requires USDA to conduct a Renewable Nitrogen Fertilizer Study to assess the feasibility of producing nitrogen fertilizer from renewable energy (and submit a report to Congress within 270 days). Provides discretionary funding of $1 million for FY2008. [Sec. 9001]	Requires a report within 1 year of appropriations on the production of fertilizer from renewable energy sources in rural areas. Report must identify challenges to commercialization of rural fertilizer production, processes and technologies and the potential impacts of renewable fertilizer on fossil fuel use and the environment. Appropriations of $1 million are authorized for FY2009. [Sec. 9003]
TITLE XI: Livestock			
Study on Bioenergy Operations			
No provision.	No provision.	No provision.	Requires USDA study on the use of animal manure as a fertilizer, and on the impact on consumers and on agricultural operations of limitations placed on the use of animal manure and the effects of increased competition for manure due to biofuel uses. [Sec. 11014]
TITLE XV: Trade and Tax Provisions; Subtitle C PART II—Energy Provisions (also see *Description of the Chairman's Mark of the "Heartland, Habitat, Harvest, and Horticulture Act of 2007"*, Joint Committee on Taxation, JCX-94-07, October 2, 2007)			
Credit for Production of Cellulosic Biofuel			
Under the American Jobs Creation Act (AJCA) of 2004, (P.L. 108-357), cellulosic ethanol, once developed, would receive the current tax credit of $0.51 per gallon available to any ethanol blended into gasoline as provided, through Dec. 31, 2010. [26 U.S.C. 40]	No provision.	Provides for a total income tax credit of $1.25 per gallon for cellulosic ethanol blended (inclusive the standard ethanol tax credit of $0.51/gal. and the small producer credit of $0.10/gal.). The 15 million gallon limitation on the $0.10 credit is waived for small cellulosic ethanol producers. Applicable only for domestically produced cellulosic ethanol. The producer credit is extended through April 1, 2015. [Sec. 12312]	Provides a fourth tax credit under 26 U.S.C. 40, the Cellulosic Biofuel Producer Credit. The credit is $1.01 per gallon less the amount of small-producer ethanol credit claimed and the alcohol mixture credit claimed for ethanol. [Sec. 15321]

(Continued)

2002 Farm Bill (FSRIA, P.L. 107-171) or Other Law (as indicated)	House-Passed Farm Bill (H.R. 2419)	Senate-Passed Farm Bill (H.R. 2419)	Enacted Farm Bill (P.L. 110-246)
Comprehensive Study of Biofuels			
No provision.	No provision.	No provision.	The Secretary of Treasury, with USDA, DOE, and EPA shall commission the National Academy of Sciences to produce a report on biofuels, including current and projected production, economic and environmental impacts, the impact of government programs, and the relative impacts of different types of biofuels. [Sec. 15322]
Alcohol Fuel: Modification of Alcohol Credit			
Any ethanol blended into gasoline is eligible for a tax credit of $0.51 per gallon as provided under current law (AJCA of 2004, P.L. 108-357) through Dec. 31, 2010. [26 U.S.C. 40]	No provision.	Reduces the ethanol blenders' tax credit of $0.51 per gallon to $0.46 per gallon beginning in the first calendar year after the year in which 7.5 billion gallons of ethanol is produced. [Sec. 12315]	Reduces the ethanol tax credit of $0.51 per gallon to $0.45 per gallon beginning in the first calendar year after the year in which 7.5 billion gallons of ethanol is produced. [Sec 15331]
Alcohol Fuel: Calculation of Volume of Alcohol for Fuel Credits			
Under current law (AJCA of 2004, P.L. 108-357) the volume of bio-alcohol counted as fuel eligible for the tax credit may include up to 5% of the volume as denaturant. [26 U.S.C. 40]	No provision.	Reduces the permissible volume of denaturant to 2% for purposes of calculating the volume of alcohol eligible for the tax credit. [Sec. 12316]	Reduces the permissible volume of denaturant to 2% for purposes of calculating the volume of alcohol eligible for the tax credit. [Sec. 15332]
Alcohol Fuel: Ethanol Tariff Extension			
Under current law (Heading 9901.00.50 of the Harmonized Tariff Schedule (HTS)) imports of ethyl alcohol are subject to a duty of 14.27¢ per liter ($0.54 per gallon) and a duty of 5.99¢ per liter (Heading 9901.00.52; HTS) on imports of ethyl tertiary-butyl ether through Dec. 31, 2008. [19 U.S.C. Chapter 18]	No provision.	Extends the tariff of 54 cents per gallon for imported ethanol or mixtures of ethanol (headings 9901.00.50 and 9901.00.52 of the HTS) through Dec. 31, 2010. [Sec. 12317]	Adopts Senate provision. [Sec 15333]

(Continued)

2002 Farm Bill (FSRIA, P.L. 107-171) or Other Law (as indicated)	House-Passed Farm Bill (H.R. 2419)	Senate-Passed Farm Bill (H.R. 2419)	Enacted Farm Bill (P.L. 110-246)
Alcohol Fuel: Limitations on, and Reductions of, Duty Drawback on Certain Imported Ethanol			
Under current law (Section 1313 of the Tariff Act of 1930, as amended) permits the refund of duty if the duty-paid good is re-exported or used to make a good that is exported. In particular, a person who manufactures gasoline with ethanol subject to the duty imposed under either HTS 9901.00.50, can export jet fuel (which does not contain ethanol) and obtain a refund of the duty paid. [19 U.S.C. Chapter 18]	No provision.	Eliminates the ability to obtain a refund of the duty imposed under HTS 9901.00.50 on imported ethanol by substitution of either ethanol not subject to the duty, or of another petroleum product, that is exported to obtain the drawback. [Sec. 12318]	Adopts the Senate provision. [Sec. 15334]

Author Contact Information

Tom Capehart
Specialist in Agricultural Economics
tcapehart@crs.loc.gov, 7-2425

End Notes

[1] For more information, see CRS Report RL32712, *Agriculture-Based Renewable Energy Production*, by Randy Schnepf. For details of previous law, see USDA, *2002 Farm Bill*, "Title IX—Energy," available at http://www.ers.usda.gov/Features/ Farmbill/titles/ titleI Xenergy.htm.

[2] For more information on agriculture and bioenergy, see CRS Report RL32712, *Agriculture-Based Renewable Energy Production*; and CRS Report RL3 3290, *Fuel Ethanol: Background and Public Policy Issues*, by Brent D. Yacobucci.

[3] For a listing of federal incentives in support of biofuels production, see CRS Report RL33572, *Biofuels Incentives: A Summary of Federal Programs*, by Brent D. Yacobucci.

[4] For more information, see CRS Report RL34294, *Energy Independence and Security Act of 2007: A Summary of Major Provisions*, by Fred Sissine.

[5] For more information, see CRS Report RL34265, *Selected Issues Related to an Expansion of the Renewable Fuel Standard (RFS)*, by Brent D. Yacobucci and Tom Capehart.

[6] In April 2008, Texas Gov. Rick Perry wrote the Environmental Protection Agency seeking a waiver from the federal ethanol mandate, noting its contribution to higher food prices and dire impact on the cattle industry. Also, see CRS Report RS22870, *Waiver Authority Under the Renewable Fuel Standard (RFS)*, by Brent D. Yacobucci.

[7] For more information see CRS Report RL32712, *Agriculture-Based Renewable Energy Production*. For a complete listing of renewable energy legislative proposals introduced during the 110th Congress, see CRS Report RL33831, *Energy Efficiency and Renewable Energy Legislation in the 110th Congress*, by Fred Sissine, Mark Gurevitz, and Lynn J. Cunningham.

[8] The conference agreement on the 2008 farm bill was originally approved by the House and the Senate as H.R. 2419 and vetoed by the President in May 2008. Both chambers overrode the veto, making the bill law (P.L. 110-234). However, the trade title was inadvertently excluded from the enrolled bill. To remedy the situation, both chambers repassed the farm bill conference agreement (including the trade title) as H.R. 6124. The President vetoed the measure in June 2008 and both chambers again overrode the veto, which made H.R. 6124 law (P.L. 110-246), and repealed P.L. 110-234.

[9] Advanced biofuels include biofuels derived from cellulosic feedstocks; sugar and starch other than corn kernel-starch; waste material including crop residue, animal, plant, or food waste; diesel fuel produced from renewable biomass including vegetable oil and animal fat; butanol or other alcohols produced through the conversion of organic matter; and other fuels derived from cellulosic biomass.

[10] For more information see CRS Report RL33928, *Ethanol and Biofuels: Agriculture, Infrastructure, and Market Constraints Related to Expanded Production*, by Brent D. Yacobucci and Randy Schnepf.

[11] Mandatory funds are authorized and provided by legislation. Discretionary funds must be appropriated annually.

CHAPTER SOURCES

The following chapters have been previously published:

Chapter 1 – This is an edited, excerpted and augmented edition of a United States Congressional Research Service publication, Report Order Code RS22131, dated September 23, 2008.

Chapter 2 – This is an edited, excerpted and augmented edition of a United States Congressional Research Service publication, Report Order Code RL34696, dated November 6, 2008.

Chapter 3 – This is an edited, excerpted and augmented edition of a United States Congressional Research Service publication, Report Order Code RL34557, dated January 13, 2009.

Chapter 4 - This is an edited, excerpted and augmented edition of a United States Congressional Research Service publication, Report Order Code RL34130, dated January 23, 2009.

INDEX

F

G

H

Q

R

U

V

W

Y